RADIO MANUFACTURERS OF THE 1920'S

RADIO MANUFACTURERS OF THE 1920'S

BY ALAN DOUGLAS

Vol. 1—A-C Dayton to J. B. Ferguson, Inc.

"Says Confucius: 'A man who, while living in the present age, reverts to the ways of antiquity, is one who will bring calamity upon himself.' "

The Vestal Press, Ltd.
Vestal, New York 13851

A note about the cover:
The MAJESTIC (actually a 1930 model) is from the collection of James Spalik of Kirkwood, New York
The RCA is owned by Harvey N. Roehl of Vestal, New York, and it came from the Murray Clark collection
The ATWATER-KENT was loaned by Joyce Demchak of Johnson City, New York

Library of Congress Cataloging-in-Publication Data

Douglas, Alan, 1943-
Radio manufacturers of the 1920's.

Includes index.
Contents: v. 1. A-C Dayton to J.B. Ferguson, Inc.
1. Radio industry and trade—United States—History.
I. Title.
HD9696.R363U6345 1987 338.7′62138418′0973 87-31088
ISBN 0-911572-67-8 (v. 1)
ISBN 0-911572-68-6 (pbk. : v. 1)

For additional copies of this book
or for our complete catalog, write to:

The Vestal Press Ltd.
PO Box 97
Vestal New York 13851-0097

Preface

By the end of 1929, an industry that had barely existed ten years earlier had sold three and a half billion dollars' worth of radios, parts, and accessories. While there have been many books written on the technical history of radio and its inventors, and still more on the broadcasting stations and personalities, there have been none on this billion-dollar segment: the companies that built the receivers.

These volumes are an attempt to fill that gap. Since the history of each company is largely a history of the radio models it made, I have particularly emphasized these, and have aimed to describe every advertised model of each company, whether common or scarce, collectible or not.

This is not a complete history of the radio industry. Even in three volumes, there is only room to sketch about 70 of the largest makers. Nor is there much coverage of "homebrew" sets, although until 1925 they outnumbered factory-made models, and by the end of 1929 accounted for one-third of the total made to that time. But almost every company in business three years or more is here, as well as many of the small companies they acquired. Several that dealt only in kits are included, and quite a few others who sold components to homebrewers in the early period, so this phase has not been entirely neglected either.

The few companies that existed before 1920 are treated in some detail for that era, but my main emphasis is on the twenties, when broadcast radio was evolving so rapidly. A clean break at December 1929 includes the entire history of most makers, while avoiding the cathedrals and midgets that revolutionized radio manufacturing in the thirties.

THE ORIGIN OF THE SINGLE CONTROL

The sheiks are responsible for the original demand for single-control sets. They could not understand why a man who needed only one arm to drive a car needed two to tune a set.

Radio World (Sept. 17, 1927)

Organization

Many stories apply to more than one company: the Neutrodyne's development, for instance, or the RCA lawsuits leading to industry-wide patent licensing in 1927. Rather than repeat these stories each time, certain details have been scattered among the individual chapters; thus while most of the Neutrodyne's history is in the Freed-Eisemann chapter, significant parts are also under Fada, Garod, etc. Similarly, references to Atwater Kent appear in the Freshman and Splitdorf chapters. For these reasons the index is necessary for retrieval of all the bits and pieces.

The length of each chapter is largely dependent on the amount of material available. A few companies left a well-marked trail, but others managed to avoid all publicity at the time and have vanished without a trace, other than the radios they made. Fortunately the companies that attracted the most publicity tend to be the important ones.

All dates given with each model are the dates of first advertising, and prices are those at introduction. Prices are list, although many sets were openly advertised for less, and are generally without tubes or accessories.

I have tried to avoid footnotes, not because the facts have no source, but because the sources are so scattered that footnotes would occupy more space than the text. Introductory dates and prices of course come from advertising, a fair sample of which is reproduced throughout these volumes. Company history comes largely from contemporary trade publications: *Radio Retailer and Jobber*, *Radio Dealer*, *Radio Retailing* and others. Some comes from radio magazines in general; some from later historical accounts by the men or companies involved. Corporate financial data can be found in yearly volumes of Poor's Manuals.

Acknowledgments

My thanks to friends who helped with information on specific radio companies: Hugh Aitken (RCA), Lee Barron (Westinghouse/RCA), Bill Biddle (Day-Fan, AC Dayton), Wendell Carlson (GE/RCA), Fred Cassens (Silver-Marshall, Zenith), H.L. Chadbourne (De Forest, Thompson, and others), J.N. Clapp (Tri-City), Bill Condon (De Forest), Dexter Deeley (Tuska), Leo Gibbs (AC Dayton, Bremer-Tully, Day-Fan, Erla, Metro, Silver-Marshall), George Harris (Splitdorf), Bob Hertzberg (Pilot), Will Jensby (De Forest, Leutz), Robert Lozier (Magnavox), Dave McKenzie (Fada, Steinite, Tri-City), Wilson Norwood (Adams-Morgan), Bob Palmer (Thompson), Don Patterson and *RADIO AGE* (Ozarka, Grebe, and others), Floyd Paul (Gilfillan, with Ed Reitan; Magnavox; Radio Shop), T.M. Pletcher, Jr. (Zenith), Dick & Ann Schamberger (Federal), Wardell Smith (Adams-Morgan), J. McWilliams Stone, Jr. & Michael Supitilov (Operadio), Toledo-Lucas County Public Library (Air-Way), Harold Wheeler (Neutrodyne licensees), John M. Williams (RCA), Ralph & Elinor Williams (Atwater Kent), John Wolkonowicz (Atwater Kent, Philco).

Thanks also to Merrill Bancroft, Rich Elskamp and Walt Sanders, to Dave Crocker for originally collecting many of the ads, to Rich Wolven for the long-term loan of vital magazines, and to Rich, and John Wolkonowicz, for years of stimulating exchange and interest that made this subject come alive. Friends who contributed photographs or lent radios to be photographed are credited under each illustration. Radios not credited are from the author's collection, past or present.

There are many others who should be thanked also, but the full list could probably fill a volume. You know who you are, and your help is deeply appreciated.

Hear the Voices in the Air

Near every large city there are now one or more powerful wireless telephone stations. Speech from anyone of these stations can be clearly heard with the Amrad Receiving Set a score or more miles away. Frequently, these stations give radio "concerts." By means of the **Amrad Receiving Set** you can easily hear these "concerts" with surprising clearness though you are miles distant and in the seclusion of your own room.

Radio telegraph messages, press reports, etc., from commercial stations and ships several hundred miles away, as well as the constant intercommunication of the amateur stations in your general vicinity can be heard any hour of the day.

Amrad ad, Radio News (Dec. 1920)

Radiophone Music.—It appears very much as though we are on the verge of a new era in radio communication, namely, radiophone music for the home. The idea, in brief, is to have radiophone stations at central points sending out concert music as well as speeches and lectures via radio, and compact and simple radio receiving sets in various homes, clubs and so on to intercept the waves. Already there are several radiophone stations in operation, and at least one wireless company has developed a receiving set made in the form of a cabinet phonograph, incorporating a concealed loud-speaking telephone unit, so that the music, speech, lecture or other radiophone transmission can be heard throughout a room. It is believed by radio men that leading manufacturers of radio equipment will see fit to maintain radiophone concert and lecture services for their patrons.

Scientific American (May 28, 1921)

RADIO CONCERTS

We are pleased to inform our readers that the DeForest Radio Tel. & Tel. Co. is giving radio concerts between 7:30 and 8:30 every evening, except Saturdays and Sundays. These concerts are given on a wavelength of 1,400 meters, and a transmitting set radiating 5¼ amperes is used. The input into the tubes is one kilowatt.

The DeForest Company would appreciate receiving reports on these concerts from amateurs, stating what kind of receiver is used and the distance at which they are heard from New York City.

Radio News (Jan. 1921)

Now Dad-Hear the tenor

"I can listen in on all the radio concerts now, Dad. Those Baldwin phones you bought me for Christmas are corkers. Just listen to that famous tenor singing over the wireless telephone. Every note records just as clearly as though it were a phonographic reproduction right here in our own home."

Actually Baldy Phones reproduce in identically the same manner as do the high grade phonographs. Instead of a heavy iron diaphragm, as in most phones, a selected grade of mica is used. This is much more susceptible to distortion and as a result responds more readily to the thousands of overtones and harmonics of the human voice or any musical instrument.

Baldy's are the most sensitive phones in the world. This is attested to by the fact that the leading radio engineers, with every facility at their command for testing the audibility and sensitiveness of every make of phone, choose Baldwin for their personal use.

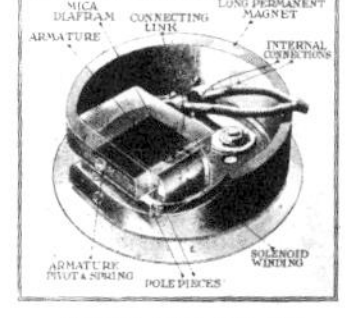

Study this X-Ray photo of a Baldwin ear-piece and you will see why they are world famed for their sensitiveness.

Our new booklet will give you some interesting facts about Baldwin Phones, in addition to prices. Ask your dealer for a copy. If he can not supply you write direct, giving his name and address.

JOHN FIRTH & CO., Inc.
18 Broadway New York

Sole Distributors of

Eldridge Electrical Instruments
Kolster Decremeter
U. S. Bureau of Standard Wavemeter
Navy Standard Lyden Jars
Brownlie Adjustable Phones

BALDY FOR LAND SEA AND IN THE AIR **PHONES**

Radio News (Dec. 1920) p. 347

One of the first advertisements to explicitly mention broadcasting.

Introduction

Broadcasting and KDKA

Although the human voice had first been transmitted by radio in 1900 (Fessenden, one mile at Cobb I., Maryland), and various experimenters had broadcast in the years following, some quite regularly, broadcasting as we know it began in 1920, and with it the radio industry. The popular accounts single out KDKA: "The sensation in the public press created by this marvelous demonstration of wireless telephony set in motion a movement that eventually encircled the globe," (Archer, *History Of Radio To 1926*) or ". . . the excitement radiating from KDKA set off a national mania," (Barnouw, *A Tower In Babel*).

Actually KDKA's November 2 broadcast of the Harding-Cox election returns created little stir. The major radio magazines published at the time (*QST*, *Radio News*, *Wireless Age*) didn't even mention KDKA. *QST*'s profile of Frank Conrad in February 1921, an issue that included amateur news up to January 23, said nothing of broadcasting; KDKA first appeared in *QST* in May. Popular weeklies were even slower: *Scientific American* devoted a paragraph to "radiophone music" in its May 28, 1921 issue, though it had regularly printed stories on other aspects of radio for two decades, while the *Literary Digest*, a leading newsmagazine which likewise had featured radio articles before, first mentioned broadcasting on July 9, 1921.

In the Beginning

Radio in 1920 and 1921, the starting point of this account, was therefore what it had been for twenty years: commercial and amateur point-to-point communications. At this time, the commercial ship-to-shore and transoceanic business was far more important than the amateur, but it approximately only doubled during the decade, and very few commercial wireless manufacturers moved successfully into broadcast radio. The broadcast industry was largely descended from the amateur-equipment makers. Indeed, if it hadn't been for the flock of "amateur and experimental" licensees under Armstrong's regeneration patent, in the early twenties when the industry was taking shape, RCA and its "radio group" allies would have had the field entirely to themselves. The group (RCA, GE, AT&T, Westinghouse, United Fruit/Wireless Specialty Apparatus) spent a great deal of their legal energies fighting off their own licensees, but by 1925 when they had succeeded in driving them out of business or into relying on some other circuit, a much more dangerous herd had their noses into the tent. Atwater Kent, Crosley, Sparton, Zenith and others who would soon take most of the market from RCA, might never have done so, or existed at all, if RCA had been able to take a tougher stance when they were smaller.

Armstrong Regenerative Sets

In 1920 Howard Armstrong's legal expenses were mounting, while he waited for one of the large electrical or wireless companies to make an offer for his regeneration and superheterodyne patents and applications. His attorneys hit on the idea of licensing the myriad makers of regenerative receivers for the amateur market, under his patent 1,113,149 (1914), for a royalty of 5% of sales price, and were soon signing them up as fast as they could spot their magazine ads. Since the ham market was considered negligibly small, issuing these licenses would not endanger the eventual sale of the patents to commercial interests. The following companies were licensed, more or less in the order listed:

- American Marconi (only for one or two stations)
- International Radio Telegraph Co.
- A.H. Grebe & Co., Inc.
- Chicago Radio Laboratory (5/15/20) (became Zenith)
- Clapp-Eastham Co. (4/18/20)
- Cutting & Washington, Inc. (7/7/20)
- Adams-Morgan Co.
- The Precision Equipment Co. (later bought by Crosley)
- Jones Radio Co. (affiliated with Kellogg)
- Mignon Mfg. Export Corp.
- Tri-City Electric Mfg. Co.
- Klitzen Radio Mfg. Co. (affiliated with Michigan)
- The Radio Shop (supposedly 9/2 to 9/4/20)
- Oard Radio Laboratories
- Pennsylvania Wireless Mfg. Co.
- The C.D. Tuska Co.
- Radio-Craft Co., Inc. (9/20/20) (later bought by De Forest)
- The Colin B. Kennedy Co.
- Eastern Radio Co.
- Chelsea Radio Co.

All these companies but the first two were small; indeed, some were nothing more than high-school boys working in their attics. And the licenses were non-transferable, so that not all continued in business. But some became substantial companies: Chicago Radio Laboratory (Zenith), Cutting & Washington (Colonial), Grebe, Kennedy, Precision Equipment (Crosley). Others were important in the early twenties: Adams-Morgan, Clapp-Eastham, Klitzen (Michigan), Radio Craft (De Forest), Radio Shop (Echophone), Tri-City (mfr. for Montgomery Ward), Tuska.

RADIO-CRAFT PRODUCTS

Detector and Two-Stage Audio Frequency Amplifier.

This instrument is made of the finest materials and workmanship, and gives perfect amplification without any interfering noises. Binding posts are provided for separate "B" battery connections, for Detector and Amplifier.

Our Detector and Amplifier line consists of:
Detector Unit $15.00
Detector and One-Stage Audio Frequency Amplifier $45.00
Detector and Two-Stage Audio Frequency Amplifier $70.00
Two-Stage Audio Frequency Amplifier $50.00

RADIO-CRAFT CO., Inc.
FRANK M. SQUIRE, Pres.
413 Third Ave., Brooklyn, N. Y.

"YOU'D BE SURPRISED"

at the distance, strength of signal and freedom from QRM you get with Ace equipment. The type T. T. Tuner is **the kind** you want for that extreme long distance work thru the worst QRM. Doc Ace has the right dope when he says "You may pay more, but you can't buy better". Type T. T. Regenerative Tuner, range 150 to 2750 meters, F. O. B. Cincinnati, **$55.**

The Precision Equipment Company
2437 Gilbert Ave., Dept. D, Cincinnati, Ohio.

INTERNATIONAL RADIO

VACUUM TUBE DETECTOR
or
SINGLE STAGE AMPLIFIER

Single tube Detector and Amplifier Units are among the most useful of the Interchangeable panel instruments in the International Experimental Apparatus Line.

READY FOR IMMEDIATE DELIVERY.
AUDION DETECTOR RTD-102-A.......... $15.00
SINGLE STAGE AMPLIFIER RTA-104-A.. $26.00
F.O.B. NEW YORK CITY
Write to Department 23 for our new completely illustrated six page bulletin index.

International Radio Telegraph Company
326 Broadway New York City

RW 4"—SIZE 12" x 9" x 6½"

Mignon "RW4"
Undamped Wave Receptors

NO LOOSE COUPLERS
LOADING COILS
VARIOMETERS
AMPLIFIERS

Strongest, most distinct signals

— IMPORTANT NOTICE —
The Mignon System Apparatus "RW1," "RW2," "RW3" and "RBD8" are no longer manufactured and are superseded by the improved "RW4," "UW1" and "BD1." These apparatus are manufactured exclusively by the MIGNON MFG. CORP., sole owners of the Mignon System Patents. (See U. S. Letters Patent No. 1329672.)
INFRINGERS WILL BE PROSECUTED.
Mignon Manufacturing Corporation - - Newark, N. J.

Damped and undamped wave apparatus for all purposes
Endorsed by Radio Dept. of D. L. & W. Ry. Co.

WRITE FOR LITERATURE
Mignon Manufacturing Corporation, Dept. "Q", Newark, N. J.
Canadian Representative—Canadian Radio Mfg. Co., Bienville, Que.

THE JONES CABINET LONG WAVE RECEIVING TRANSFORMER

For DAMPED or UNDAMPED Wave Reception

A carefully wound Receiving Transformer with an approximate Wave Length of from 200 to 8000 Meters Equipped with Variable and Fixed Condensers Rheostat and Grid Leak. Fitted to take the latest MARCONI Vacuum Tube. Bakelite Panel, Silver Plated Switches and Contacts, Mission Oak Cabinet. Entire Apparatus built within Metal frame to permit easy removal. Model "D", as illustrated, SIXTY DOLLARS, Net. Model "E" fitted with Modern "DEAD END" Switch and High Grade Hot Wire Ammeter, Eighty-Five Dollars net.

New Bulletin sent on request.

THE JONES RADIO CO., 384 Monroe Street, BROOKLYN, N. Y.

QST (June 1920)

Typical Stock for Radio Merchant

Quantity.	Assortment No. I Merchandise	Minimum	Maximum
2-10	E.R. 753 G. E. receivers @ $18......	$36.00	$180.00
1-5	AR.-1300 G. E. receivers @ $50.....	50.00	250.00
1-5	A.A.-1400 G.E. amplifiers @ $75....	75.00	575.00
2-6	S--307421 Westinghouse Aeriola, Jr., @ $25	50.00	150.00
2-6	S-319564 Westinghouse Aeriola, Sr., @ $65	130.00	390.00
1	S-307189 Westinghouse tuner, @ $68	68.00	68.00
1	S-307190 Westinghouse detector, 2-stage amplifier	70.00	70.00
2-10	S-307215 Westinghouse R. C., @ $132.50	265.00	1,325.00
1-2	DT-600 De Forest sets, @ $25......	25.00	50.00
1-2	DT-700 De Forest sets, @ $36......	36.00	72.00
1-2	DL-800 De Forest amplifier, @ $35..	35.00	70.00
1-2	"S" Paragon receivers, @ $50......	50.00	100.00
1	DA-2 Paragon amplifier, @ $65....	65.00	65.00
2-5	2596 Amrad receivers, @ $45.......	90.00	225.00
2-5	2634 Amrad amplifier and detector, @ $47.50	95.00	237.50
2-5	2626 Amrad load coils, @ $3.85.....	7.70	19.25
2-5	2962 Amrad load coils, at $3.85......	7.70	19.25
1	CR-8 Grebe, @ $80................	80.00	80.00
1	CR-3 Grebe, @ $65	65.00	65.00
1	CR-9 Grebe, @ $130	130.00	130.00
1	Rord-Grebe, @ $75	75.00	75.00
1	Rork-Grebe, @ $65	65.00	65.00
1-2	224 Tuska popular receiver, @ $35..	35.00	70.00
1	225 Tuska popular receiver and amplifier, @ $75...................	75.00	75.00
1	H.R. receiver, @ $40	40.00	40.00
1	H.Z. amplifier, @ $40.............	40.00	40.00
2-3	105 Marvel sets, @ $15............	30.00	45.00
2-6	833 DeVeau standard loud speakers, @ $35	70.00	210.00
2-6	Midget speakers, @ $15............	30.00	90.00
1	R-3 Magnavox, @ $45............	45.00	45.00
1	R-1118 Acmefone, @ $80...........	80.00	80.00
2-6	837 Victrola attachments, @ $15....	15.00	90.00
1-2	838 Grafonola attachments, @ $15...	15.00	30.00
1	839 Edison attachment, @ $16.50...	16.50	16.50
1-2	R-1131 Amplitone loud speaker, @ $12	12.00	24.00
1-2	70 Paragon V.T. controls, @ $6.....	6.00	12.00
1	A-5 G.A. standard detector, @ $5.95	5.95	5.95
1-2	A-6 G.A. standard amplifier, @ $13.50	13.50	27.00
2-6	223 Tuska parts set, @ $27.50.......	27.50	165.00
2-6	321 Horne variocouplers, @ $7......	14.00	42.00
3-12	Horne variometers, @ $8...........	2400	96.00
3-6	R-1012 Baldwin variocouplers, @ $8	24.00	48.00
3-6	R-1012 Baldwin variometers, @ $7..	21.00	42.00
3	R-1137 Bruno loose couplers, @ $10	30.00	30.00
2	7874 tuning coils, @ $3.30..........	6.60	6.60
12	829 DeVeau radio plugs, @ $1.25....	15.00	15.00
10-25	843 DeVeau head sets, @ $8........	80.00	200.00
6-12	844 DeVeau head sets, @ $10.50...	63.00	126.00
12	56-2 Murdock head sets, @ $5......	60.00	60.00
12	56-3 Murdock head sets, @ $6.......	72.00	72.00
12	102-A Fada crystal detectors, @ $2.25	27.00	27.00
24	324 Murdock crystal detectors, @ 70c	16.80	16.80
12	112 Puritche crystal detectors, @ $2.25	27.00	27.00
200	R-1121 unmounted galena crystals, @ 25c.	50.00	50.00
200	Mounted galena crystals, @ 40c....	80.00	80.00
50	Detector tubes, UV-200, @ $5......	250.00	500.00
50	UV-201 amplifier tubes, @ $6.50....	325.00	1,300.00
6-12	WD-11 Aerictron tubes, @ $7.50....	45.00	90.00
50	UP-542 sockets, @ $1...............	50.00	50.00
25	R-500 DeForest sockets, @ $1......	25.00	25.00
25	30 Paragon sockets, @ $1..........	25.00	25.00
25	F-500 DeForest rheostats, $1.20.....	30.00	30.00
6	P.R.-535 filament rheostats, @ $3...	18.00	18.00
25	56-OB Murdock rheostats, @ $1....	25.00	25.00
2	P.R.-536 potentiometers, @ $2......	4.00	4.00
6	C.V.-101 DeForest variable condenser, @ $5	30.00	30.00
6	C.V.-100 variable condensers, @ 04.75	28.50	28.50
6	C.V.-201 DeForest variable condensers, @ $5.50	33.00	33.00
6	C.V.-200 DeForest variable condensers, @ $5.25	31.50	31.50
1	Carton Amrad fixed condensers....	22.50	22.50
1	Carton 2972 Amrad grid leaks......	40.00	40.00
50	A-2 grid condensers, @ 35c.........	17.50	17.50
50	A-3 phone condensers, @ 35c.......	17.50	17.50
50	grid leaks and condensers, @ 50c...	25.00	25.00
12	R-29 fixed receiving condensers, @ 90c.	10.80	10.80
2	DL-25 DeForest honeycomb coils, @ $1.54	3.08	3.08
2	6010 Horne lightning switches, @ $6.75	13.50	13.50
10	7001 battery switches, @ 32c........	3.20	3.20
10	7000 battery switches, @ 20c........	2.00	2.00
25	15 Paragon switch levers, @ 70c....	17.50	17.50
12	10 Paragon knob and dial, @ 90c....	10.80	10.80
12	12 Paragon knob and dial, @ 1.35....	15.00	15.00
12	14 Paragon knob and dial, @ $1....	12.00	12.00
12	R.A.-15 Paragon knob and dial, @ $1.75	21.00	21.00
12	210—1 Tuska knob and dial, @ $1.65	19.80	19.80
12	212—1 Tuska knob and dial, @ $1.10	13.20	13.20
12	800-H Clapp-Eastham knob and dial, @ 75c	9.00	9.00
2	DL-50 DeForest honeycomb coils, @ $1.65	3.30	3.30
2	DL-75 DeForest honeycomb coils, @ $1.65	3.30	3.30
2	DL-100 DeForest honeycomb coils, @ $1.71	3.42	3.42
12	763-E.R. B batteries, @ 98c net....	11.76	11.76
	766 E.R. B batteries, @ $1.65.......	19.80	19.80
12	Tungar rectifier, @ $28...........	28.00	28.00
1	231-A amplifying transformers, @ $5	50.00	50.00
10	226-W. Federal amplifying transformers, @ $7	70.00	70.00
10	R-1202 Thordarson transformers, @ $4	40.00	40.00
10	R-1103 Novice antennae equipment, @ $2.50	25.00	25.00
10	100-foot coils hard drawn antennae wire, @ 25c. lb.................	50.00	100.00
50	Also standard antennae wire, No. 14		
12	6001 Horne lightning arresters, @ $2.45	29.40	29.40
100	4501 aerial insulators, @ 25c........	25.00	25.00
50	4500 aerial insulators, @ 40c........	20.00	20.00
12	G-2 electros switches, @ $1.50......	18.00	18.00
12	137F G.R. knob and dial, @ $1.85....	21.20	21.20
12	137D G.R. knob and dial, @ $1......	12.00	12.00
6	2608 Amrad knob and dial, @ 65c...	3.90	3.90
6	2069 Amrad knob and dial, @ 65c....	3.90	3.90
25	DeVeau binding posts, No. 852, @ 15c.	3.75	3.75
25	DeVeau binding posts, No. 854, @ 17c.	4.25	4.25
25	DeVeau binding posts, No. 853, @ 17c.	4.25	4.25
25	DeVeau binding posts, No. 855, @ 20c.	5.00	5.00
25	DeVeau binding posts, No. 856, @ 20c.	5.00	5.00
25	DeVeau binding posts, No. 857, @ 25c.	6.25	6.25
		$4,396.81	$9,060.16

Typical Initial Stock for Radio Merchant

Assembled Sets — Assortment No. II	List Prices
Aeriola Junior Receiving Set..................................	$25,00
Aeriola Sr. Receiving Set..................................	65.00
R. C. Short Wave Regenerative Receiver....................	132.50
Aeriola Grande Receiving Set.............................	325.00
C. R. 3 Grebe Special Short Wave Receiver..................	65.00
C. R. Grebe Short Wave Regenerative Receiver.............	80.00
C. R. 5 Grebe Intermediate Wave Receiver..................	80.00
C. R. 9 Grebe Intermediate Wave Rceiver..................	130.00
C. R. 10 Grebe Receiver (Radiotone)........................	315.00
RORK Two Stage Amplifier..............................	55.00
RORD Vacuum Tube Detector............................	75.00
Kennedy Universal Receiver..............................	250.00
Kennedy Two Stage Amplifier............................	85.00
Kennedy Intermediate Wave Receiver......................	125.00
Kennedy Short Wave Receiver............................	80.00
Kennedy Two Stage Amplifier............................	55.00
De Forest Everyman Receiving Set........................	25.00
De Forest Radiophone Receiving Set.......................	36.00
Federal Junior Receiver..................................	25.00
	$2,028.50
Add Miscellaneous Radio Accessories......................	1,000.00
	$3,028.50

Radio Merchandising published this list of recommended stock for new radio dealers on July 1, 1922.

For several years, the Armstrong circuit was far and away the best performer of all the standard ones, and the most efficient in its use of tubes (which were very expensive, as were the batteries to run them). So any company licensed to use this circuit could make radios fully as good as RCA's. Regenerative radios became less important after 1925, partly because of legal pressures, partly because they were too tricky for the average fan to handle, and caused interference to other sets when misadjusted. Only a few companies continued making them, while most switched to TRF, tuned radio frequency, in 1924. Crosley, specializing in cheap sets with few tubes, needed regeneration for maximum performance from each tube. RCA used regeneration in its low- and medium-priced models (superheterodyne for the top of the line) and did not make a TRF until 1927.

The Radio Boom

Since the amateurs were predominantly young men and boys, with no money for manufactured apparatus, they built their own; therefore in 1920 the radio industry consisted mainly of component makers (excluding commercial and government manufacturers). And at first, broadcast fans were likely to be the families of amateurs, or the amateurs themselves, who similarly used homemade receivers. It was not until late 1921 that the general public began to be drawn in, and significant numbers of manufactured receivers sold. M.B. Sleeper gives a contemporary description of the resulting boom:

THE WIRELESS WORLD AND RADIO REVIEW December 23, 1922

Distributing Problems of Radio Manufacturers.

By M. B. Sleeper.

In the following article Mr. Sleeper describes some of the difficulties which were experienced in America in the process of establishing the wireless industry on a sound footing.

ALTHOUGH the experiences of radio manufacturers in England may differ from the series of events which have taken place during the first year of broadcasting here, the general outline will probably be the same. Possibly some of the very expensive errors made by the majority of our manufacturers and dealers that have almost paralysed the industry may be worth relating.

Last Christmas the demands of the public upon the dealers were such that they were operating on the daily deliveries from the manufacturers. The latter, sensing a great increase in business, bent all their efforts to enlarge their facilities. New companies sprang into existence, in the main duplicating the products of existing concerns. The newcomers found plenty of orders, because the dealers knew that, of whatever they ordered, only a small part would be delivered. We call it "pyramiding" orders.

For example, a dealer who could sell a thousand rheostats a month ordered two thousand from each of six companies, with the idea that the total of the partial deliveries might make up that amount. Because of this phantom of demand, manufacturers lost their perspective and tried to get on an immediate delivery basis. In the meantime business developed so rapidly that, when they could deliver a thousand rheostats the dealer could sell ten thousand. Moreover, new stores opened up by the hundred, and a certain amount of goods had to be turned out to supply each one.

Jobbers also jumped into radio, bringing with them the question of jobbers' discounts. Many of them, instead of selling to the stores, took advantage of the jobbers' discounts, but sold directly to consumers. Department stores, one after another, opened radio departments, complicating matters by asking for jobbers' discounts. Moreover, machine shops and "cellar factories" made trouble by offering jobbers' discounts to the retail stores. Radio stores included, by the way, ironmongers, chemists, tailors' shops, millinery stores, shoe stores, sporting goods and clothing shops, stationery stores, garden implement houses, motor car accessory shops, garages—in fact, almost any kind of place selling to the public.

Before this time we had had no real jobbers, and no one could definitely define a radio jobber anyway. As a result many manufacturers sold in large quantities to stores whose purchases on credit were really not justified by their normal credit rating. Then however, their turnover was so rapid that they could easily pay their bills and—here is where the trouble came—they were allowed to continually increase the size of their orders.

In May the crash came. The production of established manufacturers, plus that of the newcomers, reached the demand, and the consumers' purchases fell off slightly. Immediately orders were cancelled right and left. A little later goods were returned to the manufacturers. Retailers stopped paying their bills.

The effect was first felt by new manufacturers who had ordered materials in huge quantities, for theirs were the first orders to be cancelled. As a result, they had no accounts receivable to cover their accounts payable. They had not had time to establish themselves in the minds of the public, nor had they advertised extensively. One after another closed shop in rapid succession.

During July and August the strain was felt by the older concerns, for both orders and collections fell off to almost nothing. Since they sold directly to retailers instead of putting orders through jobbers of high financial standing, they found trade connections, built up at great expense, of no further value. Many retailers, finding business poor during the summer, quickly disposed of their stocks, but showed no interest in paying their bills, because, having made money during the period of great demand, they decided to go back to their old business, that of selling medicine, clothes, shoes, or whatever it was.

This autumn, orders have been very slow at first where, ordinarily, things are in full swing by the middle of September. This was due to the fact that many stores were still stocked with last spring's purchases and the public was waiting for the end of the unreasonably warm weather we were experiencing. Some retailers could not and others would not pay bills dating back to April and May until trade resumed. In the meantime many manufacturers who kept up production right through the summer, in preparation for autumn demands, had not been able to finance themselves, and had gone under. This has put on the market a tremendous amount of merchandise at sacrifice prices. In many cases goods are offered at one-fourth their advertised prices.

All this has benefited, in a way, those who are able to stay in business, though the lessons learned have been dreadfully costly. Credit relations between material supply houses and manufacturers, as well as with the banks, have been strained, often creating a lack of confidences in the manufacturer at a time when he needs the greatest assistance.

A word about discounts may be of interest. Discounts to retailers have been established at 25 and 3c per cent., sometimes running to 33⅓ per cent. Jobbers are allowed 40 per cent., and distributors who cover the territory of several jobbers are given 50 per cent. Distributors have been necessary here because the United States is too large to be covered from one central office. In England, on the other hand, the population is sufficiently concentrated that this necessary evil may not be needed.

Comparatively little of the autumn business is being handled directly with the retailers. Such orders are now filled by shipping the order to the jobber who covers the city where that retailer is located. The goods are billed to the jobber who, of course, has demonstrated his financial standing prior to his appointment. Such distribution has been achieved that the fraction of direct mail orders from the consumer is very small. Manufacturers generally try to sidestep mail orders by advertising that their goods can be purchased in the local stores.

It has not been my intention in the foregoing paragraphs to paint a black picture of the radio situation here as a prophecy of conditions in England. On the contrary, my desire in preparing these notes was to present a rough outline of our experience with the hope that from them some helpful ideas might be obtained by the English manufacturers in whose work I have always been interested, and for whose products, particularly after my visit during the summer, I have the highest regard.

Radiola Superheterodyne AR-812 Feb. 1924 RCA
$220 without tubes or accessories.

The boom-and-bust cycle repeated itself throughout the decade (see the chart on p. xx). Radios sold well during the winter (especially at Christmas) when reception conditions were good and people were confined indoors with time to listen. They sold poorly in the summer when static overpowered the relatively weak broadcast stations, and outdoor activities lured listeners away. Generally, radio manufacturers conformed to this cycle by engineering their new models in the spring, introducing them at summer trade shows, and running their factories only during the fall. If they guessed well, they sold all their production by spring; if not, they were left with mountains of obsolete stock to be dumped to cut-price city dealers.

A different view of the boom, from the fans' perspective, can be seen in an article from *Radio Broadcast* magazine of October 1922, on the situation in Chicago (excerpted):

Radio Has Gripped Chicago

Grand opera, news expensively and quickly gathered, the words of political and religious leaders, instrumental music by great artists—all these are carried by the house-top antennas down into dingy rooms for the comfort of persons for whom such things simply did not exist a year ago.

The alacrity with which antennas have appeared on the skyline of the west side is the most dramatic and most hopeful phase of the development of radio broadcasting in Chicago. Crude homemade aerials are on one roof in ten along all the miles of bleak streets in the city's industrial zones. For thousands of families, life has acquired new savor through radio.

Chicago caught the radio fever in earnest last fall, when the Westinghouse Company established Station KYW on the roof of a downtown skyscraper. Its KDKA station, in East Pittsburgh, had then been broadcasting for nearly a year, and stations had been created or were being built at Newark, N.J., and Springfield, Mass. The Chicago staff of the company wanted to get abreast of the others. The approaching season of the Chicago Opera Company seemed an opportunity.

Now the Westinghouse people do not pretend to be philanthropists. Their broadcasting service is business, and they admit it. They manufacture radio sets. They want a market for those radio sets. To create a market they must make the sets valuable to purchasers. Hence the broadcasting. Hence, too, the excellence of the broadcasted programmes, for the better the entertainment the larger the audience.

In arranging for the opening of their Chicago station the Westinghouse radio men found a willing ally in Miss Mary Garden, then director general of the Chicago Opera Company. Efforts were then being made to enlist the public generally in support of opera. Miss Garden and her associates in the management of the company were appealing to all Chicago to back the enterprise out of civic pride.

The suggestion that opera be broadcasted by radio was welcomed. Grand opera is an exotic dish. Taste for it is not instinctive, but acquired. Miss Garden saw in the broadcasting plan a chance to instill a liking for good music in thousands of minds outside the range of any other appeal, and so the plan was adopted.

The consequences were amazing. In Chicago at the opening of the opera season were approximately 1,300 radio sets. Announcement of the fact that opera was to be broadcasted started a clamor for equipment. As the season advanced and professional critics added their praises of radio transmission to the ecstatic comments of radio enthusiasts the clamor increased. To "listen in" on the opera became

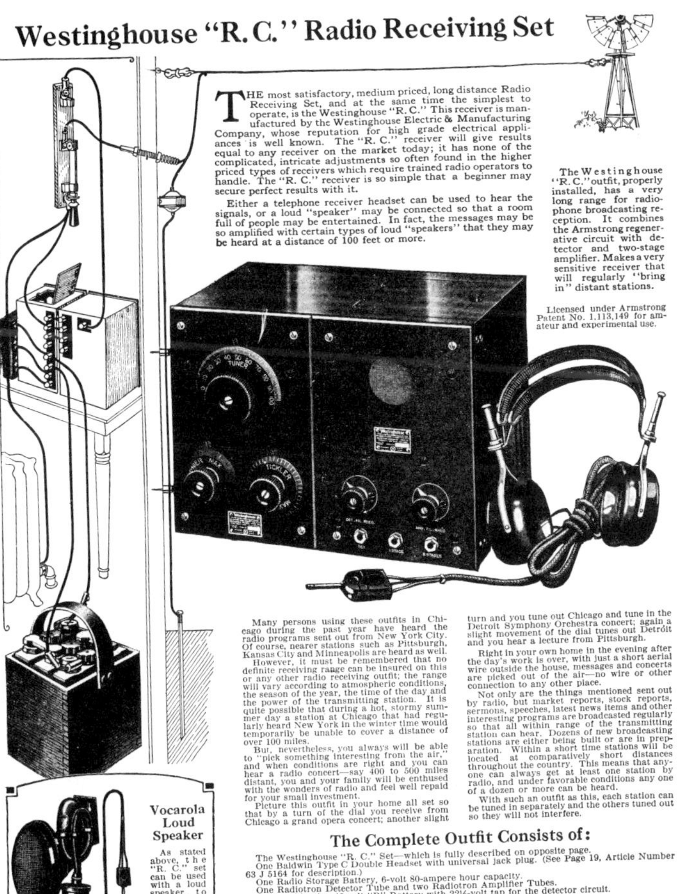

Westinghouse "R. C." Radio Receiving Set

THE most satisfactory, medium priced, long distance Radio Receiving Set, and at the same time the simplest to operate, is the Westinghouse "R. C." This receiver is manufactured by the Westinghouse Electric & Manufacturing Company, whose reputation for high grade electrical appliances is well known. The "R. C." receiver will give results equal to any receiver on the market today; it has none of the complicated, intricate adjustments so often found in the higher priced types of receivers which require trained radio operators to handle. The "R. C." receiver is so simple that a beginner may secure perfect results with it.

Either a telephone receiver headset can be used to hear the signals, or a loud "speaker" may be connected so that a room full of people may be entertained. In fact, the messages may be so amplified with certain types of loud "speakers" that they may be heard at a distance of 100 feet or more.

The Westinghouse "R. C." outfit, properly installed, has a very long range for radio-phone broadcasting reception. It combines the Armstrong regenerative circuit with detector and two-stage amplifier. Makes a very sensitive receiver that will regularly "bring in" distant stations.

Licensed under Armstrong Patent No. 1,113,149 for amateur and experimental use.

Many persons using these outfits in Chicago during the past year have heard the radio programs sent out from New York City. Of course, nearer stations such as Pittsburgh, Kansas City and Minneapolis are heard as well.

However, it must be remembered that no definite receiving range can be insured on this or any other radio receiving outfit; the range will vary according to atmospheric conditions, the season of the year, the time of the day and the power of the transmitting station. It is quite possible that during a hot, stormy summer day a station at Chicago that had regularly heard New York in the winter time would temporarily be unable to cover a distance of over 100 miles.

But, nevertheless, you always will be able to "pick something interesting from the air," and when conditions are right and you can hear a radio concert—say 400 to 500 miles distant, you and your family will be enthused with the wonders of radio and feel well repaid for your small investment.

Picture this outfit in your home all set so that by a turn of the dial you receive from Chicago a grand opera concert; another slight turn and you tune out Chicago and tune in the Detroit Symphony Orchestra concert; again a slight movement of the dial tunes out Detroit and you hear a lecture from Pittsburgh.

Right in your own home in the evening after the day's work is over, with just a short aerial wire outside the house, messages and concerts are picked out of the air—no wire or other connection to any other place.

Not only are the things mentioned sent out by radio, but market reports, stock reports, sermons, speeches, latest news items and other interesting programs are broadcasted regularly so that all within range of the transmitting station can hear. Dozens of new broadcasting stations are either being built or are in preparation. Within a short time stations will be located at comparatively short distances throughout the country. This means that anyone can always get at least one station by radio, and under favorable conditions any one of a dozen or more can be heard.

With such an outfit as this, each station can be tuned in separately and the others tuned out so they will not interfere.

Vocarola Loud Speaker

As stated above, the "R. C." set can be used with a loud speaker to entertain a room full of people, but no loud speaker is included in the outfit. Vocarola is the most satisfactory to use with this set. For entertaining larger audiences we recommend the Magnavox. See Page 18 for Loud Speakers.

The Complete Outfit Consists of:

The Westinghouse "R. C." Set—which is fully described on opposite page.
One Baldwin Type C Double Headset with universal jack plug. (See Page 19, Article Number 63 J 5164 for description.)
One Radio Storage Battery, 6-volt 80-ampere hour capacity.
One Radiotron Detector Tube and two Radiotron Amplifier Tubes.
One combination 45-volt "B" Battery with 22½-volt tap for the detector circuit.
A complete antenna equipment consisting of:

150 feet stranded aerial wire cable.
2 air-gap type, extra high grade aerial wire insulators.
1 extra high grade wall insulator (see Page 35).
1 600-volt 100-ampere radio grounding switch; 25 feet No. 4 insulated ground wire.
1 lightning arrester.
1 porcelain base switch.
8 large porcelain insulating knobs with screws.
2 screweyes.
1 ground clamp.
50 feet rubber covered connecting wire.
15 feet flexible cord to connect batteries to instruments, etc.

Shipping weight, complete, 50 pounds.

663 J 621—Complete outfit, with instructions for installation and operation **$169.00**

All merchandise shown in this Catalogue shipped from Northern Illinois — Montgomery Ward & Co.

Westinghouse D. A. and R. A. Radio Instruments

These two instruments, combined in one cabinet, make up the R. C. set shown on opposite page. They are supplied separately so that either can be used with radio instruments of other makes if desired.

Type R. A. Short Wave Regenerative Tuner

Licensed Under Armstrong Patent No. 1,113,149 for amateur and experimental use

This instrument takes the incoming radio wave collected on the antenna wire and "tunes" it so the other apparatus used can change the wave so it may be heard in the head receivers. In order that this "tuning" can be done easily by anyone, the instrument is made as simple as possible and requires but one adjustment in order to tune to the desired signal. The wave length range is from 180 to 700 meters, which means that amateur, broadcasted and commercial messages may be tuned in. May be used with a crystal or an audion detector, working alone or in conjunction with an amplifier.

How It Works

This is a single circuit tuner and the oscillating circuit consists of a condenser of variable capacity and a variometer inductance connected in series. The rotating plates of the condenser and the rotating coils of the variometer are mounted on the same shaft (controlled by the large dial), and are so balanced that rotating the one shaft changes simultaneously the inductance and capacity of the antenna circuit, thereby keeping the efficiency of the oscillating circuit practically constant throughout the entire range of the receiver. A single plate variable vernier condenser (controlled by lower left hand dial) is connected in parallel with the main condenser, and makes extremely fine tuning possible. Through the use of an adjustable tickler coil (controlled by the lower right hand dial), regenerative amplification may be obtained with a vacuum tube detector. This greatly increases sensitiveness and selectivity of the set.

Details of Construction

All connections are made at the back of the cabinet. Binding posts extend through the rear of the tuner and they are plainly marked by machine engraved insulating plates. The capacity effect of the operator's body on tuning is eliminated by means of a metal shield mounted on the back of the front panel, and connected to the ground circuit.

Panel—Micarta, dull satin finish. **Cabinet**—Height, 9½ inches; depth, 8½ inches; width, 6½ inches. Solid mahogany, varnished and polished. **Dials**—polished black Micarta with beveled edges. Markings filled in white. **Condensers**—Rotary plate type, air dielectric. **Wiring diagram** showing all connections is included, together with complete instructions for installing and operating. Net weight, 6 pounds. Shipping weight, 10 pounds.

563 J 622 . . . . . . **$68.00**

Load Coil for Use with Type R. C. Receiver or R. A. Tuner

The addition of this coil to either the R. C. or R. A. instruments, increases the receiving range, making possible the reception of signals having wavelengths from 1600 to 2800 meters. It is readily attached to two binding posts at the rear of the cabinet. Ship. wt., 1 lb.

63 J 6301 . . . . . . **$6.00**

Westinghouse D. A. Detector and Two-Stage Amplifier

This instrument can be used in conjunction with any type of tuner—loose coupler, honeycomb coils or regenerative tuner. It provides a vacuum tube detector and two stages of audio frequency amplification. The results obtained with it are vastly superior to the results obtained by a crystal detector under the same conditions, the signals as given out by the second stage of amplification being at least one hundred times louder than on a crystal detector. It is this type of instrument, used in connection with a regenerative tuner such as the R. A. set listed above, that makes possible the wonderful results obtained by modern radio. Two such instruments working together have received radio telephone messages from stations 1000 or more miles distant. (See description on opposite page.)

How It Works

The current is first passed to the detector tube, from which it passes on to an amplification transformer, which steps up the current and delivers it to the first amplifying tube. The incoming messages at this point already are greatly increased, but are further amplified by passing through another amplifying transformer, from whence the current is delivered to a second amplifying tube, and the current when delivered from this tube has been stepped up so loud and clear that messages even from far distant stations can be heard very distinctly, either in the telephone headsets or through the loud speaker. Two rheostats—one controlling the detector tube, the other controlling the two amplifier tubes—give perfect control of the filament currents. Three telephone jacks mounted on the panel enable the signal to be heard either by the detector or first or second stage of amplification by inserting the plug in the corresponding jack. All binding posts are located at the rear of this instrument, and are so arranged that direct wiring connections can be made with the R. A. tuner. The three sockets for holding any standard vacuum tube are mounted on a shockproof shelf which absorbs vibrations that otherwise would be transmitted to the tubes and introduce undesirable noises.

Panel—Micarta, dull satin finish.

The cabinet is 9½ inches high, 8½ inches deep and 6½ inches wide, solid mahogany, varnished and polished. The instrument is completely shielded on all sides, entirely eliminating capacity effects from operator's body.

Wiring diagram showing all connections is included, together with complete instructions for installation and operation.

No batteries, tubes or headphones included. Net weight, 10 pounds. Shipping weight, 13 pounds.

563 J 624 . . . . . . **$70.00**

Follow the Big Games by Radio

Only a few thousand actually will see the games, but millions will sit in grandstand seats in their own homes, and with their radios "tuned in," listen for the good word that Ruth or Hornsby or Walker has "clouted 'em" into the championship.

Last year, for the first time in history, fans miles from the scene of the battle "watched" the game, play by play, as it came to them over the wireless telephone.

This year you who own radio outfits are in the midst of the world of sports no matter where you live. Broadcasting stations all over the country will relay to you the news of the games. Not only baseball, but football, basket ball, polo, hockey, golf—all the sporting news is yours first hand.

A radio outfit keeps you "in step"—it gives you a grandstand seat for the biggest sporting events of the season, the country over.

A late 1922 Montgomery Ward catalog shows typical regenerative models *the most fashionable and popular of winter sports. Home, it seemed, couldn't be home without a radio set.*

But no radio sets were on the market!

Manufacturers and dealers had not foreseen such a demand. After the first rush nothing was left for the hundreds of frantic radio customers save "bootleg stuff"—sets rebuilt or manufactured in defiance of patent restrictions. And all the while the finest opera in America literally was wasting its fragrance on the desert air.

Came then the small boy to the rescue. He is the hero of Chicago's radio drama, the small boy is.

Just as their grandfathers fiddled with bicycles and their fathers with automobiles, the young Chicagoans of 1921 began fiddling with the radiophone. And presently the number of radio sets in KYW's field had tripled, although the dealers in electrical supplies had only one answer for customers: "We haven't a thing; perhaps next month. . ."

The tremendous enthusiasm of the youngsters threw an embarrassing burden on the public schools. Boys who had been yawning through their physics and manual training classes came suddenly to life as the enthusiasm for radio spread. "For a while I was almost ashamed to go to school in the morning," one high school instructor confessed, "because the boys were shooting over my head. I had to do the hardest sort of grinding before I could face them. They took to the business like ducks, and were speaking the lingo with the fluency of experts before the radio fever was a month old."

At the close of the opera season the number of sets in use in the city of Chicago had increased from 1,300 to something like 20,000. And you have the word of experts for it that boys of high school age were responsible for at least 75 per cent of the increase. Grown-ups occasionally took flyers in the new game, but those who made the home sets work, who contrived makeshifts to take the place of unavailable gear, were mechanics in short trousers.

The departure of the opera company left the Westinghouse company in a quandary. By broadcasting the music of Miss Garden's organization, KYW had established a radio audience of thousands. That audience wasn't going to wait ten months for another opera season; it wanted entertainment without delay. Undeniably it was KYW's move.

A musical director and a staff of performers were engaged. The newspapers, by now awake to the fact that radio was claiming as much public interest as baseball and divorce, offered cooperation. And when the curtain fell on the last operatic performance of the winter, KYW was ready with an all-day broadcasting programme. Twelve-hour service has been given daily ever since, and will be maintained.

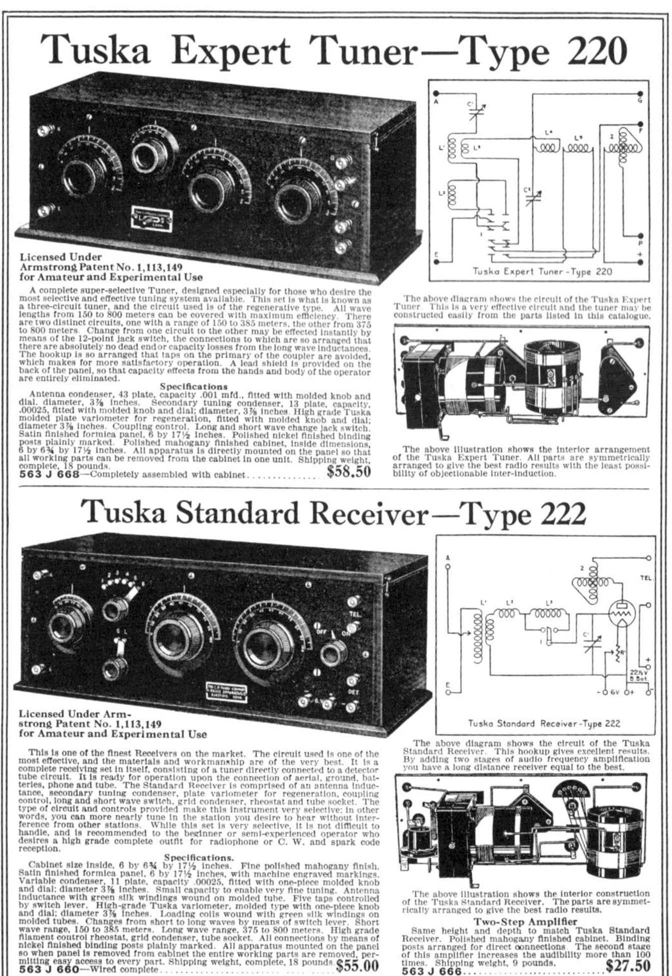

Tuska Expert Tuner—Type 220

Licensed Under Armstrong Patent No. 1,113,149 for Amateur and Experimental Use

A complete super-selective Tuner, designed especially for those who desire the most selective and effective tuning system available. This set is what is known as a three-circuit tuner, and the circuit used is of the regenerative type. All wave lengths from 150 to 800 meters can be covered with maximum efficiency. There are two distinct circuits, one with a range of 150 to 385 meters, the other from 375 to 800 meters. Change from one circuit to the other may be effected instantly by means of the 12-point jack switch, the connections to which are so arranged that there are absolutely no dead end or capacity losses from the long wave inductances. The hookup is so arranged that taps on the primary of the coupler are avoided, which makes for more satisfactory operation. A lead shield is provided on the back of the panel, so that capacity effects from the hands and body of the operator are entirely eliminated.

Specifications

Antenna condenser, 43 plate, capacity .001 mfd., fitted with molded knob and dial, diameter, 3⅞ inches. Secondary tuning condenser, 13 plate, capacity, .00025, fitted with molded knob and dial; diameter, 3⅞ inches. High grade Tuska molded plate variometer for regeneration, fitted with molded knob and dial; diameter 3⅞ inches. Coupling control. Long and short wave change jack switch. Satin finished formica panel, 6 by 17½ inches. Polished nickel finished binding posts plainly marked. Polished mahogany finished cabinet, inside dimensions, 6 by 6¾ by 17½ inches. All apparatus is directly mounted on the panel so that all working parts can be removed from the cabinet in one unit. Shipping weight, complete, 18 pounds.

563 J 668—Completely assembled with cabinet **$58.50**

The above diagram shows the circuit of the Tuska Expert Tuner. This is a very effective circuit and the tuner may be constructed easily from the parts listed in this catalogue.

The above illustration shows the interior arrangement of the Tuska Expert Tuner. All parts are symmetrically arranged to give the best radio results with the least possibility of objectionable inter-induction.

Tuska Standard Receiver—Type 222

Licensed Under Armstrong Patent No. 1,113,149 for Amateur and Experimental Use

This is one of the finest Receivers on the market. The circuit used is one of the most effective, and the materials and workmanship are of the very best. It is a complete receiving set in itself, consisting of a tuner directly connected to a detector tube circuit. It is ready for operation upon the connection of aerial, ground, batteries, phone and tube. The Standard Receiver is comprised of an antenna inductance, secondary tuning condenser, plate variometer for regeneration, coupling control, long and short wave switch, grid condenser, rheostat and tube socket. The type of circuit and controls provided make this instrument very selective; in other words, you can more nearly tune in the station you desire to hear without interference from other stations. While this set is very selective, it is not difficult to handle, and is recommended to the beginner or semi-experienced operator who desires a high grade complete outfit for radiophone or C. W. and spark code reception.

Specifications.

Cabinet size inside, 6 by 6¾ by 17½ inches. Fine polished mahogany finish. Satin finished formica panel, 6 by 17½ inches, with machine engraved markings. Variable condenser, 11 plate, capacity .00025, fitted with one-piece molded knob and dial; diameter 3⅞ inches. Small capacity to enable very fine tuning. Antenna inductance with green silk windings wound on molded tube. Five taps controlled by switch lever. High-grade Tuska variometer, molded type with one-piece knob and dial; diameter 3⅞ inches. Loading coils wound with green silk windings on molded tubes. Changes from short to long waves by means of switch lever. Short wave range, 150 to 385 meters. Long wave range, 375 to 800 meters. High grade filament control rheostat, grid condenser, tube socket. All connections by means of nickel finished binding posts plainly marked. All apparatus mounted on the panel so when panel is removed from cabinet the entire working parts are removed, permitting easy access to every part. Shipping weight, complete, 18 pounds.

563 J 660—Wired complete **$55.00**

The above diagram shows the circuit of the Tuska Standard Receiver. This hookup gives excellent results. By adding two stages of audio frequency amplification you have a long distance receiver equal to the best.

The above illustration shows the interior construction of the Tuska Standard Receiver. The parts are symmetrically arranged to give the best radio results.

Two-Step Amplifier

Same height and depth to match Tuska Standard Receiver. Polished mahogany finished cabinet. Binding posts arranged for direct connections. The second stage of this amplifier increases the audibility more than 100 times. Shipping weight, 9 pounds.

563 J 666 . . . . **$27.50**

All merchandise in this catalogue shipped from Northern Illinois — Montgomery Ward & Co.

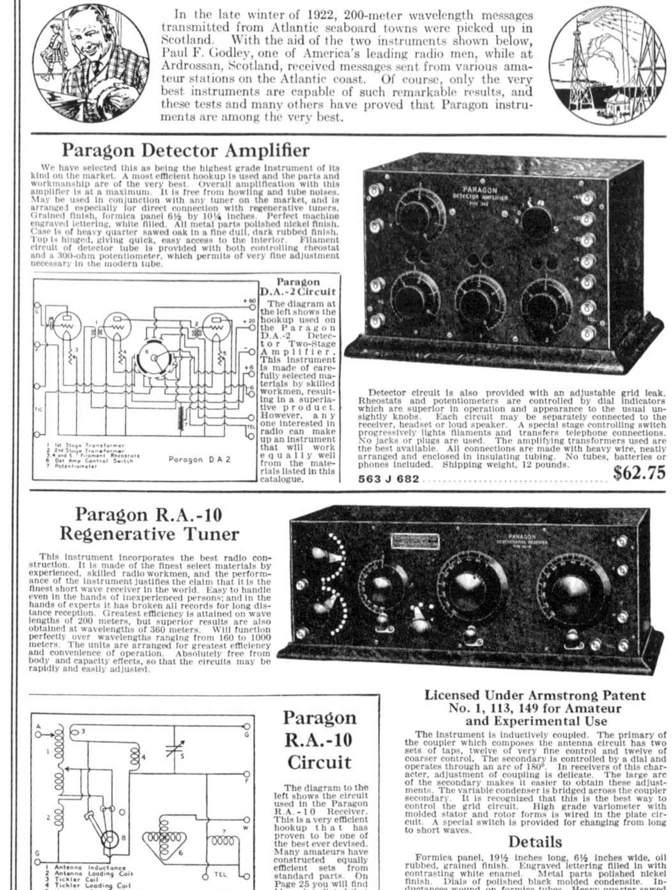

Scotland Listens to U. S. with a Paragon

In the late winter of 1922, 200-meter wavelength messages transmitted from Atlantic seaboard towns were picked up in Scotland. With the aid of the two instruments shown below, Paul F. Godley, one of America's leading radio men, while at Ardrossan, Scotland, received messages sent from various amateur stations on the Atlantic coast. Of course, only the very best instruments are capable of such remarkable results, and these tests and many others have proved that Paragon instruments are among the very best.

Paragon Detector Amplifier

We have selected this as being the highest grade instrument of its kind on the market. A most efficient hookup is used and the parts and workmanship are of the very best. Overall amplification with this amplifier is at a maximum. It is free from howling and tube noises. May be used in conjunction with any tuner on the market, and is arranged especially for direct connection with regenerative tuners. Grained finish, formica panel 6½ by 10¼ inches. Perfect machine engraved lettering, white filled. All metal parts polished nickel finish. Case is of heavy quarter sawed oak in a fine dull, dark rubbed finish. Top is hinged, giving quick, easy access to the interior. Filament circuit of detector tube is provided with both controlling rheostat and a 300-ohm potentiometer, which permits of very fine adjustment necessary in the modern tube.

Paragon D.A.-2 Circuit

The diagram at the left shows the hookup used on the Paragon D.A.-2 Detector Two-Stage Amplifier. This instrument is made of carefully selected materials by skilled workmen, resulting in a superlative product. However, any one interested in radio can make up an instrument that will work equally well from the materials listed in this catalogue.

Detector circuit is also provided with an adjustable grid leak. Rheostats and potentiometers are controlled by dial indicators which are superior in operation and appearance to the usual unsightly knobs. Each circuit may be separately connected to the receiver, headset or loud speaker. A special stage controlling switch progressively lights filaments and transfers telephone connections. No jacks or plugs are used. The amplifying transformers used are the best available. All connections are made with heavy wire, neatly arranged and enclosed in insulating tubing. No tubes, batteries or phones included. Shipping weight, 12 pounds.

563 J 682 . . . . **$62.75**

Paragon R.A.-10 Regenerative Tuner

This instrument incorporates the best radio construction. It is made of the finest select materials by experienced, skilled radio workmen, and the performance of the instrument justifies the claim that it is the finest short wave receiver in the world. Easy to handle even in the hands of inexperienced persons; and in the hands of experts it has broken all records for long distance reception. Greatest efficiency is attained on wave lengths of 200 meters, but superior results are also obtained at wavelengths of 360 meters. Will function perfectly over wavelengths ranging from 160 to 1000 meters. The units are arranged for greatest efficiency and convenience of operation. Absolutely free from body and capacity effects, so that the circuits may be rapidly and easily adjusted.

Paragon R.A.-10 Circuit

The diagram to the left shows the circuit used in the Paragon R.A.-10 Receiver. This is a very efficient hookup that has proven to be one of the best ever devised. Many amateurs have constructed equally efficient sets from standard parts. On Page 25 you will find blueprints listed that give complete information on similar hookups.

Licensed Under Armstrong Patent No. 1, 113, 149 for Amateur and Experimental Use

The instrument is inductively coupled. The primary of the coupler which composes the antenna circuit has two sets of taps, twelve of very fine control and twelve of coarser control. The secondary is controlled by a dial and operates through an arc of 180°. In receivers of this character, adjustment of coupling is delicate. The large arc of the secondary makes it easier to obtain these adjustments. The variable condenser is bridged across the coupler secondary. It is recognized that this is the best way to control the grid circuit. High grade variometer with molded stator and rotor forms is wired in the plate circuit. A special switch is provided for changing from long to short waves.

Details

Formica panel, 19½ inches long, 6½ inches wide, oil rubbed, grained finish. Engraved lettering filled in with contrasting white enamel. Metal parts polished nickel finish. Dials of polished black molded condensite. Inductances wound on formica tubes. Heavy quarter sawed oak cabinet, fine dark rubbed finish. Each dial fitted with a fine (vernier) control. Ship. wt., 20 pounds.

563 J 683 . . . . **$72.75**

At first, all broadcasting was done on the single wavelength of 360 meters (833 kHz) while stations would switch to 485 meters (619 kHz) for market reports or weather forecasts. High selectivity was therefore unnecessary. For local reception, a simple crystal set was sufficient. For greater distance the Armstrong regenerative circuit was tops, but for the unlucky manufacturers who had not gotten licenses in mid-1920 before Armstrong sold his patent to Westinghouse, RF amplification ahead of the detector stage was the next best thing. Early designs had one tuned circuit followed by several stages of RF amplification using iron-core transformers, tuned broadly to cover 360 and 485 meters. Even after the spring of 1922, when an additional wavelength of 400 meters (750 kHz) was set aside for a few of the better stations, these fixed-tuned-RF receivers were still adequate.

At this time, in fact for the entire decade, "homebrew" (Prohibition was brand new, too) radios made up a large percentage of the total in use. Statistics compiled by *Radio Retailing* (pp. xx & xxi) show roughly a million homemade sets built each year, outnumbering factory production by ten to one in 1922 and not surpassed until 1925.

By mid-decade there were 30 radio periodicals, most of which printed plans for building the latest radio circuits. Large city newspapers had their radio pages, or often, weekly radio supplements, likewise crammed with circuits to be built and advertisements for the parts necessary (and program listings, from stations coast to coast!). Imagine your present Sunday paper's TV supplement devoting two pages to program listings, half the remainder to ads, and the other half to features on how a TV works or how to build one yourself, and then imagine your teen-age son building it, and you'll have a fair idea of radio in the early twenties. Incidentally, if surviving specimens are any indication, many of these homebrew sets never worked when finished, if indeed they ever *were* finished.

20 THE RADIO GLOBE, SATURDAY, FEBRUARY 25, 1922.

Answers to Questions

(Continued From Fifteenth Page.)

the fire escape if you want to comply strictly with the rules. As you have a transmitting station, you will have to use a switch.

Radio Editor:—I have a crystal set and an aerial 100 feet long and 50 feet high. The set itself consists of a tuning coil, crystal detector, variable and fixed condenser and 3,000 ohm 'phones. Why is it that I do not hear the concerts from W D Y and W J Z?

S. L. J., Mt. Vernon.

If you are able to hear code messages you must have the set wired up correctly, and the only thing this department can suggest is for you to put in an audion bulb, as you are too far away to get the music from Newark.

Radio Editor:—With a Grebe CR 9 set installed in Long Beach, L. I., will I be able to hear Pittsburgh and Chicago? Can a third step of amplification be added to this set? Is there any broadcasting station located on Long Island? Is the Metropolitan Opera House going to erect a broadcasting station?

REGGIE DE VON DAIRE.

It is impossible to estimate the distance any set will work, as stated before in this column. You may be able to hear Pittsburgh with a suitable aerial. Any number of steps of amplification may be added, but unless they are wired very carefully all sorts of noises and squeals will be heard in the head set. Two steps are enough for any set. We have no knowledge of any broadcasting station on Long Island. While there have been rumors that the Metropolitan Opera is going to broadcast the operas, nothing has been officially given out on the subject.

RADIO MAN WANTED
With practical construction experience and ability to compose, compile and write data for radio magazine. Write stating experience, especially as to literary ability, and give references. Radio Budd, Box M 212, The Globe.

RADIO MANUFACTURING
For
Dealers, Experimenters, Students,
in any quantity, one or thousands, complete instruments or special parts. Estimates to Samples, Drawings, Descriptions. Experimentation under Customer's Supervision. Telephone "Beekman 0373"
Manufacturers' & Inventors' Electric Co., (Factory) 29 Gold St., New York.

DE FOREST RADIOPHONES
"EVERYMAN"............complete $25
"RADIOHOME"................ $36
(Less tube, phone, battery)
NOVO "B" BATTERIES
U. S. L. Distributors.
775 E. 169th St., at McKinley Sq.
Phone Bingham 4484.
Open Sundays until NOON.

"COPPERWELD"
TRADE MARK REG. U.S. PAT. OFF.
ANTENNA WIRE
The Super-Strength Copper
Better than Solid Copper or Strands.
Buy it in cartons.
100-FT. COILS 200-FT. COILS
At your Dealers.
COPPER CLAD STEEL CO.
30 CHURCH ST., NEW YORK

RadioFans!!
Bronx Headquarters
Wholesale and Retail
for Radio Equipment. We carry a full line of standard Radio supplies. Also complete receiving and transmitting sets. Estimates furnished on sets built to your own specifications.
U.V. 201—U.V. 202
Amplifying Tubes in Stock
Detectors, 2 Stage Amplifiers.
Finished Sets in Stock.
43 Plate Condensers.
Novo "B" Batteries in Stock.
IMMEDIATE DELIVERY
Dealers Supplied.
Bronx Radio Equipment Co.
687 Courtlandt Ave., at 154th St.
Melrose 1613 Radio Call 2-BXA

Radio Exchange

FOR SALE—Deforest everyman receiver, perfect condition; without phones, $15. C. W. KEPPER, 387 Webster Ave., Jersey City, N. J.

RADIO SETS—BUILT AND INSTALLED. Complete instructions for operating. Postal brings details. EWD. H. RANSON, 80 Monroe St., Brooklyn.

RADIOPHONE RECEIVING SET, ideal for local and long distance music broadcasting. Rare opportunity for beginner; act quick; will demonstrate. H. SACHS, 'phone Rhinelander 4725, 7 to 8 P. M.

LISTENING IN

Radio-chapel services seem to offer the best chance in a century that the church has had to "bring religion into the home." In this field, at least, religion and science can have no possible quarrel.

It is well to remember that an indoor aerial will not work for any distance with a crystal set, so do not attempt it. A single wire about 100 feet long and as high as it is possible to get it is the best aerial that can be used for receiving. Three- and four-wire aerials and the cage aerial will not help in receiving. Save your money and time by erecting a single wire aerial only.

Not all gas-pipes can be used as ground connections. In some houses an insulating joint is put in the pipe where it enters the house and consequently there is no connection with earth at all.

Once more this department is going to repeat the unfairest question in the radio game: "How far will this set receive?" This question cannot be answered by any one, as too much depends on the operator, local conditions and the power of the transmitting station at the other end. Readers are requested not to ask this question.

Galena comes in as many different grades as it is possible to imagine. Beginners who cannot hear any signals will do well to purchase several pieces of this mineral before complaining to this department. Sometimes by simply changing the mineral a set that has been a failure will at once become a wonderful success.

TUBING FOR WIRELESS
SET OF 6 CARDBOARD TUBES IN HEAVY MAILING TUBE FOR $1.00
Postage Prepaid.
Each Tube is 12" Long with ⅛" Wall.
Sizes are 3", 3½", 4", 4½", 5", 5½"
Any One Size 12" Long for 25 cents.
Rutherford Supply House
78 Chestnut St., Rutherford, N. J.

RADIO SETS AND WIRELESS SUPPLIES
Of the Highest Grade in Stock.
EXPERT TECHNICAL ADVICE.
Apparatus We Advise You to Purchase or Assemble Guaranteed to Give Satisfactory Results.
Buy From a Real Radio Organization.
HOYT ELECTRIC COMPANY
686 LEXINGTON AVE.
Between 56th & 57th Streets.
Phone Plaza 5866-7.

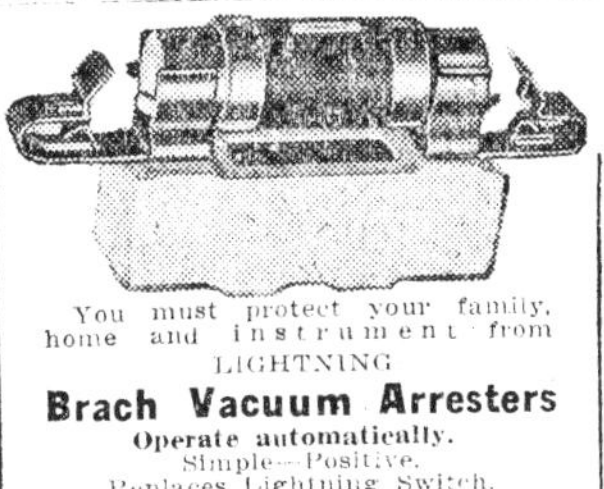

You must protect your family, home and instrument from LIGHTNING
Brach Vacuum Arresters
Operate automatically.
Simple—Positive.
Replaces Lightning Switch.
Sold by all Radio dealers.
Manufactured by
L. S. BRACH MANUF'G CO.
Newark, N. J.

PERFECT CONNECTIONS
are always sure with
Fahnestock
PATENT WIRE
Terminal

Grips like a vise.
Eliminates Vibrations
Easily Attached
Best of All for Radio Outfits.
At your dealer's, or direct from
FAHNESTOCK ELEC. CO.
East Ave. & 8th St., Long Island City.

In May 1923, however, stations began to be spread out over the band, rather than share time slots on two frequencies, and receivers needed more selectivity than one tuned circuit could provide (Armstrong regenerative sets, or sets that could be made to regenerate, could get sufficient selectivity from one tuned circuit, so some models using iron-core transformers did appear in 1923—the De Forest D10 or Federal 61, for instance). Neutrodynes came to the fore; about 100,000 were sold in 1923 and 1924, even at an average price of $150, popularizing the arrangement of three tuning dials across the panel. The following companies were licensed by Hazeltine to make Neutrodynes (for the full story of its development, see the individual chapters):

American Radio & Research Corp. (later bought by Crosley)
F.A.D. Andrea, Inc. (Fada)
Carloyd Electric & Radio Corp. (Malone-Lemmon)
Eagle Radio Corp.
Freed-Eisemann Radio Corp.
Garod Corp.
Radio Service Laboratories, Inc. (bought by Gilfillan)
Howard Mfg. Co.
Broadcast Mfrs, Inc. (acquired by King Quality Products, Inc.)
Wm. J. Murdock Co. (bought by Philco)
Stromberg-Carlson Telephone Mfg. Co.
R.E. Thompson Mfg. Co.
Ware Radio Corp.
Workrite Mfg. Co.

It didn't take Atwater Kent's engineers very long to devise a three-dial TRF that used wiring capacitance in place of physical neutralizing capacitors, a receiver he could sell profitably for just $100. He made more than 100,000 such breadboards, and nearly five times that many cabinet models, by 1925 eclipsing all the Neutrodyne makers and capturing their markets.

However, there was still plenty of room for cheaper models, and Charles Freshman moved in to fill the gap in August 1924 with his Masterpiece. By cutting corners wherever possible, he built his set to sell for only $60, and in his first seven months sold 125,000 of them. Next season, the fall of 1925, the market was flooded not only by Masterpieces, but also by dozens of knock-offs, all sporting the three large tuning dials.

AC Radios

The "newly-perfected AC sets" were headlined in 1924 . . . and 1925, 1926, and 1927. Obviously each year saw the arrival, not of perfection, but of a different expedient: first it was DC tubes running entirely from a rectifier power pack, then crude attempts at AC-heater tubes, finally the introduction of practical AC-heater tubes by RCA.

In 1924 came the first entries. While Péricaud in France had its Radio-Secteur ("power-line radio") in February; in the United States, the Dynergy was earliest, announced in June and advertised in July 1924. It was nothing more than a three-dial TRF with series-wired 01As, running on 110 VDC (common in New York where the set was sold) or on DC supplied by a power pack with Tungar rectifiers. But unlike many later entries, it *was* actually sold. Other 1924 models: Zenith's Super-X in September (advertising claims notwithstanding, made only as a battery set) or Mu-Rad's MA-20 in October. In the same class is the World, advertised in Los Angeles in February 1925.

But the first really practical AC set (one that worked well and reliably, and sold in quantity) was RCA's Radiola 30, from September 1925. Expensive and cumbersome, yes, but a number are still running today. Of the nine other AC sets featured in *Radio Retailing* in October, possibly one or two ever hit the market.

Frederick McCullough's indirectly-heated AC tube, as produced by Kellogg and advertised in January 1926, opened the way for a new generation of AC radios, although Kellogg would not use the tube in its *own* receivers until 1927. Other AC tubes followed, notably the Arcturus 15-volt series, and Sparton's Cardon tubes, but most of these died when in mid-1927 RCA began licensing the industry under its patents, because clause nine in the contract required that all radios be equipped initially with RCA tubes. Even though the clause was never enforced, few manufacturers would risk basing their product lines on a tube type that might vanish at any time, so RCA types continued to be the industry standards (Sparton made its own tubes, named Cardon after Capt. Sparks' grandsons Carter and Donald).

1927 saw the introduction of truly modern AC sets: the Radiola 17 in September, closely followed by models from most other manufacturers, all using the RCA 226 and 227 tubes.

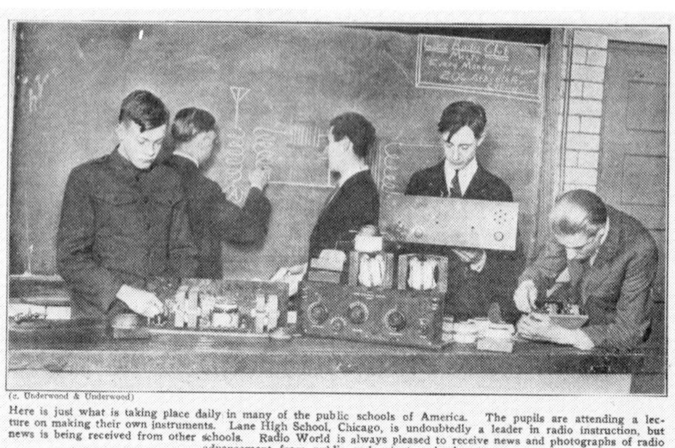
(c. Underwood & Underwood)

Here is just what is taking place daily in many of the public schools of America. The pupils are attending a lecture on making their own instruments. Lane High School, Chicago, is undoubtedly a leader in radio instruction, but news is being received from other schools. Radio World is always pleased to receive news and photographs of radio advancement from public and private schools.

Radio World (May 27, 1922), p. 17

Radio at Lane Technical High School

Superheterodynes

As early as 1922 the superheterodyne (from *super*audible *heterodyne*, transforming RF to a superaudible intermediate frequency to be amplified, before transforming again to audio) was being called the Rolls Royce method of reception, but it was far too complex and expensive for the average fan. Armstrong made the first practical application, during the war, and received patent 1,342,885 in 1920 which he sold to Westinghouse, though by 1929 had

STORE OPENS AT 9 A. M., CLOSES AT 6.00 P. M.

SUNDAY, DECEMBER 2, 1923

SNELLENBURGS

ENTIRE BLOCK—MARKET 11TH TO 12TH STREETS

What to Give the Radio Fan for Christmas

Exceptional Offerings in Our Radio Store Tomorrow

The New Atwater-Kent Radio-dyne Set Is Here at . . . $88

This set is right up to the minute—in fact it is three leaps ahead of the crowd.

The new Atwater Kent Radiodyne Set is thoroughly complete in every particular and possesses all the latest improvements that mean the utmost in radio satisfaction. In short, it is the final word in radio perfection. But come in and see it for yourself—and be convinced.

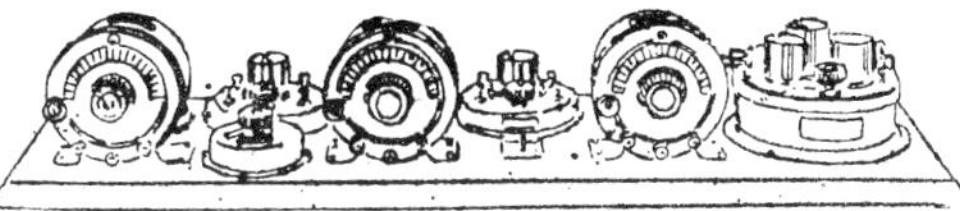

Crosley V Set, Complete, $36.75

For appearance and service this set cannot be equaled for less than twice our price.

This is the new model V, with an enlarged cabinet and up-to-the-minute improvements. All batteries are enclosed in the cabinet.

Complete with 2000-ohm head set, WD-12 tube, A battery, 1½-volt—B battery, 22½-volt, 100 ft. aerial wire, 50 ft. lead wire, 2 insulators, 2 nailet knobs, 1 porcelain tube, 1 ground clamp.

Mail and phone orders filled while they last. Sold on the Club Plan if desired.

$65.00 Westinghouse Aeriola, Sr. $39.50

Complete with WD11 tube, Brandes Superior Phone. This is the most popular single tube radio set ever built at any price. Receives over 1000 miles when properly installed and operated.

"Edison," the World's Greatest Storage Battery

Practically indestructible. Steel jars. Nickel plated.

$50.00 Batteries, Special at $22.50

10-volt, 40-ampere batteries in steel case.

$100 Batteries, Special, at $42.50

6-volt, 150-ampere.

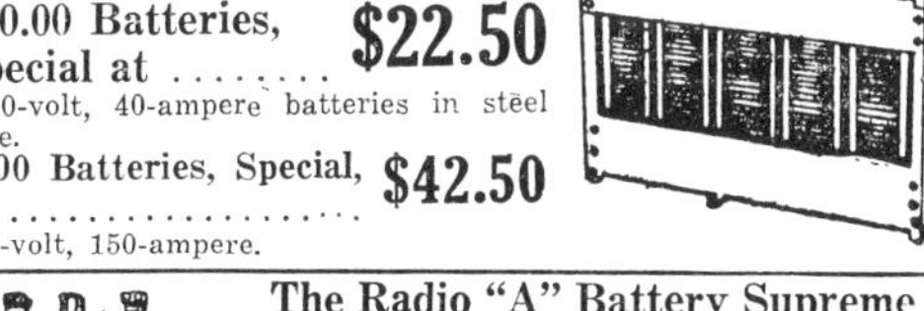

The Radio "A" Battery Supreme

A Storage Battery Without an Equal!

$1.00 Per Cell

To take the place of dry cells for low voltage tubes as well as the 6-volt tubes.

This is the lightest weight storage battery known.

The pyraline cells are clear so you can see the level of the solution at all times, thereby preventing dry plates and ruined batteries.

Each cell is 2 volts 20 amperes.

Any voltage or amperage can be made up to suit your individual needs.

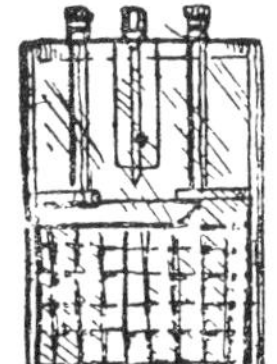

Eastern Type Crystal Radio Sets

Greatly Reduced to

$10.00

These sets are guaranteed to receive all local stations within a radius of 25 miles. Complete with 2000 ohm head set and aerial equipment. A most extraordinary value!

SNELLENBURGS Third Floor

Phila. Inquirer (Nov. 4, 1923)

Freed-Eisemann NR-5 April 1923 $150
The most popular Neutrodyne

lost every claim in it to prior inventors, notably Lucien Levy, backed by AT&T (Aitken, *The Continuous Wave*, p.467; 378 O.G. 736).

While several concerns sold superhet plans, parts, and kits from 1922 on, only RCA (cross-licensed by Westinghouse and AT&T) could legally build them. Its Radiola Superheterodynes, designed in part by Armstrong, appeared in early 1924, at which point it sued most of the kit producers and successfully clamped the lid on them. RCA eventually licensed other makers under the superhet patents in mid-1930.

One-Dial Tuning

While technically-oriented fans liked nothing better than a dozen knobs to twiddle, the public wanted something less complicated. The Armstrong regenerative circuit by its nature required careful manipulation, and little could be done to simplify it, but regens were losing favor to TRFs. TRFs were a more docile breed, that only needed three tuning dials to be set to the proper point, and one or two rheostats turned up until the music was loud enough. In spite of jokes about three-handed operators, these sets were easy to use.

Masterpiece Aug. 1924 $60

Ralph & Elinor Williams

The Radio Sun and Globe (Dec. 15, 1923)

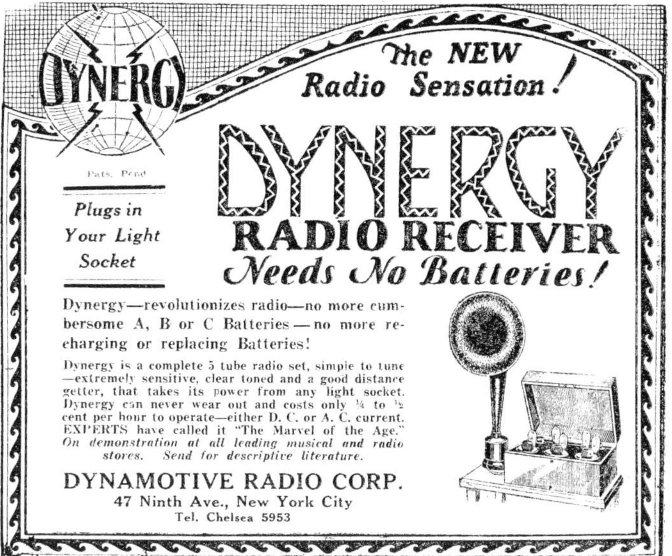

N.Y. Herald-Tribune (July 13, 1924)

Dynergy RC250 July 1924 $185 (DC) $235 (AC)

Invented by Samuel P. Levenberg of New York, who applied for a patent on Dec. 8, 1923 (1,670,893, May 22, 1928). The basic set ran from 110VDC, with a separate power pack to convert 110VAC to DC. Dynamotive failed in February 1925.

But they could be made easier. Since the second and third tuners generally tracked together, they could be combined, allowing a two-dial radio. And, with varying degrees of compromise, all three (or four) tuners could similarly be tied together. The first manufacturer to do this was Thermiodyne, in July 1924, although trimmer controls had to be provided on the panel since the four stages did not track perfectly. Magnavox did the same thing in September, hiding its trimmers inside the cabinet. Only the Mohawk, advertised in November 1924, had no trimmers, so was the first "one-dial" radio.

There was no great rush to jump on the bandwagon. The one-dial set *was* a compromise. And buyers expected to see three large tuning dials spread across the panel—it was that year's style. A year later, the 1926–27 season, when open dials were old hat, window dials or pointers came into vogue. For 1927–28, nearly everyone used one dial, behind a large ornate escutcheon. This advance was made possible partly by mechanical and electrical improvements, partly by cheaper tubes: each amplifier stage could be tuned less critically and run at a lower gain, with an additional tube making up the loss; or the extra tube could be used as an untuned antenna-coupling device with no gain, allowing all tuned stages to track closely together.

Naturally not everyone followed the trends: Zenith for instance was two years ahead in using pointers, while Atwater Kent clung to open dials ("full-vision") well into 1929!

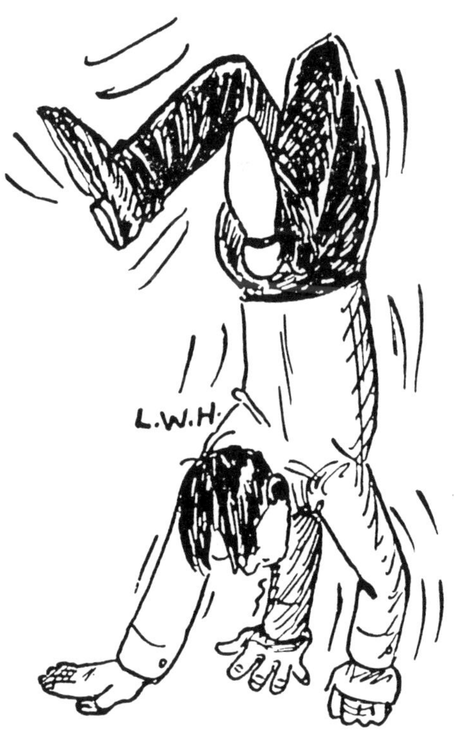

QST (May 1925)

NEUTRODYNE NEWT
THE THREE ARMED WONDER

Radiola 30 Sept. 1925
$575 with tubes

Rich Elskamp

AC TUBES

In the 1924–27 period, before AC sets were fully established, B-battery eliminators rang up more than $30 million in sales, the most popular model being the Majestic Super-B.

A-battery eliminators were less successful, as no really good rectifier was available for the high currents needed (1½ amperes for a six-tube radio).

Radio Retailing (Oct. 1925), p. 493

"Lamp Socket" Radio to Have Important Place

Many Efficient Receivers Designed Especially to Operate from A.C. or D.C. House Current Have Been Announced

On this page is a review of the leading "batteryless" radio receivers which were on exhibition at New York's radio shows last month. It is the concensus of the radio trade that battery elimination will rapidly grow in popularity as the degree of perfection in this field is reached and that radio dealers should give it serious attention this season.

Miessner Radio Corporation, Brooklyn, N. Y. A.C.—$175

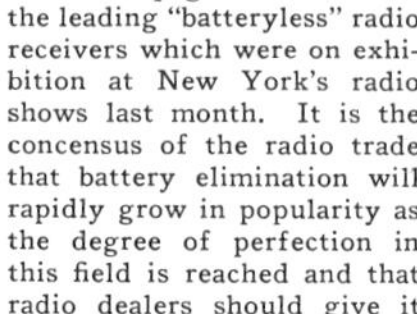

Radio Receptor Company, New York A.C.—$185 D.C.—$135

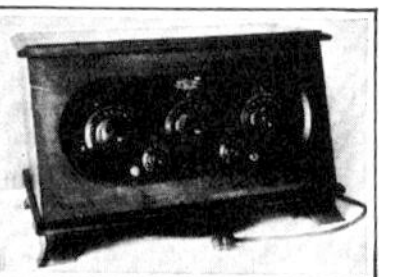

Batteryless Radio Corporation, New York, N. Y. A.C.—$200 D.C.—$140

Andrews Radio Corporation, Chicago, Ill. A.C.—$365

Argus Radio Company, New York, N. Y. A.C.—$160

San Pedro Radio Laboratories, San Pedro, Cal. A.C.—$100

*Radio Corporation of America, New York, N. Y. A.C.—$245**

Peerless Radio Corporation, Chicago, Ill. A.C.—$125

Radio Corporation of America, New York, N. Y. A.C.—$575

Powerola Radio Corporation, New York, N. Y. A.C.—$165 D.C.—$115

Music Master Corporation, Philadelphia, Pa. A.C.—$400

*This is a speaker operated from 110-volt, 60-cycle house current which, when used with a Radiola 25 or 28, replaces all batteries, and when used with any other set, supplies the B battery voltage.

Radiola 60 Sept. 1928 $175
The first modern AC superhet, using chassis construction much like the TRF Radiola 17 of a year earlier.

Photo by
Rich Wolven

Thermiodyne TF6 July 1924 $140

Mohawk 100 Nov. 1924 $100

RADIO RETAILING, *A McGraw-Hill Publication*

What's New in Radio and Where to Buy It

News of Latest Products Gathered by the Editors

This editorial section is prepared purely as a news service, to keep readers of "Radio Retailing" informed of new products on the market.

It should be noted that all announcements appearing on these pages are published without advertising considerations of any kind whatsoever.

Five-Tube Portable Receiver

Radio Retailing, July, 1925

The new circuit of the DeForest Radio Company, Central Avenue and Franklin Street, Jersey City, N. J., consists of two stages of tuned radio frequency amplification, detector with tuned input and two stages of transformer coupled audio frequency amplification. It is a non-radiating and non-oscillating circuit, with the setting of the three tuning controls practically the same for a given station. The F-5 portable incorporating this circuit is entirely self contained and uses the new DeForest cone reproducer as shown in the illustration. The dimensions are 15¼ in. x 14 in. x 11 in. when closed for carrying and weighs 37 pounds complete with batteries. 150 feet of flexible antenna wire are wound on a built-in aluminum reel in addition to 15 feet of flexible ground wire conveniently assembled in a spring clip. Intended retail price is $130. This same circuit is used in two other non-portable models which have intended retail prices of $85 and $90. Both are in walnut cabinets but the latter has B battery compartments.

Four-Tube Receiver

Radio Retailing, July, 1925

David Grimes, Incorporated, 1571 Broadway, New York, has brought out the new model of inverse duplex receiver as shown in the illustration. Included in this model are two stages of tuned radio frequency amplification, tube detector and three stages of audio frequency amplification although the set actually uses only four tubes. Straight line condensers and a new form of inductance are used in the circuit. Intended retail price, less accessories, for the type illustrated, $125. The Baby Grand model, a three tube reflex set has an intended retail price of $49.50. Both models use UV-199 tubes.

Combination Book Case and Cabinet

Radio Retailing, July, 1925

Constructed along the same lines as the ordinary sectional book case, and made in either quartered oak or birch mahogany finishes, the F. E. Hale Manufacturing Company, Herkimer, N. Y., has added a combination book case and radio cabinet to its many types of radio furniture. If it is desired to place the batteries in the basement, two book sections filled with books may be placed under the radio section instead of the battery section. The total height of the four pieces making the assembly of the radio cabinet is 43¾ in. The intended retail price is $32.

Five-Tube Receiver

Radio Retailing, July, 1925

The "Radiodyne WC-17 A" is one of the latest products to be announced by the Western Coil and Electrical Company, Racine, Wis. Two stages of tuned radio frequency are used with only two tuning dials. The receiver is furnished with an ebony effect, satin Duco finish. The panel is of highly polished Micarta, engraved and inlaid in gold. Overall dimensions are 22 in. x 7½ in. x 9 in. Either an inside, outside or lamp socket antenna may be used with the set. The intended retail price, less accessories, is $75.

Five-Tube Receiver

Radio Retailing, July, 1925

Selectivity, true-tone reproduction and ease of operation are three outstanding features of the No. 830 receiver made by the Globe Electric Company, 14 Keefe Avenue, Milwaukee, Wis. It is a five tube receiver using two stages of tuned radio frequency amplification. The cabinet is of genuine mahogany with a panel opening of 7 in. x 21 in. Throughout the receiver all construction is of bakelite, with heavy metal brackets. The intended retail price is $65.

Six-Tube Receiver

Radio Retailing, July, 1925

The R. E. Thompson Manufacturing Company, 30 Church Street, New York City, has brought out a six-tube neutrodyne receiver known as the "Concert Grand," type S-70. It employs two stages of tuned-radio frequency amplification, a detector and three stages of audio-frequency amplification. The cabinet is of mahogany with a sloping panel, and has overall dimensions of 29 in. x 14 in. x 13¾ in. Special adapters for UV-199 tubes are furnished with the set. Intended retail price is $180.

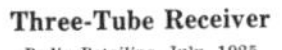

Three-Tube Receiver

Radio Retailing, July, 1925

The "Ray Isler" is a three-tube receiver made by the Ray Isler Radio Company, 1021 Park Avenue, Rochester, N. Y. The circuit uses a detector and two stages of audio frequency amplification. A feature pointed out by the maker is that the set will operate on either indoor or outdoor antenna. The cabinet, finished in walnut, has outside dimensions of 22 in. x 8 in. Intended retail price $70.

Triple Socket

Radio Retailing, July, 1925

Made of molded "Ballsite," for either panel or base mounting, with phosphor bronze double contacts, flush bottom, and highly polished, nickel plated brass tube, is the socket produced by the Ballston Insulating and Composition Company, Inc., Ballston Spa, New York. The Ballston company also makes of the same material, single sockets, dials, reamed for ¼-in. shaft and furnished with a case hardened steel set screws, in sizes from a 2-in. to a 4-in. diam., and several styles of knobs, as well as binding post tops.

Three-tube Reflex Console Type Receiver

Radio Retailing, July, 1925

The Console model 419-3 built by the Electrical Research and Manufacturing Company, Waterloo, Iowa, is a three-tube reflex receiver using either dry cell or storage battery tubes. It is non-radiating and covers a wave band frequency of from 200 to 600 meters. There are compartments for log book, magazines, head sets, etc. The "Armstrong Symphony" reproducer is a built-in feature of the solid walnut cabinet. Intended retail price, without tubes or batteries, is $250.

Five-Tube Receiver

Radio Retailing, July, 1925

The "Hollywood Five" employing two stages of tuned radio frequency, detector, and two stages of audio frequency amplification, has been placed on the market by the Hollywood Radio Company, 154 Nassau Street, New York City. Three dials for tuning are provided, as well as a rheostat and potentiometer control. The wave band frequency covered is from 200 to 570 meters. The set is encased in a mahogany cabinet and wired with flexible braided wire. Intended retail price, $45.

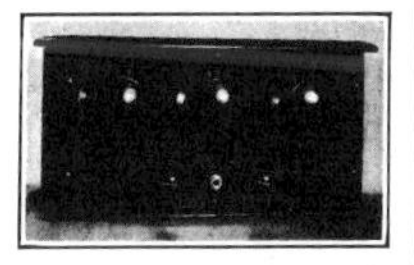

Five-Tube Receiver

Radio Retailing, July, 1925

The "Kenman-5", made by the Kenman Electric Company, 454 Merrick Road, Lynbrook, N. Y., is a five-tube receiver with low-loss construction throughout. The wave-band frequency covered is from 220 to 550 meters. Intended retail price is $37.50.

Four-Tube Portable

Radio Retailing, July, 1925

The "Ozarka" portable made by Ozarka, Inc., 855 Washington Boulevard, Chicago, Ill., is a four-tube tuned radio frequency set using UV-199 tubes. The loudspeaker is built in the top of the case, which is fabrikoid-covered. The set is furnished complete with aerial equipment, and has a weight of 25 lb. The intended retail price is $75.

Spring Bumper for Radio Cabinet

Radio Retailing, July, 1925

A self leveling spring bumper with a very heavy rubber insert, to protect any furniture upon which the cabinet may be placed, is being made by the Lakeside Supply Company, 73 West Van Buren Street, Chicago, Ill. The "Protecto-Groove" is also made by this company and is a device used to keep the grooves in the cabinet in which the panel fits from breaking out. It is arranged so that either ⅛-in., 3/16-in. or ¼-in. panels can be used, and is furnished in nickle, gold or any other color to match either the panel or the cabinet.

A, B, and C Batteries

Radio Retailing, July, 1925

Dry cell B and C batteries and storage A batteries are recent productions of the Music Master Corporation, 130 North Tenth Street, Philadelphia, Pa. The B batteries are made in 22½ and 45-volt sizes, with intended retail prices of $2 and $3.75; while the C battery is a 4½-volt battery with an intended retail price of 60c. All the batteries are unconditionally guaranteed. Two sizes of 6-volt storage batteries are also made of 90 and 110 ampere hour capacities. They are incased in mahoganized boxes having the top of the battery sealed with a moulded hard rubber lid so that no metal parts are exposed.

Radio Table

Radio Retailing, July, 1925

The Wasmuth-Goodrich Company, Peru, Ind., has recently placed on the market the radio table shown in the illustration. It is finished in duo-tone hand-rubbed varnish finish, with the top and front made of 5-ply select mahogany or American walnut. The size of the top is 14 in. x 30 in. and with both ends raised it is 47 in. long. The battery compartment is 10 in. x 24 in. x 13¼ in. Intended retail price is $25.

Combined Lamp and Reproducer

Radio Retailing, July, 1925

The "Peerlite" is a product brought out recently by the Peerless Light Company, 539 Broadway, New York, which combines a high class reading lamp with a practical reproducer. The vase comes in a number of colors, including orange, maroon, black and jade, with a stippled gold base. It is fitted with two-light cluster on each side of the fiber horn. The shade is said to be unique because of its exquisite glow when the light strikes it from the outside, due to the imported bead overlay over the hand painted surface. The intended retail price complete is $32.50.

Radio Retailing (July 1925), pp. 56–57

Typical designs of the 1925–26 season

Radio Receiving Sets Manufacturers Are Marketing for the Season 1926-27

Table Type, $38 to $85

Crosley Radio Corp., $38

Automatic Radio Mfg. Co., $45

Diamond T Radio Mfrs., $49.50

Reichmann Co., $50

Keystone Radio Labs., $55

American Specialty Co., $59.50

Amrad Corp., $60

Rhason Mfg. Co., $65

Murdock Co., W. J., $65

U. S. L. Radio, Inc., $75

Indiana Mfg. & Electric Co., $75

Malone-Lemmon Products Corp., $78

Beacon Radio Mfg. Co., $75

Neutrowound Radio Mfg. Co., $85

Table Type, $90 to $450

Cardinola Radio Corp., $90

Anylite Electric Co., $100

Roth-Downs Mfg. Co., $100

Wells Mfg. Co., $100

Radio Corp. of America, $115

A. C. Electric Mfg. Co., $135

Air-way Electric Appliance Corp., $137.50

Advance Radio Research Lab., $140

Distantone Radios, Inc., $150

Mazda Radio Mfg. Co., $175

Machen Mfg. & Distribution Co., F. E., $180

Andrea, Inc., F. A. D., $300

Ferguson, J. B., $226

Golden-Leutz, Inc., $450

Radio Retailing, A McGraw-Hill Publication — *Radio Retailing, July, 1926*

Radio Receiving Sets, 1926-27 (continued)

Console Type, $89.50 to $190

Diamond T Radio Mfg. Co., $89.50

Crosley Radio Corp., $90

United Electric Co., $100

Barty Radio Inc., $110

Gormac Electric Co., $115

Farmers Wireless Telephone Co., $115

Larkin Co., $120

Roth-Downs Mfg. Co., $130

Rhason Mfg. Co., $136

A. C. Electrical Mfg. Co., $139

International Radio Corp., $160

Reichmann Company, $185

Simplex Radio Co., $185

International Radio Corp., $190

Radio Receiving Sets, 1926-27 (continued)

Console Type, $195 to $610

Simplex Radio Co., $195

Federal Radio Corp., $200

Stromberg-Carlson, $225

R. C. A., $260

A. C. Elec. Mfg. Co., $255

Mazda Radio Mfg Co., $285

MacLaren Mfg. Co., $300

Porter, Frank R., $350

Stromberg-Carlson, $365

Federal Tel. Co., $375

Rodgers Radio Corp., $395

Garod Corp., $400

Golden-Leutz, Inc., $610

1928 Radio Receivers of Beauty and Utility

FRESHMAN G-4
Here is a new electrically operated Freshman receiver, complete and ready to operate as delivered. The price is $225.00, which includes new RCA a. c. tubes

OFTEN it is said that a radio receiver, to be fully appreciated by the feminine half of the domestic republic, must be encased in housings which are esthetically as well as technically satisfactory. It is natural and right that, as radio has become more an accepted part of the equipment of every home, women have had an increasing voice in the selection of the radio receiver. What, then, do the radio manufacturers offer to the prospective purchaser? To be brief, the radio receiver of 1927-28 is a thing of beauty as well as of utility. On these two pages, and on others in this issue, are grouped illustrations of the sets which have come from the manufacturers' designing laboratories. Various makers have grappled with the important problem of appearance, and, as these illustrations show, have met it in widely different ways. Some of these models achieve their decorative effect through the dignity of utter simplicity; others are notable for the use of ornament in the treatment of the control elements. The metal tuning and control escutcheon is supplied with some decoration, slightly reminiscent of that simplicity to be found in the grouping of instruments on present day automobile dashboards. Illumination of the control escutcheon is also a feature which has much in common with the instrument panel on the modern motor car. From the representative models shown on these pages, the housewife can gain an excellent idea of the appearance of moderate priced radio receivers which are offered to suit her taste as well as that of her husband. Her ideas of the necessary limitations of her domestic decorative scheme should blend with her husband's technical opinions.

THE STEWART-WARNER 520
The tendency towards single-control is again manifest in this six-tube receiver. Two tuned and one untuned stages of r. f., detector, and two transformer-coupled audio stages, comprise the circuit. The cabinet is of dull polished walnut veneer. The price, $125.00

BOSCH 66
Another single-control receiver employing six tubes, this one being priced under a hundred dollars. The beauty of the cabinet is a feature but the efficiency of the circuit itself has in no way been neglected. The illuminated dial is graduated into both kilocycles and arbitrary numbers

GREBE'S "SYNCHROPHASE" SEVEN
The five tuned stages (four r. f. and detector) are tuned by means of a single dial and vernier, the rigidity of the tuning condenser assembly insuring permanency of the accurate factory adjustment. The receiver has been carefully shielded, while the special coils employed enhance the selectivity and permit maximum amplification. These coils are of the binocular type and are wound with Litz wire

WORKRITE
This eight-tube receiver has all-metal chassis, and complete copper shielding. There is a single illuminated drum type dial control. Price, $160.00 without table

BELOW
The Kolster 6H consists of a six-tube receiver power cone reproducer, and B socket power unit. Space is provided for either an A socket power device or a storage battery, which is extra. Price, $265.00

THE "WARWICK"
In a cabinet of Tudor period. A six-tube Splitdorf receiver, similar to that shown on page 351, is built in. Price, $275.00 with cone and power-supply unit

ZENITH
The Model 17E electric. A single-tube 400-mA. unit supplies the necessary power. The receiver itself employs six tubes. The price is $350.00. Spinet base, $20.00 extra

THE KELLOGG 507
This is a six-tube, non oscillating, t. r. f. set, balanced and shielded. "Inductance tuning" is employed, a system which requires no variable condensers and which permits equal amplification throughout the frequency spectrum covered. The price of the receiver is $190.00

"COMMANDER"
Here is a King six-tube completely shielded t. r. f. receiver which makes use of single control tuning. There are three r. f. stages. The loop may be folded into the cabinet. Price $220.00

THE "NAVAJO"
Six tubes, shielded, single - control—these words describe this new Mohawk receiver just as they do other prominent receivers of the season, for this circuit combination predominates in 1928 models. The "Navajo" sells for $67.50. Equipment to adapt it for a. c. operation is obtainable at additional cost

MU-RAD
Two t. r. f. stages, detector, and three audio stages are included in this receiver. The audio stages, all of the double-impedance form, give uniform response between 30 and 5000 cycles, according to the manufacturers. Single-control tuning is featured. Price $98.00. In a console, equipped for complete a. c. operation, the receiver sells for $265

THE "MAYFLOWER"
Here is a five-tube completely a. c. operated receiver by Cleartone, of Cincinnati. Complete with a. c. tubes, loud speaker, and built-in power unit, the "Mayflower" lists at $250.00

A MARTI RECEIVER
Another electric receiver, this one employing six tubes, three of which constitute the audio channel, resistance coupling being featured. Tuning is accomplished by means of two dials

Servants of Your Light Socket

The Radio Receiver Powered Directly from the Light Socket is like the Automobile with a Self-Starter Pushing the Button Starts the Machine—Models on This Page are Representative of the Season's A. C. Set Offerings.

TO THE LEFT
The well-known Loftin-White circuit, six tubes, is employed in the Arborphone Model 253 receiver. Tuning is accomplished by means of a single dial calibrated in wavelengths. A Peerless cone loud speaker is included, and the receiver may be used with either batteries or socket-power devices. The cabinet is of matched walnut, curly maple and rosewood veneers, and gumwood. Price, $250.00

A NEW DEPARTURE IN POWER SUPPLY
The Day-Fan Electric Company, of Dayton, Ohio, has placed on the market receivers which are powered by motor generators, all batteries being eliminated. The efficiency of the motor generator principle of transforming alternating current to direct current has long been recognized, but its application to the powering of a radio receiver has hitherto been neglected, so far as receivers commercially available are concerned

AN ECONOMICAL A. C. RECEIVER
The manufacturers of this receiver, the Argus Radio Corporation, of New York City, state that the cost of operating this receiver is no more than one fifth of a cent per hour. The receiver is completely a. c. operated, and employs no batteries or chargers. There are six tubes in all, three of which are r. f. in a combination of tuned and untuned stages. The final audio stage employs a 210 type tube, a plate potential of 400 volts being applied to it. Provision is made so that the audio amplifier may be used in conjunction with a phonograph pick-up device. Price $195.00, less tubes and loud speaker

LEFT
This is a Freed-Eisemann electrically operated receiver, listing at $295.00. A neutrodyne circuit is used, and extraordinary selectivity is claimed by the manufacturers

AN ALL-AMERICAN HIBOY
The receiver employs six tubes, and may be obtained either to use the new a. c. tubes or for ordinary tubes, batteries or power units being required in the latter case

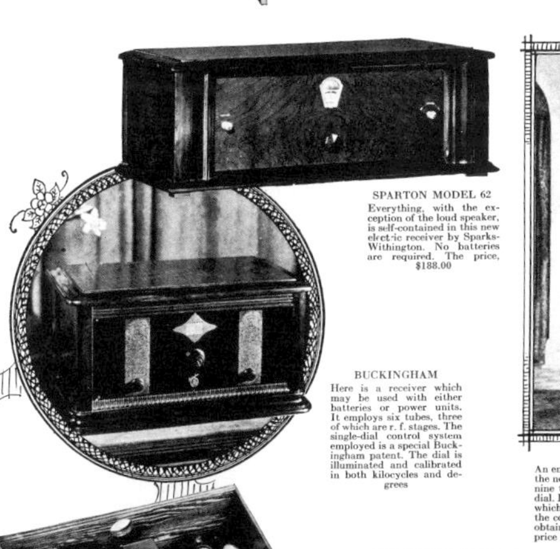

SPARTON MODEL 62
Everything, with the exception of the loud speaker, is self-contained in this new electric receiver by Sparks-Withington. No batteries are required. The price, $188.00

BUCKINGHAM
Here is a receiver which may be used with either batteries or power units. It employs six tubes, three of which are r. f. stages. The single-dial control system employed is a special Buckingham patent. The dial is illuminated and calibrated in both kilocycles and degrees

THE CASE CONSOLE
An enclosed loop, controlled from the front panel, obviates the necessity for an outdoor antenna. The receiver employs nine tubes and tuning is accomplished by means of a single dial. It is equipped with a Newcombe-Hawley loud speaker, which has an air column of seventy-two inches. The price of the console is $350.00, less accessories. The receiver is also obtainable for use with a. c. tubes, in which case the list price is $475.00 complete with tubes and ready to operate

ELKAY "SENIOR"
Complete a. c. operation with every necessary piece of equipment built in, unified tuning control, "Truphonic" audio amplification, total copper shielding, six tubes—thus can this new Elkay receiver be described. The list price is $195.00. Six McCullough a. c. tubes, and one Raytheon BH rectifier, are required as extras

TO THE LEFT
The electric receiver drawer of McMillan, Chicago. This drawer forms the nucleus of several attractive receivers by McMillan, ranging in price from $260.00 to $325.00. A table cabinet model lists at $170.00

Radio Retailing (Mar. 1928), pp. 36-37

Statistical Survey *of the* Radio

Estimated figures pertaining to the radio industry from 1922 to date, compiled by *Radio Retailing* from sources as authentic and accurate as it is possible to obtain

Number of Homes With Sets

(As of Jan. 1)

Including both factory-built and home-made after accounting for obsolescence

Year	Homes
1922	60,000
1923	1,500,000
1924	3,000,000
1925	4,000,000
1926	5,000,000
1927	6,500,000
1928	7,500,000

Radio Audience

(As of Jan. 1)

Number of people listening to sets in use

Year	Audience
1922	75,000
1923	3,000,000
1924	10,000,000
1925	15,000,000
1926	20,000,000
1927	26,000,000
1928	35,000,000

Total Radio Sales

(At retail, in numbers and dollars, during the year)

	1922	1923	1924	1925	1926	1927	Total to Date
Radio Sets, factory-built No.	100,000	250,000	1,500,000	2,000,000	1,750,000	1,350,000	6,950,000
(including furniture) $	$5,000,000	$15,000,000	$100,000,000	$165,000,000	$200,000,000	$168,750,000	$653,750,000
Speakers No.	25,000	500,000	1,500,000	2,000,000	2,000,000	1,400,000	7,425,000
$	$750,000	$12,000,000	$30,000,000	$32,000,000	$30,000,000	$28,000,000	$132,750,000
B-Power Units No.	...	...	10,000	100,000	500,000	400,000	1,010,000
$	...	...	$400,000	$4,000,000	$18,000,000	$12,000,000	$34,400,000
Storage Batteries and A-Power Units No.	...	650,000	2,000,000	1,700,000	2,100,000	...	6,450,000
$	...	$7,000,000	$25,000,000	$26,000,000	$37,000,000	...	$95,000,000
A and AB Power Units No.	...	...	...	...	...	550,000	550,000
$	...	...	...	...	...	$22,000,000	$22,000,000
Dry Batteries, B and C $	$4,500,000	$6,000,000	$55,000,000	$66,000,000	$80,000,000	$68,000,000	$279,500,000
Receiving Tubes No.	1,000,000	4,500,000	12,000,000	20,000,000	30,000,000	39,000,000	106,500,000
$	$6,000,000	$17,000,000	$36,000,000	$48,000,000	$58,000,000	$58,500,000	$223,500,000
Rectifying Tubes No.	...	...	...	...	...	2,200,000	2,200,000
$	...	...	...	...	...	$8,800,000	$8,800,000
Other Accessories (including headsets, separate cabinets, aerial equipment, etc.) $	$3,750,000	$4,000,000	$11,600,000	$24,000,000	$33,000,000	...	...
Other Accessories (not including furniture) $	...		...	...	...	$7,500,000	$83,850,000
Radio Furniture (separate) $	$...	...	...	...	...	$52,000,000	$52,000,000
Parts $	$40,000,000	$75,000,000	$100,000,000	$65,000,000	$50,000,000	$21,000,000	$351,000,000
No. Home Made Sets No.	1,000,000	1,500,000	1,750,000	1,000,000	750,000	300,000	6,300,000
Totals							
Sets $	$5,000,000	$15,000,000	$100,000,000	$165,000,000	$200,000,000	$168,750,000	$653,750,000
Parts $	$40,000,000	$75,000,000	$100,000,000	$65,000,000	$50,000,000	$21,000,000	$351,000,000
Accessories $	$15,000,000	$46,000,000	$158,000,000	$200,000,000	$256,000,000	$256,800,000	$931,800,000
$	$60,000,000	$136,000,000	$358,000,000	$430,000,000	$506,000,000	$446,550,000	$1,936,550,000

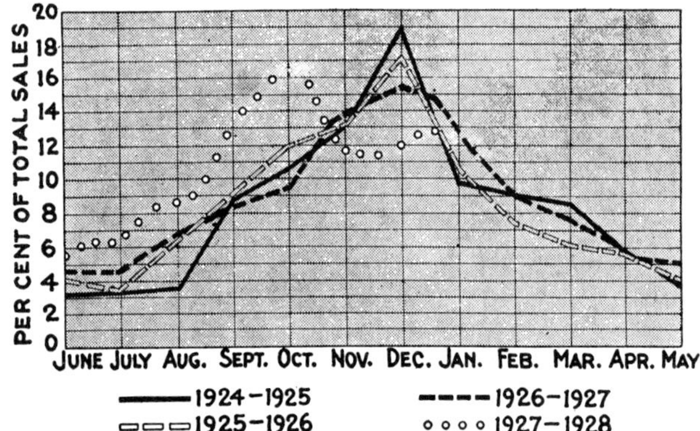

Survey made among a number of retailers shows sales curve slowly straightening out, as compared with other years.

Number Sets on Farms

(At end of year)

Year	Sets
1922	10,000
1923	145,000
1924	360,000
1925	550,000
1926	1,350,000
1927	1,600,000

FOR the past three months representatives of *Radio Retailing* have been compiling these figures by means of personal contact with leading radio manufacturers. The editors desire to acknowledge their courteous aid and co-operation.

BUSINESS, *as of* January 1, 1928

Number of Dwellings Wired and Unwired by States

States	Estimated Number of Homes	Number Wired for Electricity Service	Number Unwired
United States...	27,850,000	17,596,390	*10,559,510
Alabama	600,000	133,900	466,100
Arizona	106,000	35,700	70,300
Arkansas	454,000	107,100	346,900
California	1,030,000	1,335,900	*
Colorado	250,000	169,000	81,000
Connecticut	380,000	325,800	54,200
Delaware	58,000	26,430	31,570
District of Columbia	126,000	97,700	28,300
Florida	312,000	133,700	178,300
Georgia	746,000	135,600	610,400
Idaho	124,000	63,500	60,500
Illinois	1,710,000	1,470,740	239,260
Indiana	744,000	515,600	228,400
Iowa	576,000	328,200	247,800
Kansas	434,000	274,600	159,400
Kentucky	600,000	181,300	418,700
Louisiana	455,000	104,300	350,700
Maine	188,000	129,500	58,500
Maryland	376,000	236,400	139,600
Massachusetts	995,000	851,200	143,800
Michigan	1,040,000	856,800	183,200
Minnesota	632,000	384,750	247,250
Mississippi	426,000	51,150	374,850
Missouri	830,000	541,500	288,500
Montana	165,000	67,120	97,880
Nebraska	330,000	169,000	161,000
Nevada	19,000	13,730	5,270
New Hampshire	108,000	89,700	18,300
New Jersey	877,000	777,300	99,700
New Mexico	92,000	22,540	69,460
New York	2,690,000	2,549,900	140,100
North Carolina	676,000	161,100	514,900
North Dakota	153,000	46,400	106,600
Ohio	1,570,000	1,228,000	342,000
Oklahoma	557,000	176,950	380,050
Oregon	209,000	180,600	28,400
Pennsylvania	2,290,000	1,437,500	852,500
Rhode Island	165,000	130,100	34,900
South Carolina	434,000	89,700	344,300
South Dakota	164,000	53,900	110,100
Tennessee	586,000	164,100	421,900
Texas	1,270,000	491,900	778,100
Utah	122,000	104,600	17,400
Vermont	84,000	59,500	24,500
Virginia	596,000	167,200	428,800
Washington	364,000	330,080	33,920
West Virginia	396,000	121,000	275,000
Wisconsin	685,000	443,900	241,100
Wyoming	56,000	30,200	25,800

* The discrepancy between totals for the U. S. is due to California where the total reported for residential customers exceeds the estimated number of families.

Number of Sets Sold

Total number radio sets, home-made and factory built, sold to date	13,250,000
Homes with sets	7,500,000
Scrapped, or more than one set in a home	5,750,000

TOTAL SALES IN SIX YEARS

Total radio sales at retail, 1922 to 1927, inclusive	$1,936,550,000

Industry Census

(To date)

Manufacturers	1,200
Wholesalers and distributors	1,100
*Retailers	28,000

**Carrying full stock of sets and accessories*

Saturation Comparison

(To date)

No. Homes in U. S.	27,850,000
No. Phonographs	12,500,000
No. Passenger Autos	16,100,000
No. Telephones	18,250,000
No. Homes wired for electricity	17,596,000
No. Farms	6,500,000
No. Homes without radio sets	20,350,000
Radio Saturation	27%

Radio Exports

(At end of year)

1922	$2,800,000
1923	$3,450,000
1924	$6,000,000
1925	$9,900,000
1926	$9,500,000
1927	$9,200,000

New Table-Type Receivers for Next Season's Market

Illustrations of some of the receivers manufacturers are marketing this fall. Photos on these and the following pages are not complete presentations of manufacturers' lines and are intended only to give some idea of the new types of products.

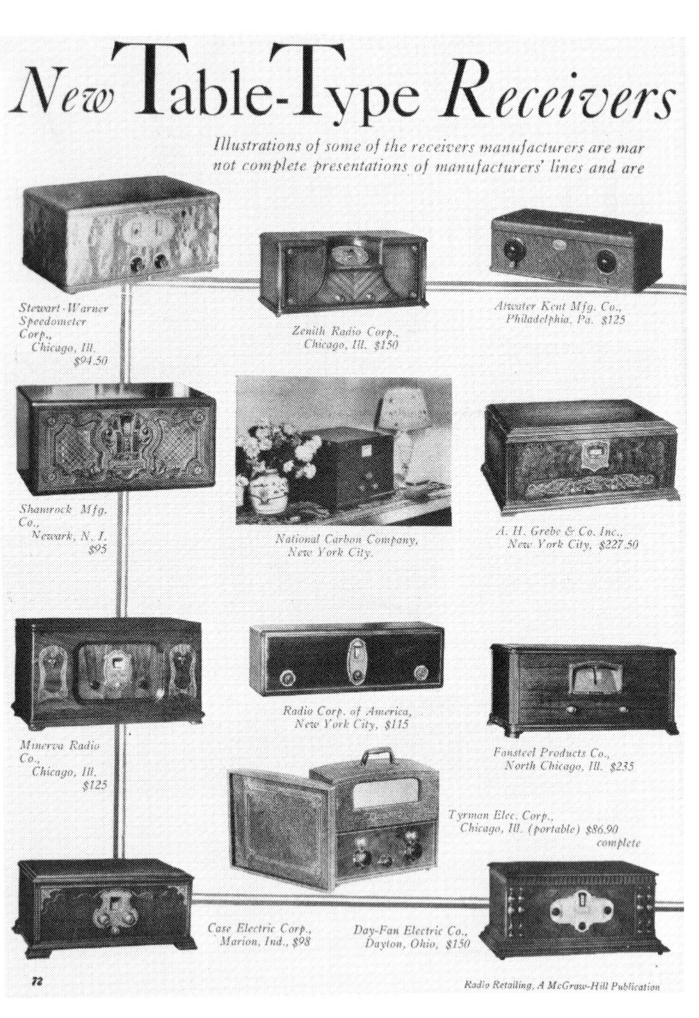

Stewart-Warner Speedometer Corp., Chicago, Ill. $94.50

Zenith Radio Corp., Chicago, Ill. $150

Atwater Kent Mfg. Co., Philadelphia, Pa. $125

Shamrock Mfg. Co., Newark, N. J. $95

National Carbon Company, New York City.

A. H. Grebe & Co. Inc., New York City, $227.50

Minerva Radio Co., Chicago, Ill. $125

Radio Corp. of America, New York City, $115

Fansteel Products Co., North Chicago, Ill. $235

Tyrman Elec. Corp., Chicago, Ill. (portable) $86.90 complete

Case Electric Corp., Marion, Ind., $98

Day-Fan Electric Co., Dayton, Ohio, $150

Bush & Lane Piano Co., Holland, Mich. $110

Crosley Radio Corp., Cincinnati, Ohio, $80

F. A. D. Andrea, Inc. Long Island City, N. Y. $110

American Bosch Magneto Corp., Springfield, Mass., $132.50

Kolster-Radio Corp., Newark, N. J. $135

Stromberg-Carlson Tel. Mfg. Co., Rochester, N. Y.

Kellogg Switchboard & Supply Co., Chicago, Ill. $169.50 complete

Philadelphia Storage Battery Co., Philadelphia, Pa. $125

Freed-Eisemann Radio Corp., Brooklyn, N. Y. $125

Bremer-Tully Mfg. Co., Chicago, Ill. $130

A-C Dayton Co., Dayton, Ohio, $98

Steinite Radio Corp., Chicago, Ill. $75

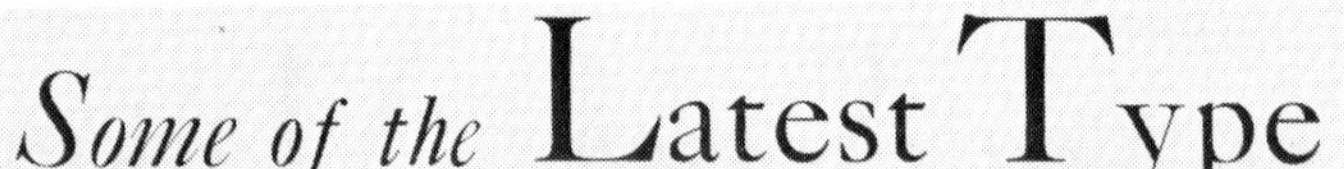

Some of the Latest Type Console Receivers for 1929

A. D. Andrea, Inc., Long Island City, N. Y. $235

Kolster Radio Corp., Newark, N. J. $350

Kellogg Switchboard & Supply Co., Chicago, Ill. $775 complete

Electrical Research Labs., Chicago, Ill. (radio-phonograph combination)

Splitdorf Radio Corp., Newark, N. J. $290

Case Electric Corp., Marion, Ind. $175

(LEFT) Argus Radio Corp., New York, N. Y. $495

Minerva Radio Co., Chicago, Ill. $225

Philadelphia Storage Battery Co., Philadelphia, Pa. $275

Day-Fan Electric Co., Dayton, Ohio, $200

Bush and Lane Piano Co., Holland, Mich. $215 complete

Marti Electric Radio Co., West Orange, N. J.

Bremer-Tully Mfg. Co., Chicago, Ill. $190

National Carbon Co., New York City.

Amrad Corporation, Medford Hillside, Mass. $475

Steinite Radio Corp., Chicago, Ill. $150

Fansteel Products Co., North Chicago, Ill. $487.50

Grigsby-Grunow Co., Chicago, Ill. $167.50

Zenith Radio Corp., Chicago, Ill. $230

Table Type Sets Shown At The Chicago Trade Show—*also Magnetic and Dynamic Speakers*

Andrea, F. A. D., Inc., Model 20, $99.50.

Audiola Radio Co., $95

Radio-Victor Corp. of America, Model 60, $175.

Stewart Warner Corp., Model 902, $89.75.

Stromberg Carlson Tel. Mfg. Co., Model 641, $155.

National Carbon Co., Model 31, $115.

Kennedy, Colin B. Corp., Model 110, $115.

National Co.

Zenith Radio Corp., Model 41, $100.

Atwater Kent Mfg. Co., Model 55, $88.

Philadelphia Storage Battery Co., $67.

U. S. Radio & Television Corp. (Apex), Model 89, $89.95.

Ware Mfg. Co., $195.

Buckingham Radio Corp., $69.50.

Utah Radio Products Co., $65.

Wright-DeCoster, Inc. $100.

Andrea, F. A. D., Inc., $72.50.

Radio-Victor Corp. of America, $30.

National Carbon Co., $50.

Stewart-Warner Corp., $19.25.

Davis Industries, Inc.

Borkman Radio Corp., $40.

Temple Corp., $39.

Trimm Radio Mfg. Co., $25.

Atwater Kent Mfg. Co., $34.

Philadelphia Storage Battery Co., $32.50.

Rola Co., $30.

COMPLETE EXHIBITORS' DIRECTORY ON PAGE 114

Expensive consoles became more and more important toward the end of the decade, luring manufacturers into abandoning their cheaper models, just in time for the Depression when nothing but cheap models would sell. Makers who were well-financed, like Atwater Kent and RCA, or were fast enough on their feet to switch to small table models, like Philco, survived. Those who based their plans on indefinitely-increasing sales of consoles, like Majestic, did not.

New Console Sets At The Chicago Trade Show *For the 1930 Season*

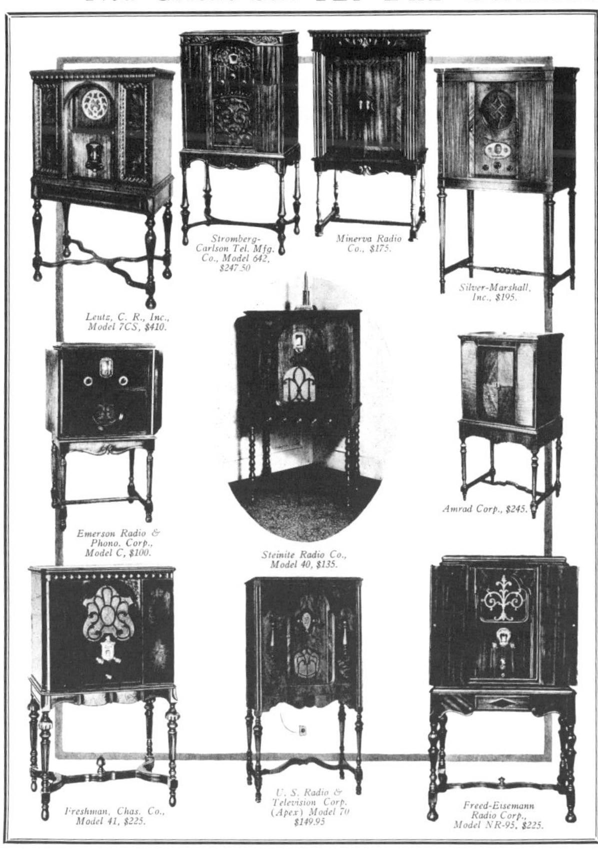

Stromberg-Carlson Tel. Mfg. Co., Model 642, $247.50

Minerva Radio Co., $175.

Silver-Marshall, Inc., $195.

Leutz, C. R., Inc., Model 7CS, $410.

Amrad Corp., $245.

Emerson Radio & Phono. Corp., Model C, $100.

Steinite Radio Co., Model 40, $135.

Freshman, Chas. Co., Model 41, $225.

U. S. Radio & Television Corp. (Apex) Model 70 $149.95

Freed-Eisemann Radio Corp., Model NR-95, $225.

Temple Corp., Model 8-80, $189.

Zenith Radio Corp., Model 42, $175.

National Carbon Co. Model 33, $210.

Bremer-Tully Mfg. Co., Model 82, $200.

Balkeit Radio Co., Model C, $175.

American Bosch Magneto Corp., $168.50.

Howard Radio Co., Model 5, $199.50.

Philadelphia Storage Battery Co., $119.50.

Kennedy, Colin B. Corp., Model 310, $197.

Consolidated Radio Corp. (Arborphone), Model 430, $119.

COMPLETE EXHIBITORS' DIRECTORY ON PAGE 114

Our Business Moves in Cycles of CAUSE and EFFECT

Showing the Two-year Relationships of Technical, Economic and Social Developments to the Progress of Radio

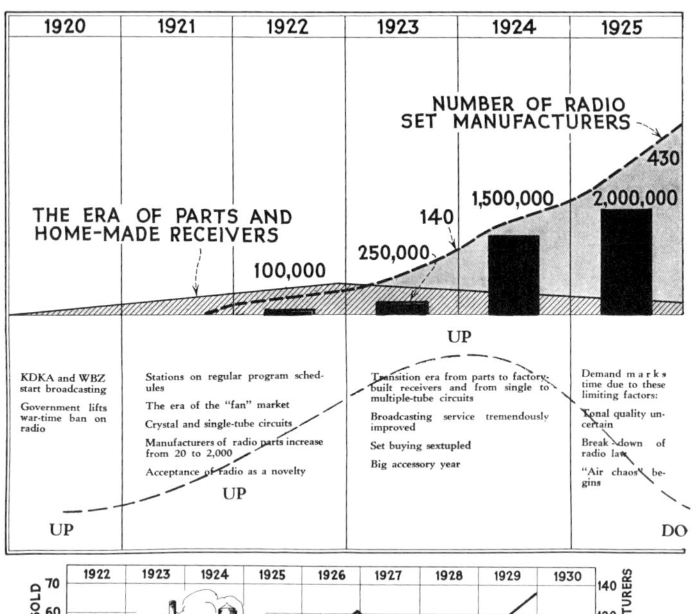

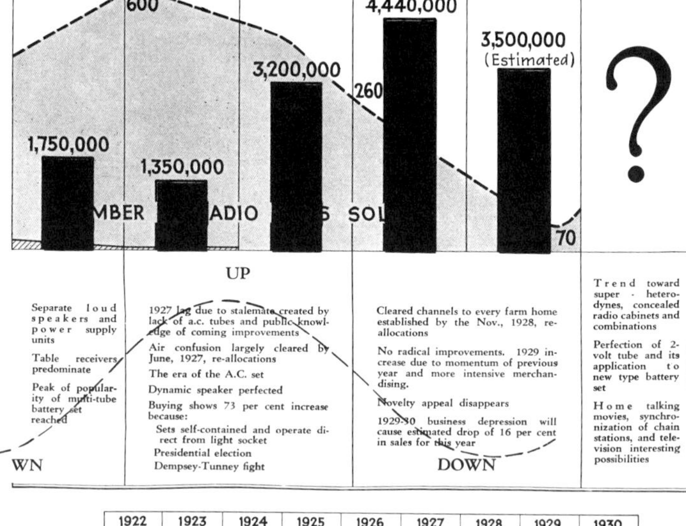

1922 1923 1924 1925 1926 1927 1928 1929 1930

MILLIONS OF TUBES SOLD — NUMBER OF TUBE MANUFACTURERS

NUMBER OF TUBE MANUFACTURERS

NUMBER OF RADIO TUBES SOLD

PERCENTAGE OF YEARLY INCREASE

1922	1923	1924	1925	1926	1927	1928	1929	1930
	200%	33⅓%	25%	30%	15%	20%	31%	14%
1,000,000	3,000,000	4,000,000	5,000,000	6,500,000	7,500,000	9,000,000	11,800,000	13,500,000

Radio Retailing (Nov. 1930), pp. 42-43.

Total Sales of Radio Products at Retail (1922-1930)

	1922	1923	1924	1925	1926	1927	1928	*1929*	*1930*	Product
Radio Sets, factory-built (including consoles and built-in reproducers)	*100,000* $5,000,000	*250,000* $15,000,000	*1,500,000* $100,000,000	*2,000,000* $165,000,000	*1,750,000* $200,000,000	*1,350,000* $168,750,000	*3,200,000* $350,000,000	*4,200,000* $525,000,000	*3,672,400* *$298,010,000*	Consoles and Midget Receivers.
Radio-Phonograph Combinations	....	....	....	....	....	....	*81,000* $38,000,000	*238,000* $67,068,000	*155,400* *$34,188,000*	Radio-Phonograph Combinations.
Tubes	*1,000,000* $6,000,000	*4,500,000* $17,000,000	*12,000,000* $36,000,000	*20,000,000* $48,000,000	*30,000,000* $58,000,000	*41,200,000* $67,300,000	*50,200,000* $110,250,000	*69,000,000* $172,500,000	*52,000,000* *$119,600,000*	Tubes.
Reproducers (excluding those in consoles and combinations)	*25,000* $750,000	*500,000* $12,000,000	*1,500,000* $30,000,000	*2,000,000* $32,000,000	*2,000,000* $30,000,000	*1,400,000* $28,000,000	*2,460,000* $66,400,000	*800,000* $16,000,000	*$3,500,000*	Speakers (excluding those already in receivers).
A-B-C (Dry) Batteries	$4,500,000	$6,000,000	$55,000,000	$66,000,000	$80,000,000	$68,000,000	$50,400,000	$30,530,000	*$21,514,000*	A-B-C (Dry) Batteries.
A-B Power Units, Storage Batteries and Chargers	....	$7,000,000	$25,400,000	$30,000,000	$55,000,000	$34,000,000	$17,500,000	$14,350,000	*$6,920,000*	A-B Power Units, Storage Batteries and Chargers.
Other accessories*	$3,750,000	$4,000,000	$11,600,000	$24,000,000	$33,000,000	$38,550,000	$46,000,000	$9,600,000	*$6,700,000*	Other Accessories*.
Parts (does not include sales to manufacturers)	$40,000,000	$75,000,000	$100,000,000	$65,000,000	$50,000,000	$21,000,000	$12,000,000	$7,500,000	*$6,000,000* (estimated)	Parts (not to manufacturers).
Totals									*4,519,500*	Automobile-Radio
Sets, plus Combinations	$5,000,000	$15,000,000	$100,000,000	$165,000,000	$200,000,000	$168,750,000	$388,000,000	$592,068,000	*$332,198,000*	Sets and Combinations.
Parts	$40,000,000	$75,000,000	$100,000,000	$65,000,000	$50,000,000	$21,000,000	$12,000,000	*7,500,000*	*6,000,000*	Parts.
Accessories	$15,000,000	$46,000,000	$158,000,000	$200,000,000	$256,000,000	$235,850,000	$290,550,000	242,980,000	*158,234,000*	Accessories (inc. tubes).
Total Sales for year	$60,000,000	$136,000,000	$358,000,000	$430,000,000	$506,000,000	$425,600,000	$690,550,000	$842,548,000	*$500,951,500*	Total Sales for Year

***Includes aerial equipment, meters, pick-ups, turntables, headsets, furniture, etc.*

Permission to quote statistics in this issue is granted if credit is given "Radio Retailing"

Radio Retailing (Mar. 1931)

Total radio sales by all companies were estimated by David Sarnoff in *Radio World*, Jan. 24, 1925, p.21: $2 million in 1920, $5 million in 1921, $60 million in 1922, $120 million in 1923, $350 million in 1924.
The number of sets sold in 1923 (250,000) was incorrectly given as 550,000 in the 1938 Broadcasting Yearbook; this error has been widely repeated since.

A-C DAYTON

A-C Electrical Mfg. Co.

The AC Electrical Manufacturing Co. was formed around 1901 to make AC motors, and incorporated about June 1919 with $100,000 capital stock, financed by C.H. "Chap" Bosler and public-utility magnate Albert Emanuel. Bosler was president and the company was located on East Fourth St. in Dayton, Ohio. In mid-1922 they were joined by Roy Stanley Copp from McCook Field Radio Laboratory, who designed a series of radio components, ready by August and advertised in the September 1922 *QST*. In its catalog and in the November QST ad, Dayton also offered complete regenerative receivers under the Ace name, apparently made for Dayton by the Precision Equipment Co. of Cincinnati.

The XL-5 appeared in September 1924, first in a long line of TRF models. Allen Apple recalled "In 1924 I worked there after school. In 1925 I wired broadcast receivers and that year I was made a foreman over a flock of women who wired sets. I had to stay after work and calibrate and test the BC sets." (1985 letter). In June 1926 AC Dayton occupied a new four-story factory.

By September 1928 Bosler had been replaced as president by Conrad Strassner, Emanuel's son-in-law. Most all large radio companies had long since signed up for an RCA patent license by this time, but AC Dayton held out, opting instead for a Technidyne license from Lester Jones, like Sparton and Slagle (Continental). The Technidyne circuit was supposed not to infringe on Alexanderson's TRF patent, the cornerstone of RCA's licensing position. Instead of cascaded RF amplifier stages, each containing a tuned circuit, the Technidyne put all the tuning at the beginning, followed by a multi-stage untuned amplifier. This circuit was used in the Navigator 1929–1930 models, designed by the new chief engineer Ford Studebaker.

Meanwhile Strassner's wife filed for divorce in late 1929, and her father Emanuel resigned from the board of directors, but he still appeared to be in control, as two days before petitioning the court for a receiver in January or February 1930, he was paid $19,000, having lent the company $80,000 earlier. Strassner was looking for a job in January (he later became a sales rep for Pilot). The creditors immediately dumped Emanuel's receiver and installed a former Dayton treasurer, and great things were forecast, but the last news item on the company appeared in May 1930.

DAYTON RADIO PRODUCTS

Are the most complete and most accurate of any Radio apparatus made by one Company in America.

Type B Vario-Coupler

Dayton Phone Receiving Set

Moulded Bakelite Variometer

VARIABLE CONDENSERS IN 4 SIZES

	List Price
Type 9 P. C. S. Square and Round Plate .0003 Mf.	$3.00
Type 17 P. C. S. Square and Round Plate .0005 Mf.	3.60
Type 31 P. C. S. Square and Round Plate .001 Mf.	4.50
Type 45 P. C. S. Square and Round Plate .0015 Mf.	5.30
Type 17 P. C. S. Condensers with Vernier attached	4.40
Type 31 P. C. S. Condensers with Vernier attached	5.30
Type A Variometer Moulded Bakelite	6.75
Type A Vario-Coupler Moulded Bakelite	7.00
Type B Variometer Bakelite Tube Type	4.25
Type B Vario-Coupler Bakelite Tube Type	4.75
Type C Vario-Coupler Bakelite Tube Type	3.75
Rheostats	1.10
Panel Switches 8 Points	.90
Phone Jacks	.65
Genuine Bakelite Knobs and Dials 3"	.75
Insulating Rubber Tubing 10 ft. Package	.45
Radio Phone Receiving Sets	75.00

Jobbers and Manufacturers write for Catalog and Discounts.

THE A-C ELECTRICAL MFG. CO., DAYTON, OHIO

MAKERS OF ELECTRICAL DEVICES FOR OVER 20 YEARS

QST (Nov. 1922), p. 112

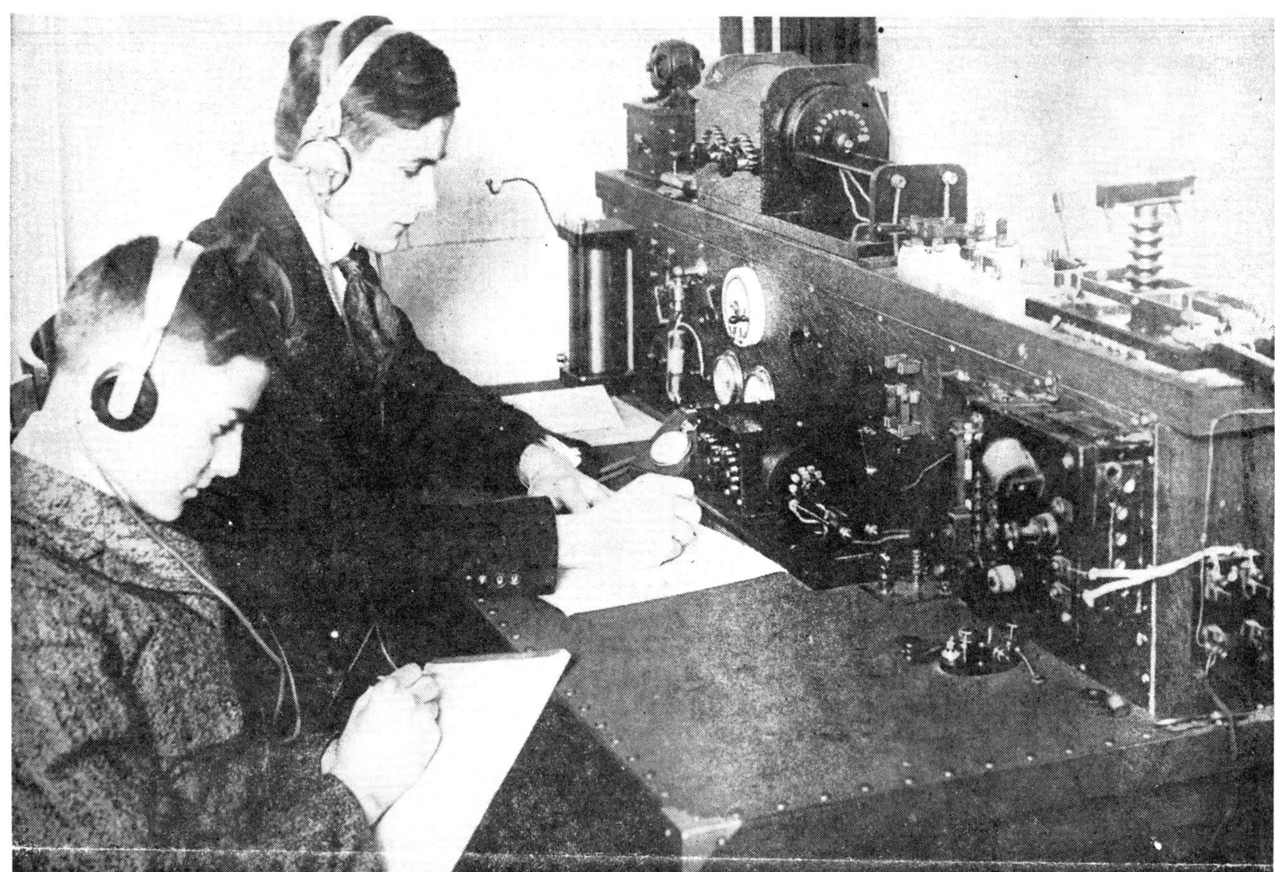

—Daily News Staff Photos.

Old photo shows Stanley Copp (R) as a boy in his radio shack with Carl Linxweiler. Picture ran in The Daily News in 1916.

A C E Radio Phone Receiving Set

(A / C / E)

Licensed under Armstrong Patent 1,113,149

This set has just been developed and is placed on the market with the one idea in mind—value for your money.

There is nothing more discouraging than trying to receive over a makeshift set. Radio broadcasting would be still more popular today if the publc did not have to contend with inferior apparatus.

It is a recognized fact that a regenerative set is the only practical set for the reception of Telephone broadcasting. Under average conditions we claim a range of 1000 miles for this set. With favorable conditions signals have been copied 1725 miles.

This receiver is built up to a standard and not down to a price. The highest grade of insulation is used throughout—bakelite.

The cabinet is genuine walnut with satin finish. The panel is black polished bakelite 15"x 6", with beautiful black bakelite knobs and dials for controlling operation of set. The tuning condenser is equipped with vernier for precision tuning. The detector tube is controlled wth vernier rheostat, assuring finest adjustment possible of the filament. All connections are made from rear of cabinet, eliminating unsightly wiring.

The circuit used in this set is the tuned plate regenerative, giving great signal strength and selectivity. Regeneration is accomplished by tuning plate circuit.

Dayton Radio Amplifier

Type A-2

This two-stage amplification unit has been designed to meet the specifications of a real, all-round audio-frequency amplifier. In building this amplifier, we have left out nothing but the howls and squeals common to many amplifying units, and we are giving you two real stages of undistorted amplification.

This unit is mounted to panel of black polished bakelite, 6"x 8", with back connections. The cabinet, 6½"x 9½", is of genuine walnut, hand rubbed to a satin finish.

All interaction is eliminated in this amplifier. The transformers are of high-ratio type. The filaments are controlled by two of our "Dayton" Rheostats operated with dials instead of knob and pointer. There are three jacks for connection to either detector, first or second stage of amplification. A canopy switch is provided for turning off the filaments.

This amplifier matches the A C E Receiver in height and appearance, and when used in conjunction with it, broadcasts can be heard throughout the room.

From an undated AC Dayton radio catalog

Broadcast Receiver — Type R-12

Employing special Copp VARIO-SELECTOR constructed on Copp Circuit No. 5.

This Receiver combines Tuned Radio Frequency with a special selector device which gives to this set the requirements necessary for perfect radio reception:

Selectivity Volume Clearness Distance Ease of Tuning

Wave Length Range — 200 to 600 meters
Distance — up to 2500 miles

Type R-12 is a 4 tube Set designed for satisfactory use at a reasonable price, both for cities where the nearness of Broadcasting Stations requires fine selectivity; as well as for suburban districts where selectivity is not as essential as clearness, volume, distance and ease of tuning.

The VARIO-SELECTOR makes this set efficient for either Sharp or Broad tuning as conditions warrant.

Broadcast Receiver Type R-12 is built of the highest grade materials using Bakelite insulation throughout, all contained in a beautiful Brown Mahogany finished Cabinet.

Copp Circuit No. 5 is the fifth proven circuit developed and perfected by our Engineering Department under the direction of Mr. R. S. Copp, who built and perfected the first single circuit regenerative Receiving Set, known as Copp Circuit No. 1.

Full information, with descriptive literature, etc., will be mailed on request.

The Price of Type R-12 . . . . $98.00
(Without Tubes)

The A-C Electrical Mfg. Co. - - Dayton, Ohio

Makers of Electrical Devices for over 20 Years

R-12 Mar. 1924 $98

COPP VARIO SELECTOR

The Copp vario selector is not a wave trap. It is an instrument designed for use with single circuit

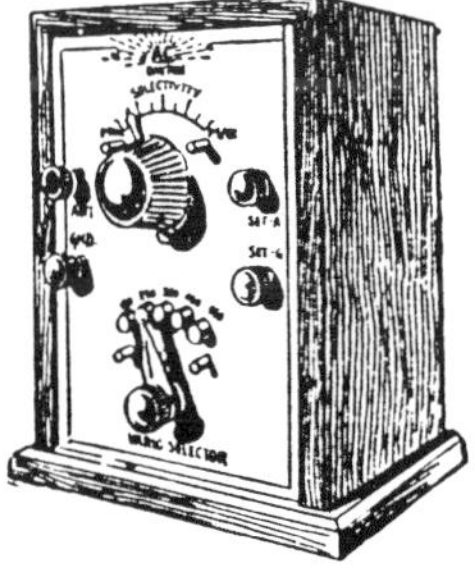

receivers and converts them into double circuit receivers. It consists of a tapped antenna or primary coil and a rotor coil of a few turns of wire inductively coupled to the antenna coil. The rotor coil connects to the antenna and ground binding posts of the single circuit receiver. With its use the single circuit receiver is made very selective. Manufactured by the A-C Electrical Mfg. Co., Dayton, Ohio.

Arrived in excellent packing.

AWARDED THE RADIO NEWS LABORATORIES CERTIFICATE OF MERIT NO. 490.

Copp Vario-Selector
Apr. 1924 $11.50

Plus Clearness

HERE is a new receiving set—designed as all fine sets, to give volume, selectivity, distance and simplicity of operation—But *PLUS* one feature that marks its superiority —*CLEARNESS OF RECEPTION.*

The A-C Dayton POLYDNE XL-5 is a super, five tube receiver that will enable you to honestly enjoy your favorite programs, without the annoyance of interference and distortion. The XL-5 receives the finest orchestral and vocal music exactly as played, with perfect clarity of modulation.

True radio enjoyment will be yours with this new receiver. Its mechanical refinements have resulted in a beautiful set, one that will fit the arrangement and decorative scheme of any room in the home.

Your radio or music dealer will gladly demonstrate the A-C Dayton XL-5 for you. Ask to see it—to hear its remarkable clearness of reception. Write for the name of the nearest dealer.

Priced at $115.00, less tubes and accessories, the XL-5 represents a wonderful value in a fine receiving set. (West of the Rockies—$120.00)

Reputable radio jobbers and dealers will be interested in a detailed description of the A-C-Dayton XL-5. Our sales plan is an attractive one. Write for complete information.

GUARANTEED WITHOUT RESERVATION

THE A-C ELECTRICAL MFG. COMPANY
DAYTON, OHIO
Makers of Electrical Devices for over Twenty-Years.

A-C DAYTON XL-5

Radio Broadcast (Sept. 1924)

John Drew

XL-5 Sept. 1924 $115
Also offered as a kit in October for $72.50. XL-10 shown below.

Ralph Thorn

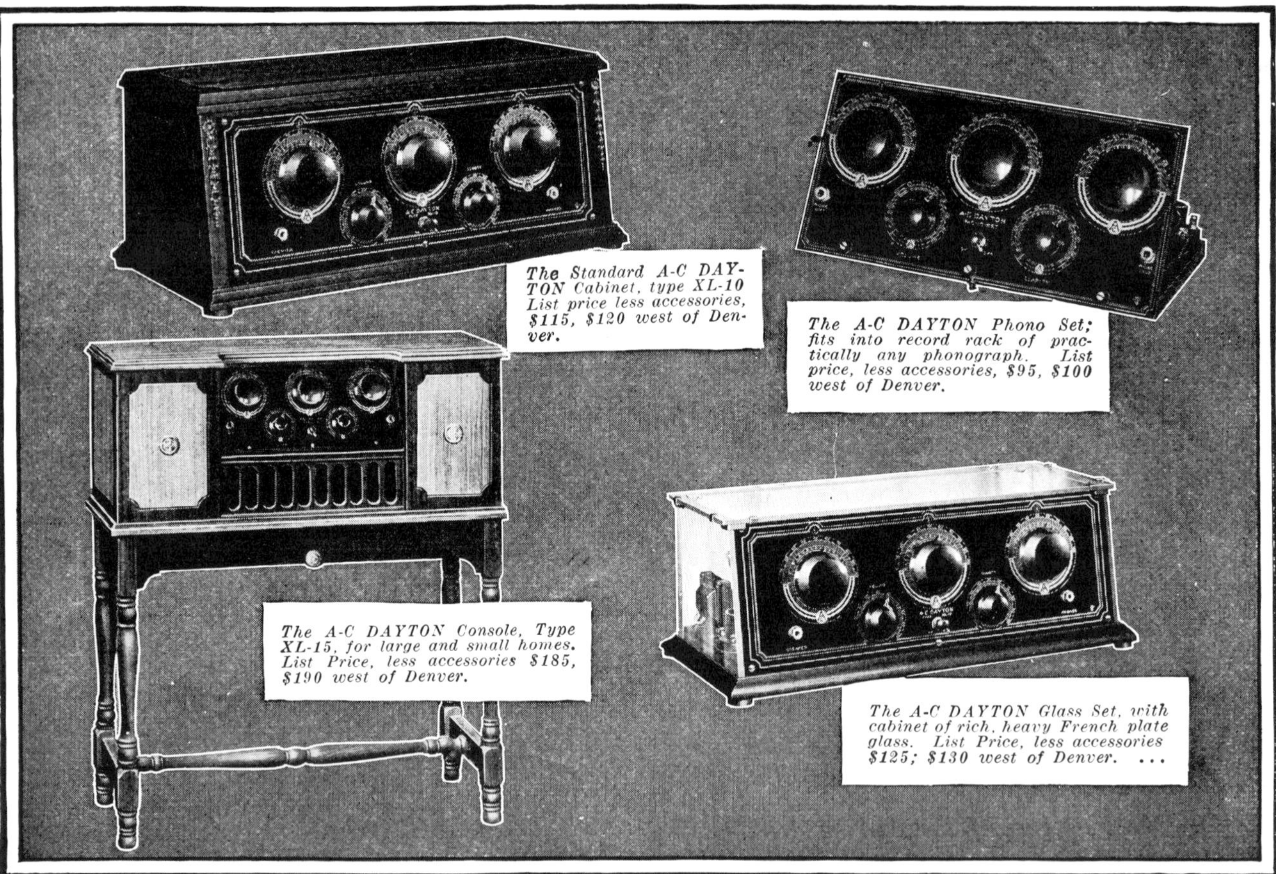

The Standard A-C DAYTON Cabinet, type XL-10 List price less accessories, $115, $120 west of Denver.

The A-C DAYTON Phono Set; fits into record rack of practically any phonograph. List price, less accessories, $95, $100 west of Denver.

The A-C DAYTON Console, Type XL-15, for large and small homes. List Price, less accessories $185, $190 west of Denver.

The A-C DAYTON Glass Set, with cabinet of rich, heavy French plate glass. List Price, less accessories $125; $130 west of Denver. . . .

Radio Retailing (Aug. 1925), p. 224

One-Dial Six-Tube Receiver

Radio Retailing, July, 1927

The A. C. Dayton Company, Dayton, O., has placed on the market the illustrated XL-60 Console Grand, 6-tube tuned radio frequency receiver. The set incorporates three stages of tuned radio frequency detector and two stages of audio frequency. The entire set is inclosed in a heavy aluminum case, the coils being individually shielded with aluminum cylinders. Provisions are made for the use of additional B and C batteries. A battery compartment provides ample room for all sizes of A, B and C batteries or power units. Overall dimensions are 43 in. wide by 41⅝ in. high by 20 in. deep. Intended retail price $285. Model XL-60 which is a table type model of the same electrical specifications has an intended retail price of $135. Model XL-25 which is a five-tube two-dial control receiver has an intended retail price of $85. Model XL-25 Console has an intended retail price of $149. Model XL-70 which is a seven-tube two-dial control receiver has an intended retail price of $165. Model XL-70 Console Grand has an intended retail price of $315.

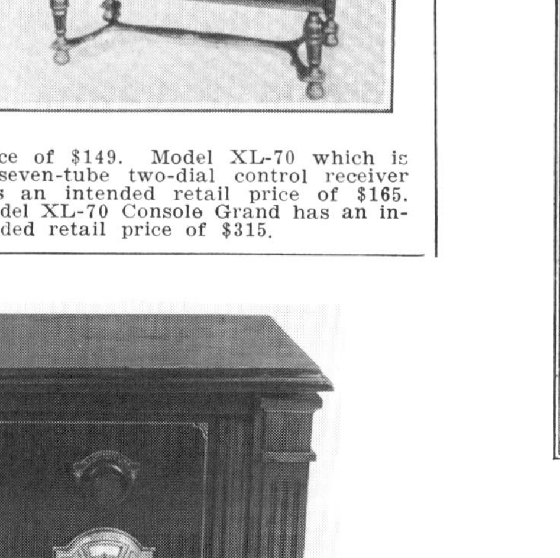

C. H. Bosler, President of the A. C. Electrical Manufacturing Company, of Dayton, Ohio, says he not only looks forward to a big year, but is right now in the midst of as busy a season as his company has yet experienced. Mr. Bosler feels that in the new models of the A-C Dayton receiving set his company has realized the utmost in five-tube tuned radio frequency perfection.

Scott MacWilliam

XL-30

Radio Dealer (Nov. 1925) p. 137)

The First News of a Tremendous Radio Story

Broad and easy or fine and selective—virtually two sets in one cabinet.

A type of set for every type of buyer.

A national advertising campaign in leading periodicals and farm papers.

Exclusive sales territory for every A-C DAYTON dealer.

Our strength protects your business. Our policy protects your price.

Radio Retailing (Aug. 1926), p. 526

ACCEPT these facts from a company of more than 20 years standing—which has been prominently identified with radio for six years and is noted for square dealing and sound management.

The A-C Dayton story concerns an original development in radio tuning, which not only establishes new standards of set performance but which supplies the direct answer to your biggest problem in selling radio receivers.

It concerns a balanced line of six models, incomparable in design, from $56 to $255 in price—and strong national advertising in leading periodicals and farm papers of more than 7,000,000 circulation.

Lastly, it concerns the building of an exclusive and permanent organization of dealers, each with his own protected territory—protected as to sales—protected as to price.

This last is tremendously important today. Time will not wait. Territories are being allotted. Regardless of your present plans or connections we urge you—get the facts in your hands immediately.

XL20 $56

XL25 $79

XL30 $135

THE A-C ELECTRICAL MANUFACTURING CO.
DAYTON, OHIO
Makers of Electrical Devices for More Than Twenty Years

Use the Coupon

To aggressive dealers of proven ability we say: Don't write us next week. Write us now—today

The A-C Electrical Mfg. Co., R-R-6
Dayton, Ohio.
Gentlemen: Send me full information about the A-C DAYTON exclusive dealer franchise, prices, advertising, etc.

Name____________________
Address____________________
City and State____________________

The A-C DAYTON line of six handsome models. Priced from $56 to $255—prices slightly higher Denver and west.

Models first advertised in June 1926.

A-C Dayton All Electric Receivers (Table Model)

This six-tube electric receiver has three stages of radio frequency, one untuned and two tuned, one tuned detector stage and two stages of audio frequency amplification. We mention this as this circuit is particularly adapted to the neat, compact, and efficiency construction of the XL-50. The tone quality of this set will satisfy the most musically trained and critical ear.

The cabinet and wooden panel are of beautifully grained walnut, and all controls are grouped on a beautifully embossed bronzed instrument plate. Dimensions are 22-in. long, 9-in. high and 9-in. deep.

For 110-120 Volt, 50-60 Cycles A.C. Current.

List Price, $175.00.

Our Price, $35.00

The XL 50 A.C. Dayton receiver installed in this beautiful console. Large orthophonic speaker gives the low, full tones.

Console built of the finest grained and matched walnut veneers.

Our Price, $43.00

CONSOLE GRAND

The XL-50 six tube AC Dayton receiver installed in the famous Duophonic Console Grand. Has two separate reproducing chambers and especially designed units.

This combination make possible the true reproduction over the entire scale.

Dimensions 43-in. wide x 41⅝-in. high x 20-in. deep.

Our Price, $49.00

Surplus ad. *Radio Retailing* (Feb. 1929), p. 129

Radio Retailing (Aug. 1928), p. 119

Radio Retailing, July, 1928 117

. . . . . The Flewelling Short-Wave Adapter Opens a Vast New Field for Immediate Radio Sales . . . .

SHORT wave reception, heretofore impossible for individual set owners, is now an engaging reality—summer programs, usually spoiled by static, are picked up with amazing clearness—International programs become daily entertainment—London and other foreign stations are heard consistently—all made possible by the A-C DAYTON-Flewelling Short-Wave Adapter.

This remarkable device, invented by E. T. Flewelling, one of radio's foremost engineers, is far and away the most important radio development announced in 1928. It was the sensation of the R. M. A. Show in Chicago, last month, attracting the attention of all who are interested in the distribution and sale of radio.

The A-C DAYTON-Flewelling Short-Wave Adapter may be instantly attached to any ordinary radio receiver without additional wiring, adapting the set for short wave reception. The retail price is only $22.50, placing it within reach of all set owners, and opening up a vast new market for immediate sales. (Already twenty-six U. S. Stations are broadcasting on short wave lengths.)

In step with A-C DAYTON'S progressive policy and immediately following Mr. Flewelling's announcement that the Short-Wave Adapter had been perfected, negotiations were consummated whereby the A-C DAYTON Company acquired exclusive manufacturing rights and Mr. Flewelling was retained as Consulting Engineer.

Alert distributors and retailers the country over have quickly sensed the significance of these announcements — every mail brings urgent inquiries for territorial rights. The coupon below is for the convenience of those who wish to send for full particulars.

THE A-C DAYTON COMPANY
DAYTON, OHIO, U. S. A.

SPECIFICATIONS

Wave length range 18-80 meters, has 3 plug-in coils for 3 wave length ranges—17-29, 27-48 and 47 to 84 meters, which are the only active broadcast bands. Additional coils for other waves can be supplied, if desired. No additional tubes or batteries required. No wiring change in your set. Adapter comes complete, housed in a handsome cabinet which will harmonize with the finest of radio furniture.

E. T. Flewelling, noted Radio Engineer and Inventor.

A-C DAYTON RADIO

The A-C DAYTON-Flewelling Short-Wave Adapter—Retail price $22.50

Distributors
Many excellent territories are yet to be assigned. Wire or use the coupon today, requesting full particulars.

Retailers
If your jobber is not already supplied, write direct, giving us his name and address.

The A-C DAYTON Company, Dayton, Ohio.
Gentlemen:
I want more information about the A-C DAYTON-Flewelling Short-Wave Adapter and the new A-C DAYTON line of receivers. Please rush your reply to
Name .... Address ....
City .... State ....
Please state whether jobber or dealer.

Flewelling Short-Wave Adapter
July 1928 $22.50 (later $15).
The Radio Products Co. of Dayton was organized in 1928 to take over production (Flewelling was president and chief engineer). These were sold into the early 1930's.

XL-61 June 1928 $65
AC-63 June 1928 $98
AC-65 June 1928 $123
AC-66 June 1928 $148 (shown in ad) Technical article in *Radio Broadcast*, Dec. 1928, p.121.

AFTER ALL *Isn't It Radio* PROFITS . . . . That Count?

By C. R. STRASSNER, *President and Treasurer*

Radio Retailing (July 1928), p. 127

For the man who believes his own ears

YOU can't be eight years in the radio manufacturing business and not learn something.

We've learned what the permanent kind of radio dealer wants—and we've learned to want that kind of dealer. Our proposition is planned for him.

While we have been making money in radio for eight years we have also been seeing that our dealers made it too. That's why our company and our dealer organization are both strong today.

Point for point, you'll find the A-C DAYTON franchise a mutual profit maker, based on a four model line that will go out and sell on straight competitive demonstration. It starts at $65 list, has an up-to-the-minute all electric set for $98, offers power amplifier performance for $123 and ends with a de luxe model that amplifies 3200 times. That's exactly six times average efficiency.

Our distributors cover their territories closely, see you often and fill orders promptly. All our advertising is concentrated on our dealers' immediate markets - newspapers, sectional farm papers, displays and literature that go directly and specifically to the people you can do business with.

There's no guess-work about this. We've proved it out since 1921. *You'll find it more profitable to sell A-C DAYTON than to sell against it.*

Let us lay our full proposition before you. Clip the coupon to your letterhead and send it today. You'll get an immediate answer to consider and decide on.

A-C DAYTON RADIO

THE A-C DAYTON COMPANY, Dayton, Ohio 2

Gentlemen: Please send complete details about your dealer's franchise.

Name..

Address..

AC-63, SELF CONTAINED ALL-ELECTRIC: a 6-tube receiver complete for light socket operation except for tubes and speaker. May be used with any type speaker. Power tube takes full rated voltage. A reliable electric set listing $98, except Canada and West. Model XL-61 listing at $65, is same as AC-63, except that it is battery operated.

AC-65, at right, has same characteristics as AC-63 except that it uses 210 power amplifier in last stage, giving electrical power amplifier results. $123, except Canada and West.

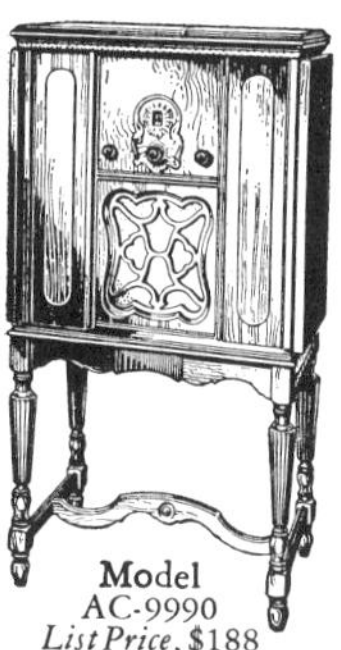

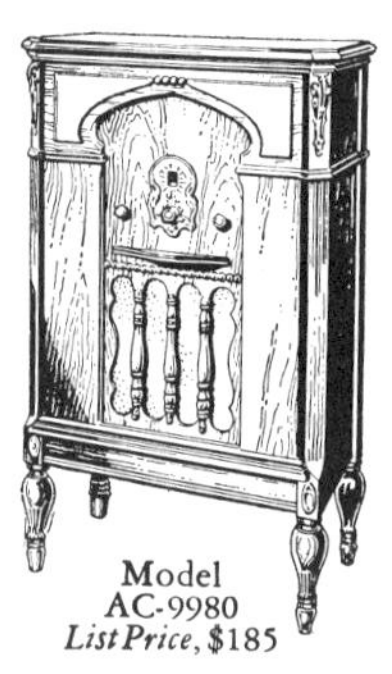

Radio Retailing (Aug. 1929), p. 119

Navigator models, 1929, using the Technidyne circuit. Technical article in *Citizens Callbook*, vol.10 no.4, Nov. 1929 p.94.

ADAMS-MORGAN/ /PARAGON

Adams-Morgan Company

Alfred Powell Morgan was born in 1889. A prolific writer, he turned out more than a dozen books, from *Wireless Telegraph Construction for Amateurs* in 1910 and *The Boy Electrician* in 1913 (which went through countless editions) to *The Pageant of Electricity* in 1943. While still in school, around 1909–1910, he formed the partnership of Adams-Morgan, which was soon doing a substantial mail-order business in wireless components.

Enter Paul Forman Godley. The same age as Morgan, Godley had been installing wireless stations in the Amazon for the Brazilian government in 1913. On his return, he met Howard Armstrong at a Radio Club of America meeting in 1914 and was amazed to learn that Armstrong had been consistently able to hear these Brazilian stations. The secret of course was the regenerative circuit. Armstrong had been unable to make his circuit work on short waves, where the amateurs were, and in any case was more interested in the commercial applications. Godley set to work, tuned his grid and plate circuits with self-resonant variometers, and the Paragon receiver was born.

Godley bought into Adams-Morgan as a one-third partner toward the end of 1915 to commercialize his design. The RA-6 was ready by October 1916, cost $35, and was quickly found in many of the better ham stations. It was last advertised in May 1917, when Adams-Morgan was busy with wartime contracts. Godley joined American Marconi in Aldene, New Jersey for the duration, having charge of receiver design.

After the war, Godley went to Adams-Morgan (according to Wardell Smith, he had not actually been there before that time) where RA-6s were again produced; later ads also hint at a "Universal Range, RA-200." He designed the RA-10, announced in October 1920, and distributed exclusively by Continental Radio & Electric Co., recently formed by well-placed radio men, which also held an RCA franchise in New York City. A January 1921 ad stated "1000 RA-10s are now ready." Like the RA-6, the RA-10 was a tuner only; the user had to provide the vacuum-tube detector and amplifiers. Adams-Morgan did make a small detector or amplifier unit containing the tube socket, grid leak and condenser, also available in January 1921, for $6.

Morgan and Godley had never gotten along too well, but the first of several "last straws" came in late 1921, when Godley was chosen by the ARRL, as the world's expert in short-wave reception, to travel to Scotland for the transatlantic tests. While he was gone, Morgan bought out his silent partner Adams and acquired majority control of the company, despite an agreement that no partner would sell without notifying the others. Paragon's history of technical innovation essentially ended at this point.

In March 1922 the DA-2 was ready, probably at the insistence of Continental, to cater to the hordes of broadcast fans who wanted complete receivers, and no doubt they sold as many as Adams-Morgan could produce, until May when the radio boom collapsed. November brought the RD-5 and A-2, nearly the same circuit but with the detector tube in the receiver cabinet as was more conventional by that time. However, the RD-5 and A-2 proved not nearly as popular as the earlier models, even when the 10R RF amplifier was added six months later (the 10R was also frequently used with the RA-10).

Around July 1923 Godley took his case to the New Jersey Chancery Court and as a result, he and Morgan settled their differences. He returned to the company long enough to design a new series of receivers, the RB2 and RB2A, III and IIIA, which appeared at the end of 1923. Clever but not especially innovative, these sets were not successful; Haynes-Griffin's New York store was offering them at a tremendous discount in May 1924. Godley left by the summer of 1924, this time for good.

By October Adams-Morgan had reorganized, acquired new capital, and introduced three new models, the Two, Three and Four. But even a $75,000 advertising outlay couldn't sell them. By the end of 1925, Paragon had essentially disappeared. In January 1926 a new "Model One" was shown in the "What's New" section of Radio Broadcast, but on January 22, 1926, receivers were appointed for the Adams-Morgan Co., Inc., liabilities $80,000, assets $197,751. Gimbels sold off much of the inventory in April. The remaining stock of models One, Two and Three went to a Brooklyn discounter in January 1927.

In October 1926 Adams-Morgan reorganized as the Paragon Electric Corp., with the financial backing of C.S. Phillips of Montclair and under the direction of Peter A. Petroff, formerly factory manager and before that the head of Millimeter Machine Works of New York. In June 1927, the new "1928 Paragon" line was announced, but by December Petroff had decided to abandon receiver production to concentrate on components then being supplied to Westinghouse, Western Union and others. The following June, even that modest plan had gone by the boards, and Paragon became a manufacturers' test laboratory at 200 Varick St., New York City. Nothing further was heard of the company.

(C. Keystone Co., Inc., of N. Y.)

PAUL FORMAN GODLEY

Paul Forman Godley was born September 25, 1889, at Garden City, Kansas. His interest in radio began when he entered Defiance College, Ohio. Being interested in communication, it was quite natural for him to become enthusiastic about radio and he studied all the available literature on radio communication published. In 1908, a commercial wireless station was built in Chicago, to which Mr. Godley was assigned as operator. In 1913, he was in the "Amazon-to-the-Andes" radio service for the Brazilian government. In 1914, he returned to his home, Leonia, N. J., and developed the short-wave regenerative receiver. In 1915, he opened a transmitting station, 2 ZE, and made many exceptional distance-records. During the World War, he served as designing engineer at the Marconi Wireless Company of America's factory. He was chosen by the American Radio Relay League to conduct its transoceanic radio tests, journeying to Scotland for this purpose

Radio World (Aug. 16, 1922), p. 8

PAUL FORMAN GODLEY (1889-1973)

Radio Topics (Sept. 1921), p. 53

ALFRED POWELL MORGAN
Author, Inventor, Manufacturer
1889–1972

Paragon Instruments Have Set New Standards

They are in a distinct class by themselves. There are no other instruments which can **EQUAL THEM IN ANY WAY**—regardless of price. **WE CAN PROVE THIS ASSERTION TO THE SATISFACTION OF ANYONE.**

It was designed **especially** and **solely** for reception of **AMATEUR WAVE LENGTHS** and its development has been carried on over a period of two years. It was the **first** and is the **only** worthy adaptation of the Armstrong circuit to short wave reception. The antenna inductance is arranged in steps. **ASIDE FROM THIS THERE ARE NO SWITCHES.** Continuously variable inductances—carefully designed variometers—are used in the closed circuits. **HIGH RESISTANCE CONTACTS, the capacity of switch points and leads, end-turn losses and the necessity for a variable tuning capacity are thus ENTIRELY DONE AWAY WITH.** The antenna and closed circuits are INDUCTIVELY COUPLED, and the COUPLING IS VARIABLE. The com-

R. A.—6—PARAGON AMPLIFYING SHORT WAVE RECEIVER, $35.00
Range 180 to 580 Meters

THIS INSTRUMENT IS SUPER-EFFICIENT, SUPER-SELECTIVE AND SUPER-SENSITIVE.

ponent parts of the instrument are not crowded into a small cabinet. The fact that ALL of these things are of extreme importance has been proven by the here-to-fore unheard of **SELECTIVELY and AMPLIFICATION** obtained by owners of this instrument. Signals may be read from stations at extreme distances or through heavy static and interference with this instrument long after other receivers have failed and WEAK SIGNALS MAY BE AMPLIFIED UP TO **ONE HUNDRED TIMES USING ONE AUDION ONLY.** We would be glad to furnish the names of prominent members of the American Radio Relay League who have offered unsolicited testimony as to the operation of the R. A.—6—, the amplification obtainable, and the distances covered both day and night.

UNQUESTIONABLY SUPERIOR TO ANYTHING ELSE ON THE MARKET

The methods employed in winding the coils eliminate leakage due to coloring matter in the insulation, put an end to the presence of moisture in the varnish, insulation and tube. The coils of the Paragon "No-End-Loss" transformers are divided into sections and fitted with **self cleaning, positive action** end turn switches which connect and disconnect the winding as required, **entirely cutting off from the circuit** unused portions of the inductance and **completely eliminating end turn effects on all wave lengths.** These switches are enclosed and are automatically controlled by the primary and secondary inductance switches respectively.

Panels, housings, switch heads, etc., are of polished black **FORMICA, which is superior in every way to hard rubber and costs more. All metal parts are of gold lacquered brass.** These instruments are adapted to extremely close tuning and due to the absence end-losses are particularly recommended as the only receiving transformers on the market suited to the reception of amateur wave lengths or for use in conjunction with the AUDION DETECTOR.

PARAGON RECEIVING TRANSFORMER

TYPE "L" $22.50 TYPE "S" $30.00 TYPE "X" $35.00

SEND STAMP TODAY FOR BULLETIN "O" WHICH DESCRIBES A VARIETY OF ENTIRELY NEW TRANSMITTING AND RECEIVING SPECIALTIES OF PARTICULAR INTEREST TO THOSE WHO DESIRE THE BEST.

SEND 6c. FOR OUR 232 pp. CATALOG

Our No. 7 Catalog shows several hundred different parts and also sets of materials for building your own apparatus. We do all the difficult work in our factory and give you the benefit of machinery and equipment

Contains complete description and prices of all the latest Wireless and Electrical Goods.

OUR PRICES WILL SAVE YOU MONEY AND OUR PROMPT DELIVERIES WILL SAVE YOU TIME.

THE BEST CATALOG OF ITS KIND IN AMERICA

Adams-Morgan Co., Sixteen Alvin Place Upper Montclair, N. J.

The factory as it appeared in 1981, when it was occupied by a theater group (with a more modern adjoining building). According to Wilson Norwood, who worked here from June 1922 to Aug. 1923, there were between 50 and 100 employees. On the first floor was a large machine shop; upstairs about 20 girls did assembly work, while wiring was done in a garage across the street (since torn down). Most everything but transformer coils was made on the premises.

The model designation "CR" indicates a Wireless Specialty Apparatus Co. design; apparently Adams-Morgan was a second source. A-M is also stated by Wardell Smith to have made the SE143 and one other receiver. Smith, a life-long friend of Morgan's, had hung around the factory since he was ten, before the war, and worked there as long as the company existed.

RA-6 Oct. 1916 $35

New England Wireless & Steam Museum

RA-10 Oct. 1920 $85
DA-2 Mar. 1922 $65

A matching transmitter, model 2-5-U, appeared in Nov. 1921, at $70.

RA 10 Interior

DA 2 Interior

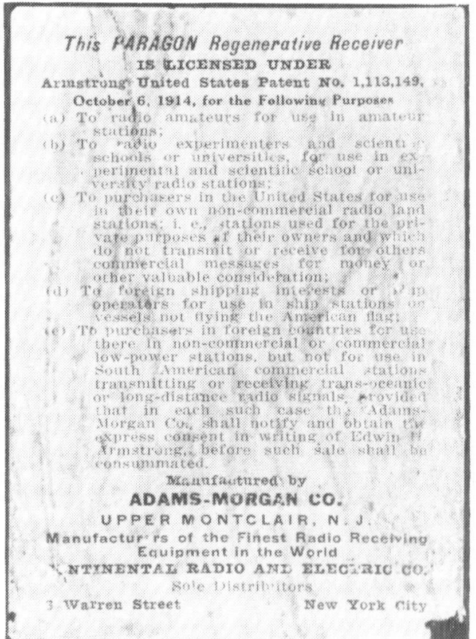

This PARAGON Regenerative Receiver
IS LICENSED UNDER
Armstrong United States Patent No. 1,113,149,
October 6, 1914, for the Following Purposes

(a) To radio amateurs for use in amateur stations;
(b) To radio experimenters and scientific schools or universities, for use in experimental and scientific school or university radio stations;
(c) To purchasers in the United States for use in their own non-commercial radio land stations; i. e., stations used for the private purposes of their owners and which do not transmit or receive for others commercial messages for money or other valuable consideration;
(d) To foreign shipping interests or ship operators for use in ship stations on vessels not flying the American flag;
(e) To purchasers in foreign countries for use there in non-commercial or commercial low-power stations, but not for use in South American commercial stations transmitting or receiving trans-oceanic or long-distance radio signals, provided that in each such case the Adams-Morgan Co. shall notify and obtain the express consent in writing of Edwin H. Armstrong, before such sale shall be consummated.

Manufactured by
ADAMS-MORGAN CO.
UPPER MONTCLAIR, N. J.
Manufacturers of the Finest Radio Receiving Equipment in the World
CONTINENTAL RADIO AND ELECTRIC CO.
Sole Distributors
3 Warren Street New York City

Label glued to bottom of RA 10 cabinet

VT Control Jan. 1921 $6

10R June 1923 $40

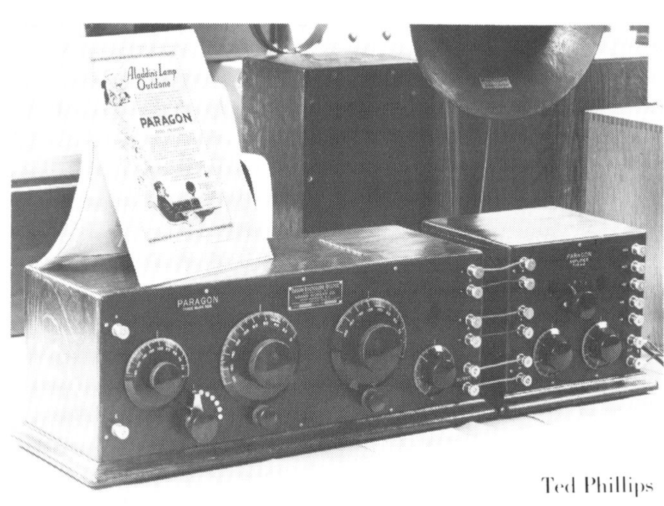

Ted Phillips

RD-5, A-2 Nov. 1922 $75, $50

RAS Mike Feher

RAS Mike Feher

RB2 Bob Wallace

Rick Weibezahl

RB2

RB2A John Williams

RB2, RB2A Dec. 1923 $135

New England Wireless & Steam Museum

III, IIIA Mar. 1924 $175
A repackaged version of the RB2.

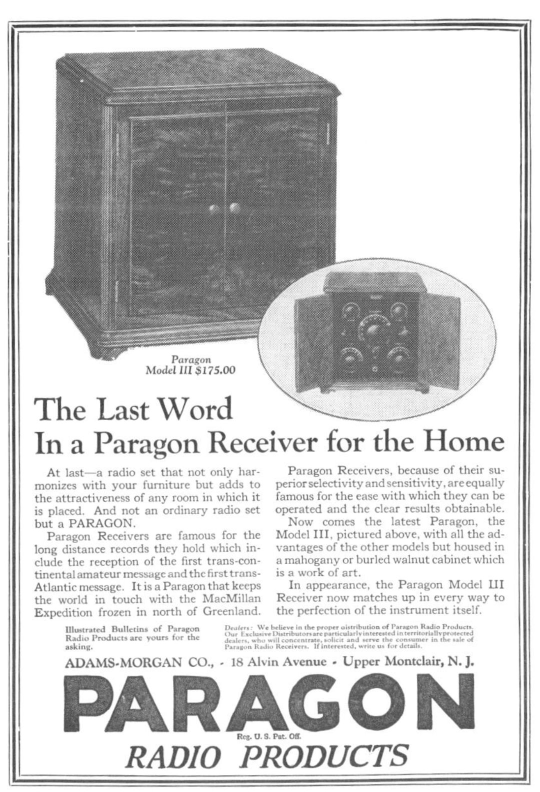

Popular Science Monthly (Mar. 1924), p. 125

The Music of the Immortals in Your Own Home
People sometimes sit hours in hot, stuffy opera houses to hear our famous opera singers. Yet the immortal notes of those same stars are broadcasted from radio stations and are listened to by thousands in the comfort and seclusion of their own homes.
Perhaps you would rather follow an athletic contest play by play, or listen to a speech by some famous man. Or perhaps the market, crop and weather reports would prove more interesting to you.
PARAGON
Reg. U.S. Pat. Off.
RADIO PRODUCTS
With the new PARAGON Three-Circuit Receiver Type RB-2 you can pick out the program you wish to hear and hear it clearly from beginning to end. For the greatest enjoyment of radio, for complete satisfaction, you should listen in with a Paragon set.
In appearance it is an addition to any home. All cabinet work is of mahogany with a brown mahogany finish. All metal parts showing, inside as well as outside of the cabinet, are nickel plated. 98% of the wiring is invisible. The whole outfit is compact, neat, solidly built, and finely finished. It is the ideal Radio receiver and the ideal holiday gift.
Illustrated Bulletins on Paragon Radio Products Are Yours for the Asking.
DEALERS: We believe in the proper distribution of Paragon Radio Products. Our exclusive Distributors are particularly interested in territorially protected dealers, who will concentrate, solicit and serve the consumer in the sale of Paragon Radio Receivers. If interested, write us for details.
ADAMS-MORGAN COMPANY
20 Alvin Ave., Upper Montclair, N. J.
Illustration shows the new Paragon RB-2 Regenerative Receiver with two-stage tone amplifier
Price $135.00
(Licensed under Armstrong Patent No. 1,113,149)

Flyer (Oct. 1924)

The New Paragon Four $65.

Range practically unlimited. Capable of clear, strong loudspeaker tone over long distance. Employs the new Paradyne circuit—non-radiating. New type SINGLE DIAL control. Will do all that sets costing three or four times as much have been able to do. Priced within everyone's means. May be used as a 3 or 4 tube set. Handsome solid mahogany cabinet. 20⅞ by 7⅞ by 8 in.

The New Paragon Two $27.50

A two-tube receiver of excellent tone and volume on loudspeaker from stations within moderate radius. Range almost unlimited for phone reception. New type SINGLE DIAL control. Mahogany finished cabinet, 11 inches long.

The New Paragon Three $48.50

Exceptionally sensitive, selective, fine-toned. With loudspeaker the volume over long distance range is amazing. Three tubes. New type SINGLE DIAL control, very simple to operate. Solid mahogany cabinet. 16¾ inches long.

THE PARAGON LINE

Oct. 1924

John Drew

Three Oct. 1924 $48.50

Wally Worth

Four Oct. 1924 $65 Technical articles in *Radio Engineering*, Apr. 1925, pp. 200–201, and *N. Y. Evening World*, Apr. 1, 1925 p. 17.

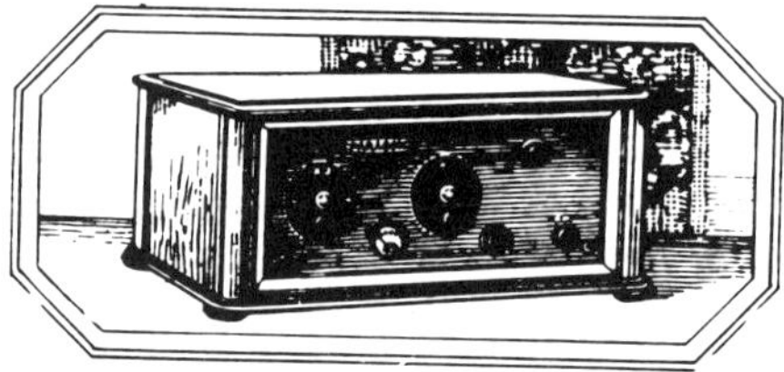

THE PARAGON MODEL THREE

The Revolutionary Improvements of the Paragon Line—The Double Impedance Circuit—the Latest, Patented Development — Equal to POWER AMPLIFICATION—

A T. R. F. Resistance-Coupled Receiver—an Average-Priced Product for the Buyer of Average Means—Nothing Like it at the Cost—

A T. R. F.-6 LONG-DISTANCE TRANSFORMER-COUPLED RECEIVER—Especially Designed and Recommended for Use in Rural Districts, Where Sensitivity is So Essential to Get DX Stations—A Set of Six Tubes—

A New Feature—A Chassis Department That Will Solve Your Problems **Your** Way—Our CHASSIS Offerings Will Lower the Price of Quality Radio This Season—With Them You Can Select Your Own Cabinet or Let the Consuming Buyer Do It—

MODEL ONE OF PARAGON

The Paramount Paragon Receivers Are Manufactured, Guaranteed, Serviced And Sold By That Pioneer And Yet Pre-Eminent Concern —

THE PARAGON MODEL TWO

Another Feature—All PARAGON SETS Can Be Supplied With Perfectly-Matched R. C. A. Tubes and With Other Equipment.

Dealers: The **New Paragon Receivers** Will Appeal to the Discriminating Public as Will No Others—It Will Pay You to Push the Most Popular **Paragons,** and You'll Positively Be Protected in **your** Territory.

Besides—There is Greater Profit in the **PARAGON**—Quick **TURN-OVER** is Assured—Prompt Deliveries Are Guaranteed By **PARAGON** Producers—Duplicated Sales Stimulated By Endorsements of Pleased Patrons of Yesterday Are Assured to You—

Again—**Paragon Receivers** Are Electrically-Equipped at Your Option—Operating Right From the Light Circuit—"B" Battery Current if More Practicable or Preferred.

THE PARAGON ELECTRIC CORPORATION

(An original Licensee Under the Patents Covering the Double Impedance Amplification.)

Upper Montclair, New Jersey, U. S. A.

Phone Montclair, 4228

Radio Retailer & Jobber (Nov. 1926)

One, $275. Two, $200. Three, $175 Nov. 1926

160 *Radio Retailing, A McGraw-Hill Publication*

The Lincoln
$175.00 to $195.00
For Battery or A.C. Operation

The Congress
Metal Cabinet **$80.00**
Walnut Cabinet, **$95**
For Battery or A.C. Operation

PARAGON SELLS
DIRECT-TO-DEALER
Exclusively

The Monroe
$125.00 to $135.00
For Battery or A.C. Operation.

Sept. 1927

Radio Retailing (Sept. 1927), p. 160

Air-Way Electric Appliance Corp.

The Air-Way Company of Toledo, Ohio made vacuum cleaners in 1919, featuring inventor D. Ben Replogle's designs: a paper bag to catch the dust, and a valve to divert the suction through the hollow handle to accessories.

The Air-Way Electric Appliance Corp. was formed in July 1920, also absorbing the Toledo Screw Products Co., which had made shells during the war, and Arrow Manufacturing Co., suppliers of Air-Way's motors. Both companies were owned by the Tracy family, prominent Toledo lawyers and industrialists: Thomas H. Tracy and his sons Pratt E. Tracy, Newton A. Tracy, and Thomas H. Tracy, Jr. Pratt Tracy was president of Air-Way, Thomas Jr. vice-president.

Besides the "Air-Way Sanitary System," the company made fractional-horsepower motors, bell-ringing transformers, and then radios. But when profits disappeared from radio, Air-Way dropped out in June 1926 to return to its first love, sanitary vacuum cleaners, whose sales were steadily growing.

The Tracys all died in the 1930s. Surviving ten years of losses in the Depression, Air-Way made vacuum cleaners until 1957, becoming Lamb Industries in 1958.

Pratt Tracy

F Nov. 1923 $50

Four tubes: 1 RF, detector, 2 audio. Early ads show two front-panel rheostats; after Mar. 1924, three.

Ken Sleger

Five Tubes: 2 RF, detector, 2 audio.

Ken Sleger

G Dec. 1923 $100

Radio News (Oct. 1922), p. 699

Each Instrument Guaranteed

AIR-WAY MODEL "C" RECEIVING SET
Here is a sound, practical set with detector and two stage amplification. It is remarkable for its long-range reception and its fineness of tuning.

AIR-WAY MODEL "B" RECEIVING SET
This set contains so many outstanding features that radio enthusiasts have wondered at the range and accuracy of reproduction. Detector and one stage amplifier with typical "AIR-WAY" fineness of workmanship.

AIR-WAY MOULDED VARIOMETER
A combination of the finest known insulating material with precise assembly and accurate winding—careful AIR-WAY manufacture doubles effectiveness and reduces interference.

AIR-WAY GREEN SEAL FILAMENT RHEOSTAT
A compact Rheostat wound with non-corrosive resistance wire over a substantial insulating ring. Trim, workmanlike, effective and durable.

AIR-WAY genuine molded rubber composition with neat white accurate graduations and numbers.

A sturdy frame with molded hard rubber endplates; uniformly mounted to assure maintained accuracy.

Air-Way
GREEN SEAL RADIO EQUIPMENT

The aim of the Air-Way engineers and the policy of the Air-Way Corporation are united in the production of Radio equipment which bears the unmistakable imprint of intelligent design, expert craftsmanship, and genuine quality throughout. In the Air-Way factory there is no compromise between quality and cost. Scientifically organized production by men skilled in volume manufacture of fine electric instruments and equipment is entirely responsible for the attractive prices at which Air-Way Radio parts and complete receiving sets are offered to the public. The Air-Way Green Seal Guarantee Tag attached to each Radio instrument is a symbol of quality that is known and preferred by experienced Radio buyers.

Air-Way Electric Appliance Corporation
TOLEDO, OHIO

Air-Way instruments appeal most to those dealers who are wisely preparing to merit a successful and growing radio business by selling products of reputation and genuine quality at fair prices. Write for Air-Way Radio Bulletin.

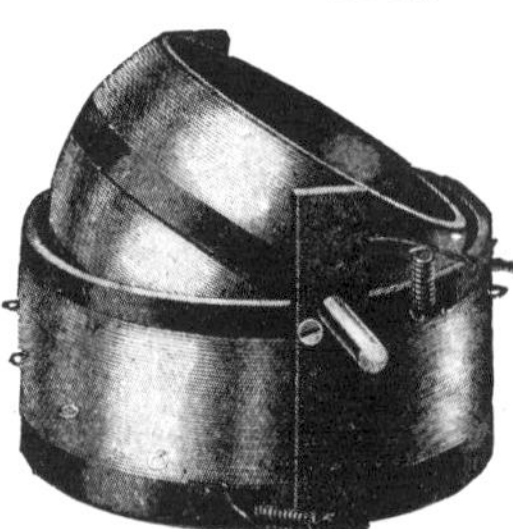

No professional operator would ask for a more precise and enduring instrument than the AIR-WAY Green Seal Variocoupler. It is built right and stays right.

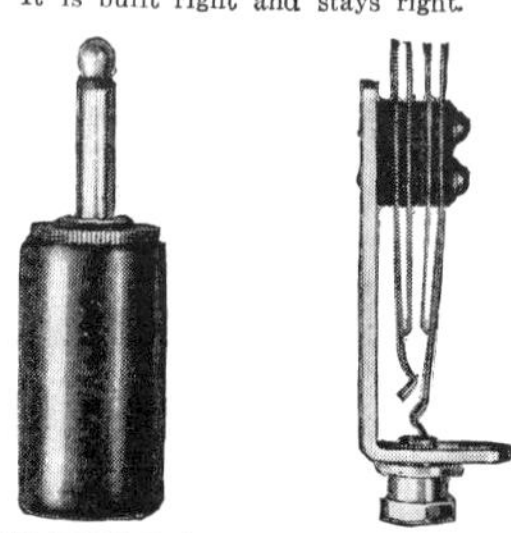

AIR-WAY Jacks and Plugs are built especially for the finer radio apparatus to do away with the loose connections for which plugs and jacks are frequently responsible.

AIR-WAY new, light - weight, positive contact tube sockets save bulbs and maintain perfect connections. New design.

AIR-WAY Amplifying Transformers do more to eliminate the whines and shrieks than any other similar piece of radio equipment. They prove their own merit by comparison.

John Wolkonowicz

41 Dec. 1924 $65

The Best in Radio Equipment

SWEEP THE AIR With Air-Way

No matter how modest the ideas of the radio beginner, he soon begins to search for distant stations.

Then is when he appreciates an AIR-WAY Receiver.

All distant signals come to any set, but they will not fight their way in through unnecessary losses and high resistances.

AIR-WAY Receivers are the last word in LOW LOSS construction and tuned radio-frequency amplification and build up the weakest signals to pleasing audibility.

Oscillation is perfectly controlled and all extraneous noises eliminated without neutralizers or complicated adjustments.

AIR-WAY No. 41, 4-Tube

We claim without reservation that AIR-WAY Model 41 is superior in every quality of radio reception to any other four-tube set ever built, and unequaled by any set at less than nearly twice the price.

The selective qualities are unexcelled in any set, operating on an outside aerial.

The dignified design of the solid walnut case and the workmanship and finish of the panel equipment give it an outward appearance in keeping with the operative quality.

A set that meets all market conditions and all individual requirements; one that the Dealer may sell to the inexperienced user or the most discriminating expert and be sure that either will attain results satisfactory in every way. Price $65.00.

AIR-WAY No. 51

The latest development in tuned radio frequency with two stages of radio frequency amplification, detector and two stages transformer coupled audio frequency amplification.

Offered without reservation as a set that will give general satisfaction to all broadcast listeners regardless of previous radio experience.

Price, as illustrated, $125.00.

Also furnished in handsome Console type cabinet of solid walnut. Price, $375.00.

AIR-WAY Apparatus is the result of several years' study and development by skilled radio engineers, and is strictly up to the minute in radio design.

Operation is simplified to the limits of the radio novice, and quality throughout is developed to meet the demands of the most discriminating of radio experts.

AIR-WAY Apparatus is distributed through established Jobbers and Dealers only. Write our Sales Department for Catalog of the complete line.

AIR-WAY ELECTRIC APPLIANCE CORP
TOLEDO OHIO

Sales Department
The Zinke Company
1323 S. Michigan Blvd
Chicago, Ill

Export Department
220 Broadway, St. Paul Bldg.
New York, N. Y.
Cable Address, Airwayvac New York

All apparatus advertised in this magazine has been tested and approved by POPULAR RADIO LABORATORY

(Jan. 1925)

51 Nov. 1924 $125
52 console $375

November 7, 1925 THE SATURDAY EVENING POST

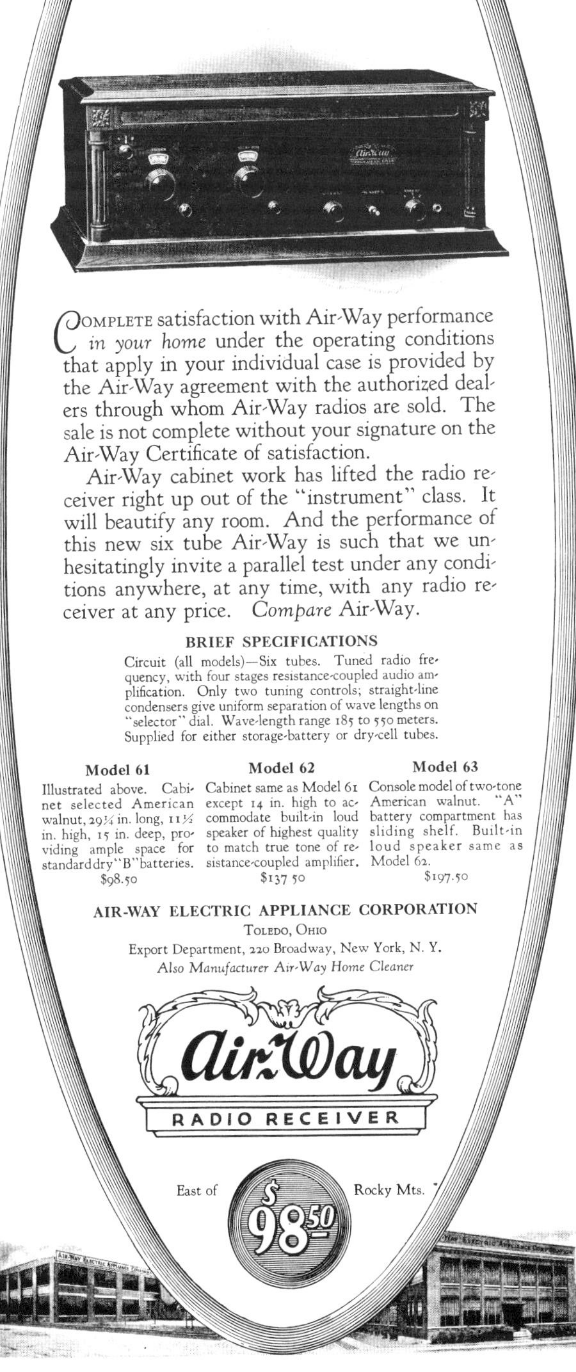

61 July 1925 $98.50
62 with speaker Aug. 1925 $137.50
63 console Aug. 1925 $197.50

61D, 62D, 63D dry-cell models also available after Nov. 1925, same prices.

ALL-AMERICAN MOHAWK

All-American Mohawk Corp.

All-American Mohawk was a merger of two Chicago companies, both started in 1920. All-American Electrical Manufacturers was formed early in that year by E.N. Rauland and a partner. Rauland, who had been interested in radio since 1909 and in charge of the coil-winding department of a large Chicago wire and coil maker, named his company after the 82nd "All-American" division in which he had served with the Signal Corps during the war. The company was incorporated and the name changed in May 1922 to All-American Radio Corp. when a public stock offering was made. Up to this point Rauland had made only components, and two reflex kits, the All-Amax Jr. and Sr., but a factory-wired set was announced in October 1925 and advertised in November.

The Mohawk Electric Corporation (later Mohawk Corp. of Illinois) was begun in 1920 as Electrical Dealers Supply House by brothers Louis and Gustav Frankel, changing its name in September 1924. By November Paul A. Chamberlain and Douglas De Mare had created a one-dial receiver that in some respects was years ahead of the industry. Chamberlain patented the idea of placing several variable condensers on a common shaft perpendicular to the front panel, and although this arrangement was unsuited to the long, narrow cabinets then in vogue, eventually it became universal. However, as Harrison has pointed out* this elegance of design did not extend to the remainder of the set's layout or wiring.

Nonetheless the 100 sold well, remaining Mohawk's basic model for nearly two years. By June 1927 Mohawk had tied up with Wurlitzer, a large and prestigious manufacturer and retailer of musical instruments. Wurlitzer made Mohawk's cabinets, while its mid-1926 radio catalog featured several one-dial models probably made by Mohawk. In May 1927 ground was broken for a factory addition of 24,000 sq. ft., giving Mohawk a total of 76,000 sq. ft., and by June, when production started on the 1927–1928 models, the factory capacity was 850 sets per day.

Mohawk stopped advertising after December 1927, and was acquired in March 1928 by All-American, which then changed its name to All-American Mohawk Corp. Wurlitzer's arrangement with Mohawk carried over to the combination, resulting in an exclusive sales contract for its 40 retail stores; Wurlitzer also leased the Neutrodyne license it had bought with Eagle Radio, to All-American for $30,000 per year. In October 1928 Wurlitzer went even further, by acquiring a large interest in the company, whose products were thereafter sold under the "Lyric" tradename (registered in 1929, use claimed since January 10, 1927. In June 1928 Mohawk lost a patent-office interference over its trademark, to a Mohawk company in Newark, New Jersey).

At the August 1929 board of directors meeting, it was decided that Wurlitzer would build All-American's cabinets (selling price $13 to $20) and also the speakers, Wurlitzer ordering 130 tons of magnet wire. By November, economic pressures forced All-American to move its entire Chicago plant to Wurlitzer's factory at North Tonawanda, New York. In April 1931 Wurlitzer took over radio production, in January 1933 marketing also, leaving All-American little to do but declare bankruptcy in November 1934. Tired of losing money on its radios year after year, Wurlitzer made its last Lyric in 1937, to concentrate on its rapidly-increasing juke-box business.

E.N. Rauland, having left All-American Mohawk in 1929, bought out its transformer business and machinery to establish the Rauland Corporation. This was purchased by Zenith in December 1948 and run as a picture-tube-making subsidiary for many years.

*Arthur P. Harrison, "Single-Control Tuning, an Analysis of an Innovation," Technology and Culture, April 1979, pp.296–321. Also IEEE Spectrum, Feb. 1983, pp.67–71.

Radio Engineering (June 1925), p. 300

Income account, All-American:

year ending	Dec. 1923	Dec. 1924	Dec. 1925	Mar. 1927	Mar. 1928	Mar. 1929
gross income				1,265,000	1,474,000	547,000
net income (loss)	96,000	365,000	47,000	(261,000)	(170,000)	(169,000)

ALL-AMERICAN

Solves Every Gift Problem

If He Has a Radio Set Already—

It can be made a better one by installing genuine ALL-AMERICAN Audio Transformers. Two of these instruments, fitted into any set not already equipped with them, will give the receiver greater loud-speaker volume with remarkable purity of tone. ALL-AMERICAN Transformers are so designed that they amplify fundamentals and harmonics equally, throughout practically the entire audible range. Hence, voice and tones are reproduced *faithfully*.

Give him ALL-AMERICANS, the Audio Transformers which, through sheer merit, have become the largest selling transformers in the world. **3 to 1 Ratio, $4.50; 5 to 1 Ratio, $4.75; 10 to 1 Ratio, $4.75.**

If He Needs a High Grade Receiver—

Give him ALL-AMERICAN Super-Fine Parts, and he can build an intermediate-frequency receiver embodying all the most advanced features known in Radio. His set will be the envy of "distance" fans, as well as of his musical friends.

Super-Fine Parts are easily installed. No critical adjustments are necessary. Operation is smooth and flawless. And every part is ALL-AMERICAN—ask any Fan what that means in Radio! Sets built with Super-Fine Parts are unsurpassed for selectivity, range, volume, and tone quality. They represent in a very real sense *the ultimate* in radio broadcast reception.

Price, $26.00

ALL-AMAX JUNIOR

An All-American One-Tube Reflex

This is the ideal gift for the youthful beginner in Radio. It comes completely mounted on panel and baseboard, and can be easily wired in one delightful evening with the aid of clear photographs and a 48-page instruction book. Easy to tune—as selective as a multi-tube set—has "crystal" tone quality—volume enough for Speaker operation. It brings in far-distant stations, and *tunes out the locals.*

Price, complete (semi-finished) $22.00

The Radio Key Book

Will help anyone to hear farther and better. Contains practical hints for the set builder—tested hookups—diagrams of All-Amax and other circuits. Sent for 10 cents, coin or stamps

RAULAND MFG. CO.

2646 Coyne St., Chicago

Pioneers in the Industry

ALL-AMAX SENIOR

An All-American Three-Tube Reflex

The set to give if you are looking for a complete receiver of the highest type. Great range and selectivity are provided by three stages of tuned and SELF-TUNED Radio Frequency Amplification. A crystal detector and two stages of Standard ALL-AMERICAN-equipped Audio, insure Speaker volume on distant stations, with undistorted tone quality. Completely assembled—full directions for wiring.

Price, complete (semi-finished) $42.00

Largest Selling Transformers in the World

Radio News (Jan. 1925), p. 1255

The Manufacture of Modern Radio Receivers

The recent trend in radio set manufacture has been toward making all the component parts, as well as the receiver itself, under one roof.

Radio News (Aug. 1926), p. 128

Above is a "testing lane" in one large Western factory. Every coil, condenser, transformer, etc., must undergo a series of rigid tests before it is selected as suitable for incorporation in a receiver.

Above we see a big assembly room, where the units of parts from the various stamping, winding, and other machines are brought together, and united into complete apparatus by deft operators. Long stretches of benches can be seen in well-lighted rooms, where one operation follows another in quick succession. The suitable routing of assembly work is one of the greatest factors in high-quality and low-cost manufacture today.

At the left is shown a testing room for completed receivers. Each set, after it is assembled, must be tested over the broadcast range wave-lengths; actual local and distant reception is obtained to assure that it has suitable sensitivity and volume for distance getting.

Below, a typical punch press department, where stampings are made of all sizes, up to receiver frames and shields.

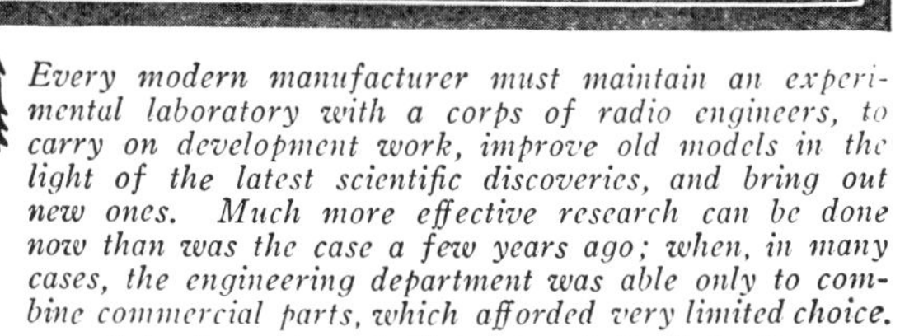

Every modern manufacturer must maintain an experimental laboratory with a corps of radio engineers, to carry on development work, improve old models in the light of the latest scientific discoveries, and bring out new ones. Much more effective research can be done now than was the case a few years ago; when, in many cases, the engineering department was able only to combine commercial parts, which afforded very limited choice.

Photos by courtesy of All-American Radio Corporation.

Radio Retailing (Aug. 1926), p. 32

RAULAND "SOVEREIGN"
(*7-tube*)
Blended Walnut Finish Base
Cabinet—Blended Walnut Brown Crackle
Parchment Finish on Gold Background
Embossed Floral Decorations
List Price—$435.00

RAULAND "LORRAINE"
(*7-tube*)
Blended Walnut Finish
List Price—$335.00
Blended Green Crackle Lacquer Finish
List Price—$335.00

RAULAND "FORTE"
(*7-tube*)
Blended Walnut Finish
List Price—$210.00

ALL-AMERICAN CONSTANT-B
Battery Eliminator
Has Tap for Power Tube
List Price—$37.50
(*Complete with Raytheon Tube*)

All-American Offers a Real Opportunity to Radio Dealers

You know that the best and soundest way to build a permanent, profitable business is to be identified with easy-to-sell goods of known high quality.

Here is one of the genuine *opportunities* in radio retailing—the splendid new line of Rauland Receivers. The franchise to sell these beautiful sets is exactly the kind of asset every substantial merchant likes to control.

You know All-American. The name is associated by those who know radio, with fine quality in those parts which really *make* radio reception. It is natural and logical that All-American should bring out as fine a line of receivers as engineering talent and manufacturing skill can possibly produce.

Rauland Receivers are as good as money can buy. We know; because we make them complete—using only units of All-American quality.

Wide-awake merchants will recognize in this line just the opportunity they have been seeking; to help make their own businesses better, as well as bigger. We'd be glad to hear from such concerns—those interested in "Radio for the years to come."

ALL-AMERICAN RADIO CORPORATION
4223 Belmont Avenue CHICAGO

ALL-AMERICAN REPRODUCER
Combines Cone and Sounding Chamber
Walnut Finish
List Price—$25.00

RAULAND "DUET"
(*6-tube*)
Blended Walnut Finish
List Price—$115.00
Blended Chinese Red Lacquer Finish
List Price—$120.00

ALL-AMERICAN MODEL R CABINET
(*5-tube*)
Blended Walnut Finish
List Price—$80.00
Blended Green Crackle Lacquer Finish
List Price—$85.00

ALL-AMERICAN MODEL R HI-BOY
(*5-tube*)
Two-Tone Walnut Finish
List Price—$115.00

ALL-AMERICAN MODEL R COMPARTMENT CABINET
(*5-tube*)
Two-Tone Walnut Finish
List Price—$90.00

RAULAND "SEXTET"
(*6-tube*)
Blended Walnut Finish
List Price—$175.00

R	Nov. 1925	$90.	R	Aug. 1926	$80,$85	Hi-boy	Aug. 1926	$115.
Duet	Aug. 1926	$115,$120.	Sextet	Aug. 1926	$175.			
Forte	Aug. 1926	$210.	Lorraine	Aug. 1926	$335.	Sovereign	Aug. 1926	$435.

Forte technical article in *Radio News*, May 1927, p. 1326. By Sept. 1927, several of these models had been converted to AC:

80,90,Hi-boy	('99 tubes)	$135, $145, $170
Duet, Sextet	('99 tubes)	$160, $220
Forte, Lorraine, Sovereign	(AC tubes)	$210, $360, $460

55,88

Radio Broadcast (Oct. 1927), p. 357

44

Flea Market

66,99

Radio Retailing (Sept. 1927), p. 195

71,253. RADIO RECEIVER CABINET. **Einar N. Rauland,** River Forest, Ill. Filed Aug. 25, 1926. Serial No. 18,858. Term of patent 3½ years.

Sept. 1927: 6 models, in 3 cabinet styles, using the same chassis:

battery:	44,	$70.	55,	$125.	66,	$200.
AC:	77,	$150.	88,	$210.	99,	$225.

PATENT OFFICE

The ornamental design for a radio receiver cabinet as shown.

Rauland also patented five other cabinet styles and a speaker at the same time.

E. N. Rauland, president of the All-American Radio Corporation station WENR, Chicago, adjusting the controls of the transmitter

Radio Age (June 1926), p. 25

Radio News (May 1928), p. 16

ALL-AMERICAN ELECTRIC. Model 98. made by All-American Radio Corp., Chicago; 6-tube; carved walnut console, 48″ high, 28″ wide, 18″ deep. List price with tubes, $196. **Dealer Helps:** displays, signs, literature, mats and cuts.

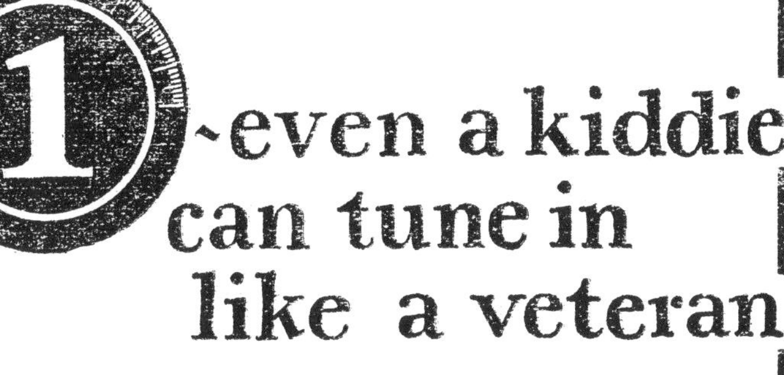

The Mohawk 5 Tube Receiver

Even your baby can tune it perfectly! What greater test can there be for the greatest radio value ever produced?

Here is a powerful 5 TUBE SET that can step neatly in and out of stations all over the country and all you do is turn ONE DIAL! The tuning is remarkably sharp. A tiny twist gets you what you want. Why you don't have to know a thing about radio. ONE simple DIAL manages the whole works. And the same number gives you the same station every time.

Every part in The MOHAWK is carefully built and adjusted. The set itself is a charming piece of craftsmanship. Note the beauty of its lines.

Get a MOHAWK for Christmas.

Distributed by

THE SUNBEAM RADIO SALES CO., Inc.,
1834 Broadway, N. Y.

THE SPARTAN ELECTRIC CORPORATION,
99 Chambers St., N. Y.

N. Y. Herald-Tribune (Dec. 21, 1924), p. 21

100 (VA) Nov. 1924 $100 110 (X) with speaker $250 (later $175) and 115 (XII) console $300 (later $225) available in Dec. 1924.

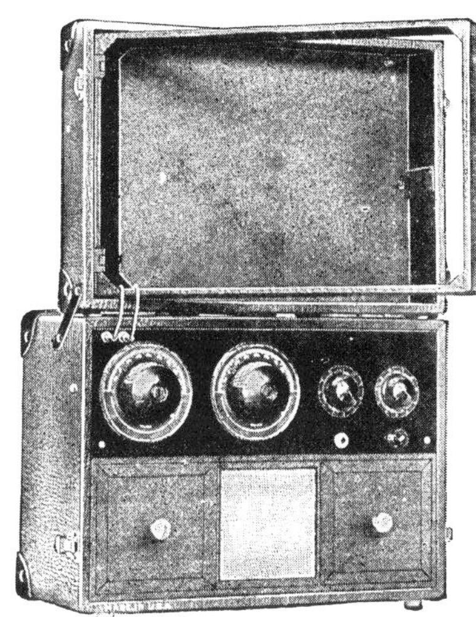

Radio Dealer (Dec. 1925), p. 152

MOHAWK SIX TUBE PORTABLE RADIO RECEIVER No. 105. Manufactured by the Mohawk Corp., of Illinois, 2222 Diversey Parkway, Chicago, Ill. Six tube receiving set using one stage tuned, R. F., two stages untuned R. F., detector and two stages, A. F. Designed for use with 199 tubes. Installed in California pine case covered with Dupont imitation shark grain leather. Cowhide leather corners; gun metal finish. Silver plated brass trimmings with carrying handle. Two 8:1 vernier tuning dials. Bakelite panel. Weighs 35 lbs. fully equipped. 11¾" high, 10¼" wide, 17¼" long. Self-contained loud speaker with wooden tone chamber. List price $135.00.

Radio Dealer (Nov. 1925), p. 135

Inventor of the well-known Mohawk one-dial receiving set, Paul A. Chamberlain has nevertheless but recently allowed the general radio public to hear of the great things he has been doing. In the obscurity of the laboratory, Mr. Chamberlain has been content to carry on his experiments, realizing the ambitions of his youth, when he first manifested the mechanical and electrical bent that later was to bring him to his well earned fame with the Mohawk Electric Corporation of Chicago.

Radio Dealer (Oct. 1925), p. 100

A Message to Dealers Who Think

(With apologies to Clarence Darrow)

Look at this radio. It has three dials . . . *Where is the three-handed operator?*

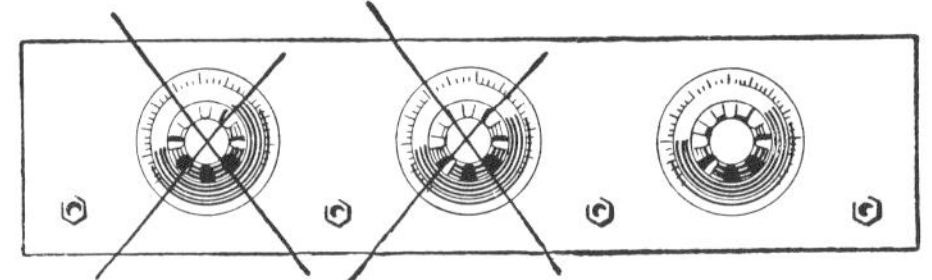

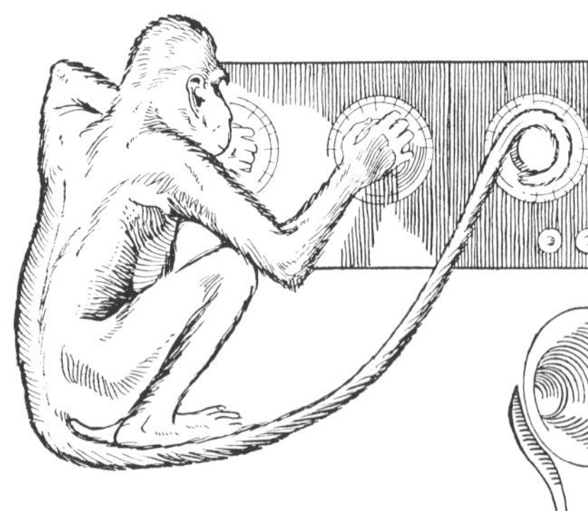

Now look again. The secret is out. Here is the operator of the three-dial radio. He has two hands and a useful tail. He is the operator.

Most people, however, do not have three-handed operators in their homes. So they prefer a radio they can operate themselves. Here it is —the Mohawk. It has five tubes—but only *one* dial to tune.

What, you ask, will this one-dial radio *do?* Is it selective? Does it give distance? Volume? Free range of the air? Beauty of tone? Undistorted, life-like reproduction?

Our answer is this. The Mohawk gives *three-dial* results with the use of only *one* dial. It has a patented, three-in-line *balanced* condenser. That is the reason. Compare results. Judge for yourself.

Other things being equal, everyone prefers a one-dial radio—a radio made for *women* as well as men. That explains the tremendous popularity of the Mohawk, the pioneer one-dial radio. That also explains the many announcements of one-dial sets now in the papers. Mohawk success has awakened the entire industry!

But only Mohawk has the patented *balanced* condenser. Only Mohawk can give three-dial results with just one dial. Sell the Mohawk and you will increase your radio sales this season. Write today for literature and list of Mohawk Jobbers.

Manufacturers
MOHAWK CORPORATION OF ILLINOIS
Independently organized in 1924
Chicago, Ill.

Sales Department
THE ZINKE COMPANY
1323 So. Michigan Ave.
Chicago, Ill.

Mohawk Model 100, five tubes just one dial to tune. Retail price, without accessories, $100.

Mohawk Consolette, Model 110, retail price without accessories, $175.

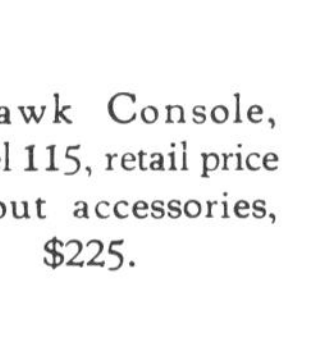

Mohawk Console, Model 115, retail price without accessories, $225.

Mohawk No. KU51 Kit contains all parts, including cabinet, for assembling a Mohawk Radio. Retail price, $75.

The Mohawk was selected from among 47 radios as standard equipment on the Pan American, crack train on the L & N Railroad. A test will tell you why.

Manufactured by Mohawk Corporation of Illinois
Independently Organized in 1924

5 tubes—just ONE dial to tune

Look at these prices!

LOOK at the prices on these pages! Study the console models that Mohawk dealers will sell at these prices in 1926-27! Do you know of values even remotely approaching these? Have you ever seen a line more obviously salable, more certain to bring quick, big profits? The Mohawk dealer of all dealers is surest of success in the coming season. Who else can sell a one-dial, 6-tube, shielded set for $65.00? Who else can sell table and console models of beauty comparable to the splendid One-Dial Mohawk line at the prices displayed here? There's only one answer — no one! There's only one move to make — write, or preferably wire, *today*, for full details of the Mohawk dealer proposition for 1926!

Mohawk Corporation of Illinois

Established 1920—Independently Organized in 1924

2220 Diversey, at Logan Boulevard, Chicago

To Distributors: A few, very few, jobbing territories remain open. What is said here to the dealer applies to you. A wire will bring our representative if your territory is one of those still open.

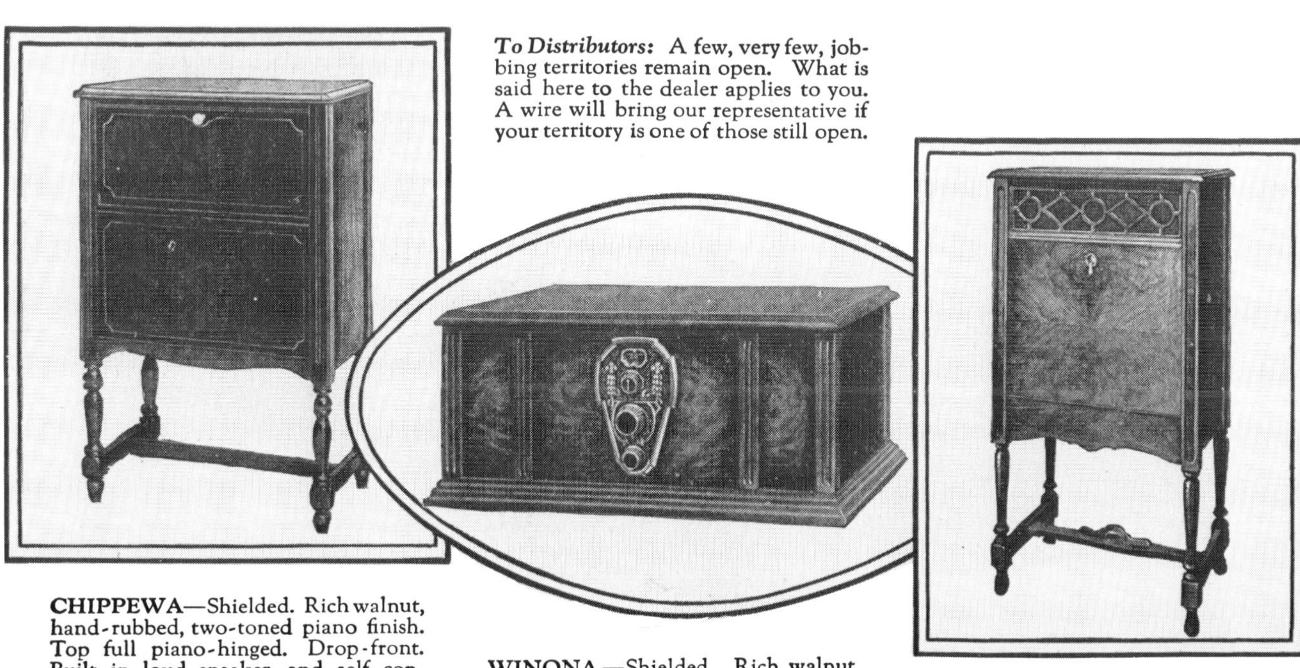

CHIPPEWA—Shielded. Rich walnut, hand-rubbed, two-toned piano finish. Top full piano-hinged. Drop-front. Built-in loud speaker and self-contained battery compartment, 40¾ inches high, 13½ inches deep, 27¼ inches wide. List price $110

WINONA—Shielded. Rich walnut, hand-rubbed piano finish. Full piano-hinged. 10¾ inches high, 13¼ inches deep, 24 inches long. List price $80

PONTIAC — Shielded. Rich walnut, hand-rubbed piano finish, with burl walnut drop front with invisible hinges. Built-in loud speaker. Self-contained battery compartment. 46 ins. high 15½ ins. deep, 25¾ ins. wide. List price . . $140

Mohawk Corporation of Illinois

GENEVA—Shielded. Rich walnut, hand-rubbed piano finish. Front full burl walnut, inlaid. Loud speaker built into dome. Self-contained battery compartment. 44 inches high, 16½ inches deep, 32 inches wide. List price—

$185

POCAHONTAS—Shielded. Colonial design in burl walnut inlaid and rich two-toned, hand rubbed piano finish walnut. Built-in loud speaker with 5-foot horn. Self-contained battery compartment. 45½ inches high, 29 inches wide, 20¾ inches deep. List price **$300**

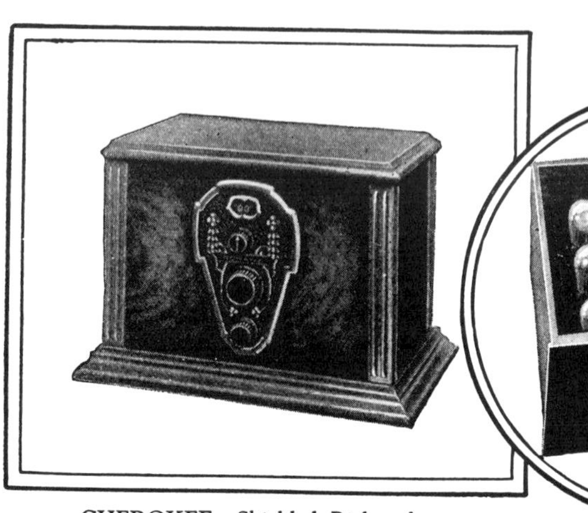

CHEROKEE—Shielded. Rich walnut hand-rubbed piano finish. Full piano-hinged. 10¾ inches high, 13¼ inches deep, 15½ inches long. List price **$65**

Prices west of the Rockies slightly higher. Canadian prices 40% higher.

SENECA — Drawer. Mohawk one-dial, six-tube shielded radio set in walnut drawer, interchangeable in all Mohawk console models. 8⅜ inches high, 12⅜ inches wide, 10 11/16 inches deep. List price **$57.50**

Established 1920, Independently Organized, 1924, Chicago, Ill.

Boston Traveler (Dec. 16, 1927)

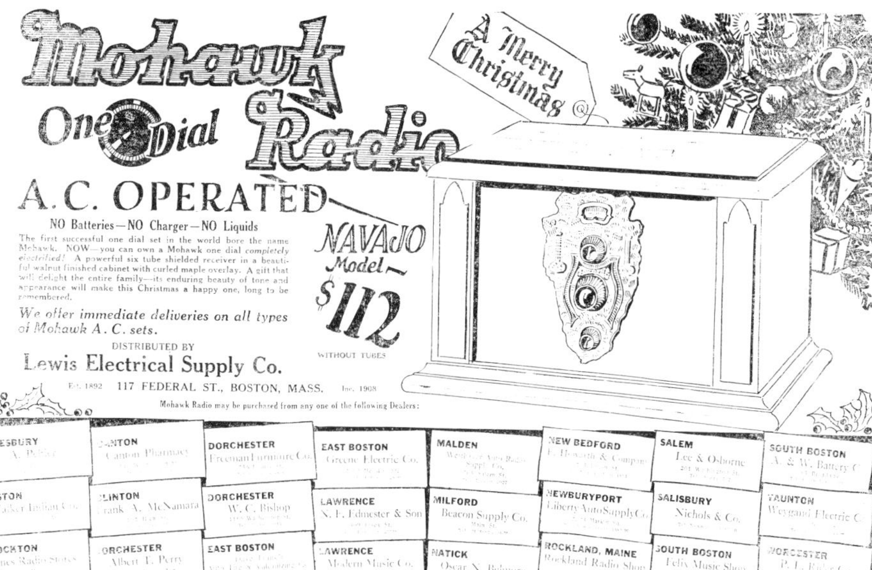

Navajo Robert Enemark

Cherokee June 1927 $65 (later Navajo, $67.50). 5 console cabinets available. AC model with Kellogg tubes, $110 extra. After Nov. 1927, AC models used RCA tubes, $72 extra.

Mohawk Corporation Officials Confer

Pictured above are the officials of the Mohawk Corporation of Illinois in executive conference, while formulating ideas for the merchandising of new Mohawk single-dial receivers. Left to right: Louis Frankel, treasurer, Otto N. Frankfort, general sales manager, Gustave Frankel, president and Douglas De Mare, chief engineer in charge of production.

Radio Retailing (Mar. 1927), p. 77

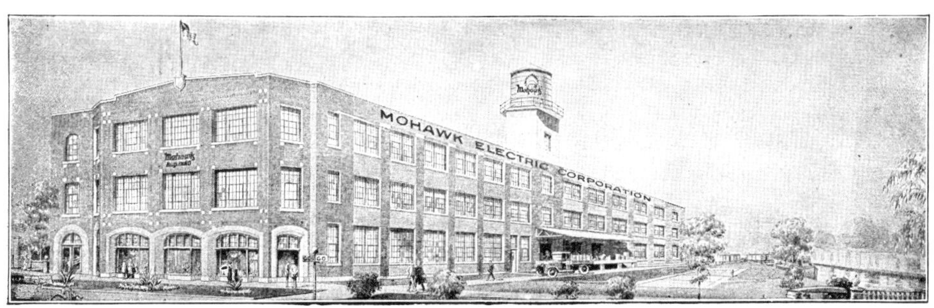

Above is shown the factory of the Mohawk Electric Corporation, of Chicago, where the famous one-dial set is made. The factory is one of the largest and most up-to-date buildings of its kind now in use for the manufacture of radio sets. Louis Frankel, Secretary of the Mohawk Corporation, says that even with the excellent manufacturing facilities at his command, the great demand for Mohawk sets keeps a full force working here all the time.

Radio Dealer (Oct. 1925), p. 155

A.C. *Receivers*

Both console and table type receivers are listed among the new products of the All-American Mohawk Corporation, 4201 Belmont Avenue, Chicago.

The illustrated table model is an eight-tube set for either battery or A.C. operation, size 20 in. by 13½ in. by 9 in. The intended retail price for A.C. operation is $127.50, battery operation, $95. Model 60, a six-tube set, slightly smaller in size, is $65 for battery operation and $92.50 for A.C.

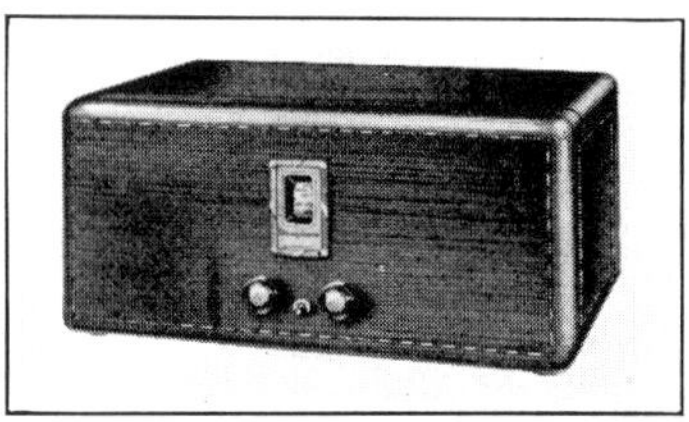

The console models include model 83, an eight-tube set with built-in magnetic reproducer for use on 110 volts A.C., 60 cycle, size 48 in. by 25½ in. by 16¾ in., price, $250 A.C., and $217.50 battery operated; the illustrated model 86, which is also an eight-tube set with built-in magnetic reproducer, size 52 in. by 27½ in. by 16 in., is $235 A.C., and $202.50 battery; model 85, same specifications as above and about the same size, $195 A.C., and $162.50 battery; model 61, which has six tube chassis No. 60 and built-in magnetic reproducer, slightly smaller than model 86, $165 A.C., and $137.50 battery; model 62, a highboy, has the six tube No. 60 chassis and built-in magnetic reproducer, $172.50

A.C. and $145 battery; and model 65, with six tube No. 60 chassis and built-in magnetic reproducer, size 40¾ in. by 19½ in. by 14½ in. is $137.50 A.C. and $110 battery operated.—*Radio Retailing*, July, 1928.

All models sold under the "Lyric" name after August, and 2 radio-phonograph models added: 66, $245 ($280 with dynamic speaker); 88, $425.

New Lyric Line

Three new sets are now being made by the All-American Mohawk Corporation, 4201 Belmont Avenue, Chicago, Ill. Each has three stages of R. F. and two A.F. stages, utilizing four 226 type tubes, two 227's, one 280 and one 250.

Model 70, a table set, is incased in a cabinet of five-ply walnut veneer. The intended retail price, with dynamic speaker in a cabinet to match, is $150.

Model 73, a console, with dynamic speaker, in a five-ply walnut veneer cabinet, is $157.50.

Model 75, a console with dynamic speaker, in a walnut veneer cabinet, 51 in. by 25⅞ in. by 14⅝ in., is $185.—*Radio Retailing*, March, 1929.

Model 93

New Lyric Models

The new series recently announced by the All-American Mohawk Corporation, 4201 Belmont Avenue, Chicago, Ill., includes three console models.

Model 93, illustrated, has ten tubes, including rectifier, in a Hazeltine neutrodyne circuit. It has double push-pull amplification and a dynamic speaker. The cabinet is made of five-ply walnut with matched walnut veneer above arch and on the apron. The intended retail price is $169.50, complete with tubes.

Model 95 has the same chassis and is known as the de luxe console. Grained walnut is used on all exposed surfaces with burl overlay on the arch. The controls are enclosed by a door which acts as an arm rest when open. The finish is natural walnut. Price, complete, $199.50.

Model SG-1, illustrated, is a seven tube set using three 224's. It has five tuned circuits, including tuned detector, three stages of r.f. and one tuned selector circuit. The doors and control panel are made of selected matched walnut. It has Oriental walnut overlays on two vertical panels. Price, complete, $187.50. *Radio Retailing*, August, 1929.

Model SG-1

Model 97

New Lyric Models

The screen-grid or the 10 tube Lyric chassis may now be had in a choice of three Italian Renaissance period cabinets, all similar as to carving on front, style of doors, etc., the only difference being in the length of the legs, says the announcement from the All-American Mohawk Corporation, 4201 Belmont Ave., Chicago, Ill. The dial panel and inside of the door panels are made of butt walnut. A coat of flat lacquer is applied over the usual lacquer coats, giving the effect of a wax finish.

Model 97 is a lowboy, 41 inches high. With the 10 tube chassis, the intended retail price is $225; with the screen-grid chassis, $230.

Model 98 in the highboy style is 48½ in. high. It retails at $235, with the 10-tube chassis; and at $240, with the screen-grid chassis.

Model 99 is the same as Model 98 but is one inch higher and has matched walnut backing from the lower part of the cabinet proper to a shelf on the bottom. The intended retail price with 10-tube chassis is $245; with screen grid chassis, $250.—*Radio Retailing*, December, 1929.

Model 99

Radio Retailing (June 1928), p. 196

WURLITZER

TEL-O-AIR
CONSOLE
PERIOD MODEL

SIX TUBES

Single Dial Control

Price **$250.00**
Less Accessories

PRICES SLIGHTLY HIGHER
WEST OF THE
ROCKY MOUNTAINS

MODEL ONE
(Open View)

(Closed View)

The Wurlitzer presentation of period models bring greater satisfaction to every radio enthusiast. Model one is an example of artistry and finished craftsmanship that commands unstinted admiration. Handsomely carved and finished in American walnut or mahogany with the grace of line and beauty of construction that make it an attractive addition to any home.

Its distinguished appearance is enhanced by its unfailing performance. Its receiving set is described on page 2. There is ample room in compartments for the Wurlitzer "A" and "B" power supply or batteries. Loud speaker and unit is built in with an artistic grille for covering.

MODEL FOUR

WURLITZER

TEL-O-AIR TABLE MODEL

Six Tube Single Dial Control

Price **$100.00** Less Accessories

This table Cabinet Model is a distinctive and efficient set with the standard Tel-O-Air Receiver as described on page 2. Attractively made and finished. Top and ends are five ply. Face veneers are selected, figured and Matched Stump Genuine American Black Walnut. Also furnished in Mahogany.

The front is solid walnut paneled effect.

The base is heavily moulded and artistically embossed.

The panel is 7″ x 18″ with $10\frac{1}{2}$″ depth behind panel. Groove arranged for either panel of metal or $\frac{3}{16}$″ Bakelite.

PRICES SLIGHTLY HIGHER WEST OF ROCKY MOUNTAINS

MODELS NINE AND TEN

WURLITZER

TABLE MODEL—Resistance Coupled

SIX TUBES, THREE DIALS Price **$49.00** Less Accessories

MODEL TEN—5 TUBES, TRANSFORMER COUPLED—THREE DIALS

Price **$39.00** Less Accessories

Another Wurlitzer table model so pleasing to look at, so enjoyable to hear. This cabinet enables the set to be moved in different rooms with convenience. The cabinet is finished either in walnut or mahogany. Handsomely carved and decorated. The panel is artistically engraved and highly finished. This model is a beautiful piece of work from an artistic as well as a perfect radio receiving set. This model has a six-tube, resistance coupled, amplifier and three dial control.

Model ten is identically the same with the exception of having only five tubes, and is transformer coupled.

PRICES SLIGHTLY HIGHER WEST OF ROCKY MOUNTAINS

MODEL SIX—*(Closed View)*

WURLITZER

DUO-CONTROL TABLE MODEL

SIX TUBES, TRANSFORMER COUPLED

Price **$85.00** Less Accessories

There is a class that prefers to have a table model radio receiving set. And Wurlitzer has designed and built this model six to satisfy this demand. It is neat and compact, can easily be placed on any living room table adding charm to the room as well as pleasure in performance. It has a six tube, tuned radio frequency with dual control. The cabinet is very artistically decorated.

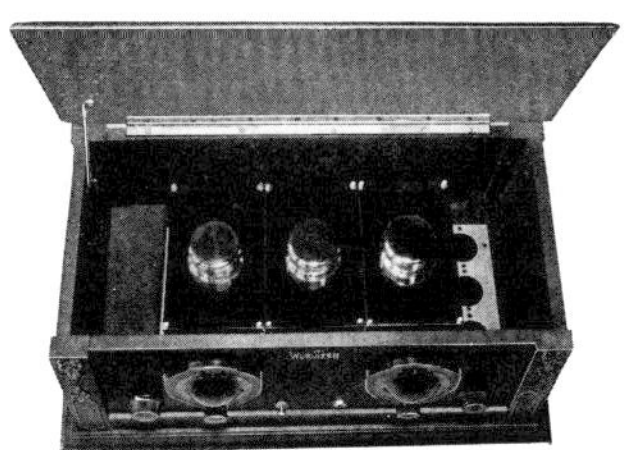

(Open View)

The interior view gives some idea of the construction, note that set is shielded, thereby adding to its selectivity. It is properly wired to give reception clearly and without the hum or distortion.

PRICES SLIGHTLY HIGHER WEST OF ROCKY MOUNTAINS

AMERICAN BOSCH

American Bosch Magneto Corp.

American Bosch Magneto Corp. was the outgrowth of a business established by Robert Bosch in Stuttgart, Germany in 1885. In 1906, Robert Bosch and Otto Heins organized an American sales agency under the name of Robert Bosch New York, Inc. which in 1912 changed its name to Bosch Magneto Co. upon construction of a plant in Springfield, Massachusetts. Bosch and Heins, being German subjects, returned to Germany at the outbreak of the war; in 1918 the Alien Property Custodian seized the company's assets and sold them to businessmen who formed the American Bosch Magneto Co. in 1919.

Robert Bosch, having formed Robert Bosch A.G., Stuttgart, in 1917 and an American sales agency Robert Bosch Magneto Co., Inc. in 1921, fought a series of lawsuits with American Bosch during the twenties for possession of the "Bosch" trademark, finally winning a decision by the Commissioner of Patents in May 1929. The two companies combined in December 1930 as United American Bosch Corp., and while the U.S. company was said to have acquired the German branch, it is a fact that by 1940 three-fourths of the company was owned by, or in the name of, Swedish interests.

Bosch announced its entry into the radio field in December 1924. In keeping with the orderly methods of a large company ($10.5 million in sales for 1926), Bosch designed the set early in 1925, produced a certain number in mid- to late-1925, and began advertising in September for the winter selling season. Apparently Bosch forecast very well, as it had sold its entire production by January 1926, and claimed to have made $1.1 million in radio sales in 1925. Indulging in a bit of speculation, we could estimate that 12,000 Amborolas were made, at a wholesale price of $90 each (assuming a 40% discount off the list price of $150) which is a reasonable production figure. Leslie F. Curtis (b. 1888) was chief engineer from January 1924 to at least 1937.

Since its magneto and auto-accessory business was shrinking in the late 1920s, Bosch found it profitable to manufacture radios for other companies too. It made several models for Sonora in mid-1927, until Sonora joined with Arborphone and got its own manufacturing plant. Sonora's place was filled by the National Carbon Co. who wanted to introduce radios under the Eveready name to take up the slack in its dwindling battery production, now that battery-powered sets were giving way to AC models. In 1928 and 1929 Bosch made a line of Eveready sets, using chassis almost identical to its own models (National Carbon also joined with Raytheon in making tubes in June 1929, an arrangement that lasted until 1933). Eveready dropped out in late-1929 but Bosch continued making radios in the 1930s. A subsidiary, Essex Radio Co. of Springfield, was created around 1934 and dissolved in January 1939.

The company was again taken over by the Alien Property Custodian during WWII. Later it merged with Arma Corp. and added military equipment to its former line of diesel fuel-injection components and magnetos and other electrical engine accessories. American Bosch, still in Springfield, is now a division of United Technologies.

Radio Dealer (July 1926), p. 44

Rich Elskamp

16 Amborola Sept. 1925 $150

The Cruiser 5 tubes—$100.

35 Cruiser July 1926 $100
A technical article and description of the factory assembly process is in *QST*, Jan. 1927, pp. 22–28.

BOSCH announces The LITTLE SIX

Those who have seen it, heard it and operated it have named this model Bosch Radio the "Little Wonder Six." They have been amazed that a six tube, single dial radio receiver with such tonal quality and perfect performance could be purchased for so little as $68.50. Consider the features which make the Bosch Little Six an outstanding radio investment at its low price of $68.50. It is space-saving —but sixteen inches long. It has a Single Station Selector, electrically lighted; six tubes, vibration proof mounted; aluminum chassis, light and strong; the Bosch Clarifier, the Bosch Volume Control and, best of all, the famous Bosch tonal accuracy. The cabinet is walnut finished and its colonial simplicity of design adds to its richness of appearance. Bosch precision workmanship and Bosch radio engineering have been so blended in the Little Six it is a revelation in performance in its class. The Bosch Little Six will appeal instantly to those who have waited for a space-saving, six tube receiver with power and Bosch tonal quality at a low price. A full description of the Bosch Little Six will be mailed on request.

All these Bosch Radio Models—ready for Bosch Socket Power Units—the Nobattry "A" and Nobattry "B"—both totaling . . $100.00

Model 66—Six tube, Single Station Selector, Table Type, wired for battery or socket power . $99.50.

Model 87—Seven tube, Single Station Selector, Table Type, loop operated, wired for battery or socket power . . . . . . $195.

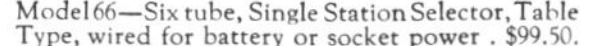

Model 57—Seven tube, Single Station Selector, Cabinet Type, concealed loop, built-in reproducer wired for battery or socket power operation $340.

Model 76—Six tube, Single Station Selector, Cabinet Type, wired for battery or socket power—with built-in speaker . . . . . . $195.
Without built-in speaker . . . $175.

AMERICAN BOSCH MAGNETO CORPORATION
SPRINGFIELD, MASS. Branches: NEW YORK CHICAGO DETROIT SAN FRANCISCO

Bosch Radio Receivers are licensed only for Radio Amateur, Experimental and Broadcast Reception. They are manufactured under patent applications of American Bosch Magneto Corp. and are licensed under patent applications and patents of Radio Corp. of America and under applications of Radio Frequency Laboratories, Inc.

first advertising dates:

57, 66, 76 June 1927
46, 87 Sept. 1927

The first digit is a serial number, the last digit the number of tubes.

Five-Tube Unified Control Receiver

Radio Retailing, March, 1927

The American Bosch Magneto Company, Springfield, Mass., has added to its line the "Imperial Cruiser" which is a cabinet type receiver. Simplicity of design is demonstrated in its colonial style which has two full length doors opening to the radio controls at arm chair height. A removable panel gives access to all batteries and power units. The cabinet is walnut throughout of matched grain and artistically highlighted. Complete with the Library Ambotone reproducer, the intended retail price is $147.50.

35 Imperial Cruiser Mar. 1927 $147.50

27 Amborada July 1926 $310

Technical article in Popular Radio, Oct. 1926, pp. 528, 570–577.

Five-Tube Table Type Receiver

Radio Retailing, March, 1927

The "Royal Cruiser" is the newest addition to the line of radio receivers made by the American Bosch Magneto Corporation, Springfield, Mass. This model is encased in a solid walnut cabinet has five tubes and the unified control which provides single dial simplicity of operation with two dial advantages. The intended retail price is $110.

35 Royal Cruiser Mar. 1927 $110

46 Bob Wallace

66 Ralph & Elinor Williams

Saturday Evening Post (Dec. 17, 1927), p. 119

BOSCH
offers five perfected AC MODELS

THE BEST IN RADIO
BOSCH RADIO

Model 107—Completely self-contained 7-tube receiver, AC tube operated, with reproducer, all tubes, loop—nothing else to buy. $440.00

Model 96—Completely self-contained 6-tube receiver, AC tube operated, with reproducer, all tubes—nothing else to buy. $295.00

Just think of a radio receiver, so perfected that it may be brought into your home, a single plug inserted in the wall socket and a world of music is at your command. These Bosch Radio models are designed as alternating current, socket power operated receivers, requiring no batteries, chargers, water or acids. Never before has radio been so simple to own and operate. There are models requiring no antenna; there are models with the reproducer in the cabinet; there are table type models—whatever may be your fancy, there is a Bosch Radio Model to please it. The range of selection in Bosch Radio includes five AC tube models and five standard tube models. All are well engineered, precision built, beautifully designed and perfectly finished. You owe it to yourself to see the Bosch Cabinets and hear Bosch Radio before buying any radio. There is an authorized Bosch dealer near you whose name we will supply if you wish.

Model 116—Completely self-contained 6-tube table type AC tube receiver, including B eliminator tube but less AC tubes. $170.00

Model 66AC—Six tube AC operated two unit model—Receiver and A & B power—complete with B eliminator tube, but less AC tubes. . . $155.00

Model 126—Six tubes completely self-contained table type AC tube receiver with B eliminator tube, but less AC tubes. $135.00

All prices slightly higher in Canada.

AMERICAN BOSCH MAGNETO CORPORATION

SPRINGFIELD, MASS. Branches: NEW YORK CHICAGO DETROIT SAN FRANCISCO

Bosch Radio Receivers are licensed only for Radio Amateur, Experimental and Broadcast Reception. They are manufactured under patent applications of American Bosch Magneto Corp. and are licensed under patent applications and patents of Radio Corp. of America and under applications of Radio Frequency Laboratories, Inc.

John Bayusik

28 June 1928 $132.50 (later $110)
This RF chassis went into various console models: the 28A, $197.50 (later $170, or $195 with dynamic speaker); finally in Mar. 1929 the 29D, $225.
Short technical articles in *Radio Broadcast*, Oct. 1928, p.369, and *Citizens Callbook*, vol. 10 no.4, Nov. 1929, p.91.

Model H

Model 16

48-H 48-16

Two Bosch Receivers

Models 16 and H have been added to the line made by the American Bosch Magneto Corporation, Springfield, Mass. The chassis in these receivers uses three 224's, one 227, two 245's and a 280 rectifier.

Model 16 is highboy, 48¼ in. high and 28¾ in. wide, with sliding doors. Price, $198.50.

Model H comes in a Tudor design walnut cabinet, with attractive walnut veneer decorations. Price, $198.50.—*Radio Retailing*, October, 1929.

Radio Retailing (June 1929), p. 172

The Bosch Radio Combination Receiver and Speaker Console will be in great demand because it embodies Screen Grid quality radio in an inexpensive combination of charming individuality. Perfectly chosen woods and veneers combined with rich carving make this console one of beauty. It has the Bosch Dynamic type speaker. List price, less tubes, $168.50

48 June 1929 $119.50 consoles $168.50, $198.50, $240. Technical articles in *Radio Broadcast*, Sept. 1929, pp.290–292, and *Radio*, Jan. 1930, pp. 52–53.
Other unadvertised Bosch models can be found in the service literature or Langley/McMahon's *Radio Collectors Guide*.

The Country Gentleman (October 1928), p. 74

Radio Sets

Built to a Fidelity Curve that proves their faithfulness of reproduction

The Eveready Fidelity Curve

is the standard of Eveready Radio Reproduction

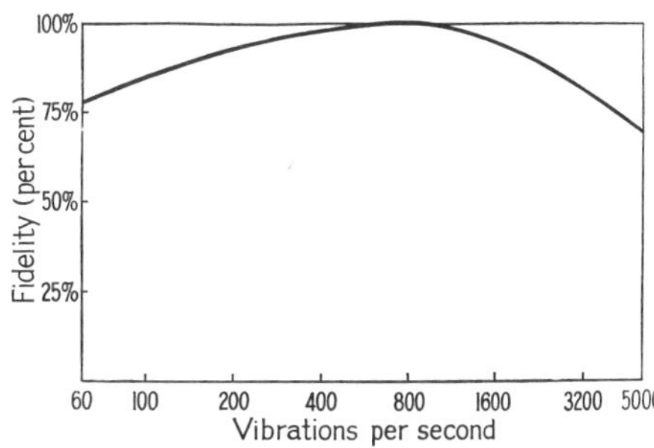

Music and speech are vibration, and by scientific measurement it has been found that the important sounds in voice and music are included within a scale of 60 to 5000 vibrations per second. This includes not only the fundamental notes but also most of the important harmonics of these notes. See on the Eveready Fidelity Curve (above) that from 60 to 4000 vibrations per second are reproduced with an unusually high degree of fidelity. The weakening of the notes above 4000 minimizes the disagreeable effects of static and other high-pitched noises. Very few notes go below 100 vibrations per second, and broadcast transmitters themselves do not put on the air lower than 60 vibrations per second.

This curve shows the faithfulness with which the Eveready Radio Set delivers speech and music to the speaker. It is essential that the speaker possess a high degree of faithfulness, for to the extent to which it may be deficient will the full measure of Eveready Fidelity be diminished. The Eveready Speaker is recommended.

Below is a list of all the principal instruments and voices, with their lowest and highest notes in vibrations per second.

Violin	192 to 3856	Saxophones	52 to 1024
Viola	256 to 1280	French Horn	60 to 682
Cello	64 to 854	Cornets	140 to 960
Bass	32 to 427	Trombone	80 to 320
Harp	30 to 3072	Tuba	42 to 341
Flute	256 to 2048	Kettle Drums	85 to 170
Piccolo	576 to 4096	Piano	78 to 4096
Oboe	240 to 1365	Soprano	256 to 1024
English Horn	160 to 960	Contralto	170 to 682
Clarinet	144 to 1920	Tenor	144 to 427
Bassoon	60 to 640	Baritone	107 to 341
DoubleBassoon	30 to 320	Bass	80 to 288

The new Eveready AC Set, Model No. 2, in die-cast aluminum cabinet, lacquered in green with striping in natural aluminum.

This receiver has seven radio tubes and one rectifier, eight in all. All power is taken from the light socket. Table model without tubes, $155.

Die-cast aluminum legs, finished to match the set, may be had to convert this into an end table, as shown above, at $20 extra.

The new Eveready AC Set, Model No. 1, in solid gumwood cabinet, antique maple finish. A cabinet that will harmonize especially well with Colonial and Early American interiors.

Same radio chassis as in the die-cast aluminum set. Price, table model without tubes, $145.

Legs in same wood finish, to convert table model into an end table, as shown above, may be had for $10 extra.

Die-Cast Aluminum Cabinets

The most modern of radio cabinets—Die-Cast Aluminum. Cabinet is everlasting, light, strong and is lacquered in green in a modern design with striping that reveals the aluminum like burnished silver; the whole protected by a transparent lacquer that prevents discoloration and resists scratching. This unusually beautiful and entirely modern cabinet is exclusive with Eveready.

The new Eveready Battery Set, Model No. 20, in solid gumwood cabinet, antique maple finish. Six tubes.

Has the lowest "B" battery drain of any 6-tube receiver yet produced, as five of its tubes are "High-Mu," combining great amplifying power with minimum current.

Price, table model without tubes, $85. Eveready speaker to match, $30. Pedestal with ample battery space, $15 extra.

At left, the new Eveready Loud Speaker in die-cast aluminum housing, decorated in green lacquer, with natural aluminum striping, to match the Eveready AC Set Model No. 2. Price, $35.

At right, the new Eveready Loud Speaker in solid gumwood cabinet, antique maple finish, to match the maple cabinets of either the AC Set No. 1 or the Eveready Battery Set Model No. 20. Price, $30.

Licensed under patents and applications of RCA and RFL

20, 21 June 1928 $85 1, 3 June 1928 $145 Same chassis as Bosch 28.
2 June 1928 $155 Cabinet design by Walter D. Teague
(patents 77,513 & 77,514 filed Oct. 18, 1928, issued Jan. 15, 1929)

2

Radio Retailing (June 1929), p. 66

Model 33, ***a larger and more luxurious console, in walnut finish with decorative carvings. Same all-electric chassis as*** **Model 32.** ***All cabinet designs are exclusive with Eveready.***
LIST $210 ***without tubes***

Model 31, ***table type. Cabinet in rich walnut finish with contrasting carved grill. Same radio chassis as the consoles. Will operate either dynamic or magnetic speaker.***
LIST $115 ***without tubes***

Model 32, ***console, in a cabinet of rich walnut finish that fits in any decorative scheme whatever. Dynamic speaker built in. Chassis has eight tubes, including rectifier.***
LIST $175 ***without tubes***

Radio Retailing (June 1929), p. 67

31 June 1929 $115 Consoles: 32, $175. 33, $210 34, $225.

By Aug., 42, 43, 44 available (same cabinets, '45 audio tubes) at $5 extra; and 52, 53, 54 (screen-grid RF) $10 more.

Technical article on the 50-series chassis in *Radio*, Dec. 1929, pp. 53–54.

Model 54

The American Radio and Research Corporation

What's in a name? "American Radio and Research Corp." was perhaps no more grandiose than "International Radio Telegraph" or "Federal Telephone & Telegraph" but, as a venture financed by the banking house of J. Pierpoint Morgan which had previously created U.S. Steel and General Electric, the intent was evident. Certainly, putting the word "research" into the title was indicative of future plans; if radio was not much of a business in 1915, it might become one in time, and "Amrad" would be ready.

However, J.P. Morgan had died in 1913, and it was his son and namesake "Jack" who was now running the empire, and lacking his father's aggressive instincts, was letting it slowly waste away. He entrusted Amrad to the wireless operator on his yacht, a boy just out of Tufts, an amateur since 1905 and a ship operator since 1907, with plenty of ambition but no business experience: Harold J. Power.

"Jimmy" immediately became "H.J." and began spending Morgan's money, building a research laboratory on the edge of the Tufts campus, and surrounding himself with personnel. From a 300-foot tower he first broadcast a few phonograph records on March 4, 1916 and on March 18 he transmitted a three-hour program to J.P. Morgan Jr. on the liner Philadelphia off Cape Cod. And added another "favorite son" to the endless debate over who was the first broadcaster.

Research might have continued to be Amrad's only activity, had not World War I occurred. But in June 1916 the Signal Corps ordered eight "cart sets," each made of a transmitter and receiver and a gasoline-engine-driven generator, mounted on two field-gun caissons. These sets, being too large for the basement machine shop, were made in New Jersey to Amrad designs. An addition was made to the rear of the laboratory, a machine shop on the first floor and a woodworking shop on the partial basement floor, but before it could be completed, in April 1917 the Navy ordered a number of motor boat transmitters, to go on recently-commandeered power boats and launches. Renting the Tufts machine shop, Amrad rushed production and shipped samples to Washington, where they were tested with generators from another company. When the first three generators burned out as soon as they were connected, the Navy cancelled the entire order, leaving Amrad with a stockroom full of sets. But fortunately other orders were received, to keep the plant humming.

About this time, Frederick W. Sammis was hired away from American Marconi, where he had long been chief engineer. He was paid $77 per week, nearly double what anyone else was making (the machinists earned $15 to $20), and his job may have been to pilot Amrad through the shoals of government contracting.

In early 1918 the Signal Corps wanted several thousand SCR74A trench transmitters, immediately, and a second floor had to be added to the building, while an empty pumping station was rented in nearby Somerville for production of accessories. Before the Armistice, Amrad had completed its half of an order for 100 battleship and cruiser transmitters. And 100 submarine detectors designed by consultant Vannevar Bush had been completed, with three installed on British subchasers, where they were technically successful in locating British submarines, but too late to find any German ones.

At this point Amrad employed about 75, but with the war's end, government contracts vanished and the company took on any sort of production work ("egg beaters and cigar lighters") to keep busy. This should have been a warning to Power that total reliance on government work was dangerous, but there was little else available, and up to now it had been quite lucrative: $700,000 worth by the end of 1921. So Amrad waited out the slump, and later in 1919 secured a Navy contract for 400 SE1420 receivers. Once the hams were back on the air in October, Amrad began advertising a few components, but its main effort there was to unload spark coils and quenched gaps left over from wartime contracts. It made no complete receivers, other than under Navy contract, and therefore when Armstrong approached Amrad early in 1920 with an offer of a half interest in his regeneration patent for $500, Power offhandedly countered with $250 and told the inventor to "sleep on it."* Armstrong at that time was in need of an ally to help pay his legal bills, until his attorneys hit upon the idea of licensing all the small manufacturers of ham gear, to bring in revenue, while waiting for one of the large electrical companies to notice the patent (Westinghouse bought it in late 1920).

But Amrad had no need of Armstrong. Government business looked good and Morgan was willing to finance a gigantic new plant of 200,000 sq. ft., of which 30,000 sq. ft. was completed by August 1920 and housed 150 workers. Part of it was devoted to manufacture of "Twin-R" fractional horsepower motors.

For the amateur market, Amrad did design a series of "units" that could be assembled in any combination to make receivers. Although the drawings were made in October 1919, it was nearly a year before samples went out to

*Story related in 1980 by Eunice Thompson, who said it could not at that time be verified, but had been common knowledge among Amrad executives.

the dealers, and January 1921 before *Radio News* carried the ads. The crystal set was one of these units, and by good fortune was ready at just the time when broadcasting began to interest the general public. 1XE had of course been broadcasting since 1916, and more or less regularly since October 1919, but no one at Amrad seemed to grasp the possibilities until KDKA and 8MK (Detroit News, later WWJ) had gathered momentum in late 1920. It's not quite true that broadcasting took the country by storm during the Harding-Cox election—KDKA's work didn't even rate a paragraph in *Wireless Age*, *Radio News* or *QST*—but someone at Amrad was rudely awakened, as the December issue of *Radio News* carried an ad for the Amrad crystal set outfit, aimed at the broadcast fan (since the set was not pictured there, one might logically assume that the ad was prepared in too great a rush for a cut to be included). Amrad also staged a pre-Christmas demonstration at Filene's, a large Boston department store, and expanded its broadcasting hours to promote crystal-set sales, but did not transmit daily until May 20, 1921.

The "unit" sets, while fine in theory, did not sell well; broadcast fans didn't want to be bothered connecting up six or more separate boxes when they could buy a complete receiver, tested and working, for less money from someone else. So Amrad engineers (Howard Tyzzer, actually) came up with a two-section receiver: a tuner in one cabinet and a three-tube detector-amplifier above. This was first shown at the September 1921 ARRL Convention and advertised in the October *QST* and November *Radio News*. Incredibly, it was advertised as a regenerative receiver, although Amrad obviously had no Armstrong license. When Westinghouse instantly laid down the law, the next month's ads were for a "short-wave" receiver (same model number).

This short-wave "double-decker" should have been Amrad's salvation, as it was ready just in time for the radio boom. Beginning around Christmas 1921, no dealer could keep enough stock to satisfy the demand. Dealers placed ten times as many orders with manufacturers as they could handle, knowing they would receive only a fraction of what they ordered, and that they could sell anything they could get. Enticed by this phantom demand, manufacturers began crash programs to increase their output, and this was where H.J.'s perfect (on paper) organization fell apart. Amrad was slow. It had too many executives, managers and directors and its policy was set by committees, all controlled by H.J. So by the time it got into production, summer had come, and the boom had collapsed back in May. Now no one wanted radios at any price, and when they finally did, in the fall, Amrad's model was a year out of date.

By late 1922 Amrad was in the hole. Morgan was said to have sunk $800,000 and was understandably reluctant to throw good money after bad. The slow-moving inventory of Twin-R motors was used as collateral for a bank loan. Sales Promotion Manager G.K. Thompson stated in 1980 that the backlog in late 1922 approached half a million dollars, none of which could be delivered. Probably these orders were for the new RF-amplifier model 3380. In an attempt to make the old double-deckers into RF models, an add-on RF-amplifier 3071 was offered in October, and by January 1923 the double-deckers themselves were modified to make them saleable. Once the Christmas season was over, though, the half-million in unfilled orders were cancelled as quickly as they had been placed. Executive dismissals began in February 1923. First to go was G.K. Thompson, whose pay, indexed to his sales volume, was higher than Power's.

Probably because it did a substantial business in crystal sets and felt threatened by Wireless Specialty Apparatus Co.'s aggressive patent policy, Amrad joined the Independent Radio Manufacturers, Inc. in March 1923. This was the group of small New York makers who had banded together to fight WSA, and were encouraging Prof. Hazeltine to develop his Neutrodyne invention. Accordingly, when licenses were issued in April to the original members, Amrad got one.

Amrad might have done well to introduce a Neutrodyne in 1923, but it did not; instead, Tyzzer designed an inductively-tuned four-tube set in August 1923, the Inductrole. It was exhibited at the New York Radio Exposition in October 1923, but true to Amrad form, could not be produced until March 1924. Up to February, meanwhile, numerous ads in Philadelphia newspapers offered model 35-U "double-deckers" made up of six unit panels—a holdover from the 1919 design! A Neutrodyne was finally produced in December 1924 but was too little, too late; Amrad went into receivership in April 1925.

By 1925 the Neutrodyne trademark had become a bit shopworn, but still signified to the public a reliable, easy-to-use radio. Armstrong regenerative sets, on the other hand, were definitely outmoded and were under attack from many quarters for their oscillating/interfering tendencies. Powel Crosley had made a great success of regenerative models but he needed something new, and he wanted a Neutrodyne license. So it was natural that he should purchase the remains of Amrad. He probably considered his $39,000 well-spent for the Neutrodyne license alone, but he also got the complete Amrad factory in the deal, as well as Mershon condenser licenses which he used extensively later.

Many of the old personnel returned to work in late 1925. Their first job was the repair of factory-reject or returned Crosley models, probably to keep them busy while Powel figured what to do with the place. Soon they were busy producing the new Amrad Neutrodynes, with an unmistakable Crosley aura about them. In early 1927 Power left to form his own company to make B-eliminators; Major J.E. Hahn of DeForest-Crosley in Toronto took over the presidency. The plan was to make a higher-priced, quality line, while Crosley kept the cheaper models; if it hadn't been for the stock-market crash and Depression, Amrad probably would have survived. However, it closed down in 1930, and the factory equipment and many personnel went to Cincinnati, while the Mershon division was sold to Magnavox and moved to Ft. Wayne.

HAROLD J. POWER

Radio World (May 6, 1922)

1915 laboratory

New England Wireless & Steam Museum

2331 E Dec. 1920 $25

Rear view, showing 1916 and 1918 additions. Later used by Tufts and known as North Hall, it burned in 1972.

Rich Elskamp

2575 Apr. 1922 $21.50

This was also advertised with the magnetic detector (p. 46).

New England Wireless & Steam Museum

SE1420C 1920

Building transmitter-receivers, 1917.

Machining rotary spark gaps, 1917.

Testing quenched-gap transmitters, about 1919.

Two views of the laboratory, about 1919

Photos from Mel Dunbrack W1 BHD

SCR74 antennas and accessories in the Somerville pumping station, 1918.

Assembly of SCR74 "trench" transmitters, 1918

Office, Sept. 1920

Office, Aug. 1919.

Rear view of newly-completed factory, Sept. 1920.

1920 factory, later a Tufts gymnasium.

Eunice Randall, Ken Thompson and Howard Tyzzer at the New York Amateur Show, Mar. 7–11, 1922.

Paul Corrette

2596 & 2634 Sept. 1921 $45 & $47.50

Known among collectors as the "double-decker." Originally the "grid load" variometer was marked "plate" until Westinghouse forced Amrad to stop advertising this as a regenerative set. However, the binding posts remained in the same positions in back, allowing the dealer or owner to reconnect this variometer into the detector plate circuit with two short jumpers, making it regenerative once again. Virtually all of them were modified in this way.

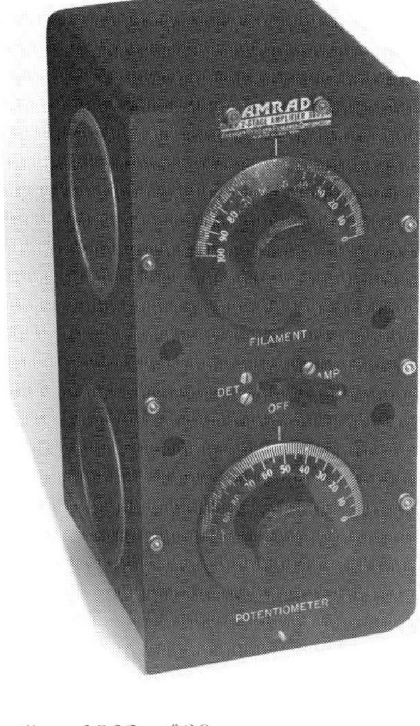

3071 Oct. 1922 $30
An add-on to the double-decker, providing RF amplification, an attempt to make up for the lack of regeneration.

Phila. Inquirer (Dec. 2, 1923

(Amrad Bulletin B.)

3380 (3108 & 2634) Oct. 1922 $125

Rich Elskamp

3500-1 (3475 & 2634) Jan. 1923
Apparently a modification of existing double-deckers, since the dial legends are on little plates pinned over the originals.

New England Wireless & Steam Museum

3500-2 (3730 & 2634)
A modification of the 3380, changing the 3108 into a 3730 with a variocoupler instead of two switch-selected variometers.

THE SATURDAY EVENING POST
March 10, 1923

AMRAD

"THE VOICE OF THE AIR"

RADIO can give to *your* household the entertainment, education and happiness that enrich family life.

In your living room, by your own fireside, celebrated artists sing and play for you—famous men of affairs talk to you—wonderful orchestras perform for you—the stock market sends you its latest news—eminent instructors share their knowledge with you—sporting events are reported play by play.

AMRAD Sets, the standard of Radio, are extremely simple to operate, as untechnical as the combination of a safe. The result of seven years of development, they are ahead of the times.

If you want to enjoy long distant radio broadcasting at the lowest possible cost, *demand* the Amrad REFLEX Receiver illustrated below. This REFLEX has twice the sensitivity of an ordinary receiver and will bring in *consistently* distant broadcasting from 50 to 300 miles. Users frequently report much greater distances up to 1000 miles. Beginners obtain astonishing results.

Ask your Dealer for a Demonstration of AMRAD Radio in Your Home and see for yourself the pleasure and real enjoyment the whole family will obtain with this up-to-date installation.

Write us for the interesting Booklet "The Voice of the Air", free on request.

Amrad Reflex Receiver, $40.00. (Accessory equipment, $23.55 or $42.50 extra, according to type of installation)

AMERICAN RADIO and RESEARCH CORPORATION
MEDFORD HILLSIDE, MASS.
NEW YORK CHICAGO

3366 Oct. 1922 $40

A relatively popular model, this one-tube reflex was designed between June 1 and Aug. 29, 1922. Eunice Randall patented the crystal detector, which used a permanent magnet to hold the "catwhisker" against the crystal.

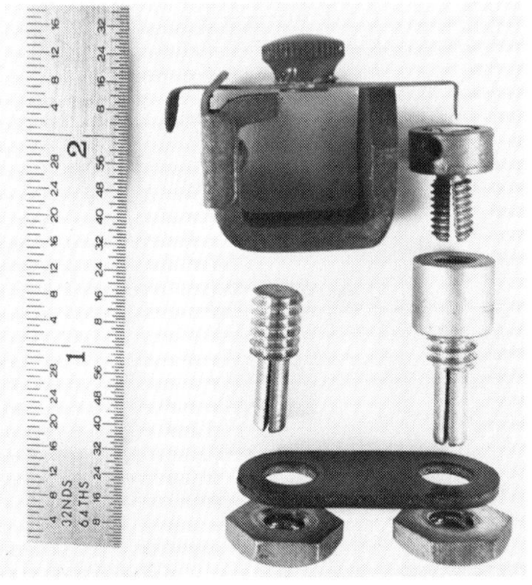

New England Wireless & Steam Museum

Doug & Jan Furney

3670 Duo Aug. 1923

Designed in May 1923 but introduced rather late for the summer trade. The covers are held in place with a strap.

Neutrodyne Dec. 1924 $85

Andrew Mooradian

3500-U

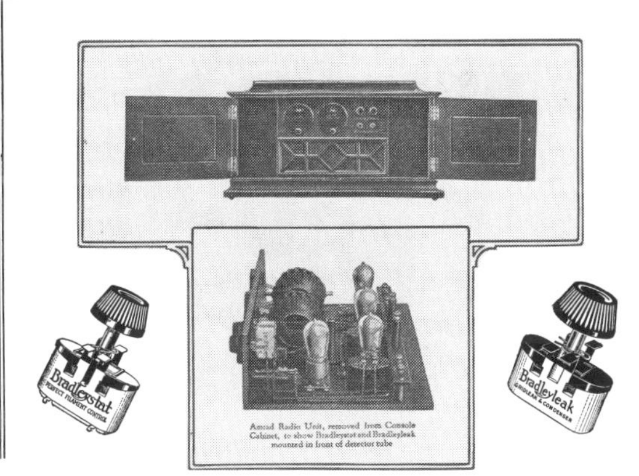

Amrad Adopts Both Bradleystat and Bradleyleak!

The Amrad Jewel—Italian Renaissance Period Art Model
Is Now Equipped With Ultra-Fine Tuning Control

ALL of the more expensive Amrad receiving sets, including the beautiful Jewel Console models, are now equipped with Bradleystats and Bradleyleaks! The noisy wire rheostats have given way to the noiseless Bradleystat. The old type of grid leak is replaced by the stepless Bradleyleak. The perfect filament control of the Bradleystat means greater range and louder reception. The stepless grid leak adjustment of the Bradleyleak, from 1/4 to 10 megohms, means higher tube efficiency.

The Console models, with self-contained loud speaker, battery compartment, and highly perfected tuner, are made more selective with the ultra-fine filament and grid control, so essential for long range reception.

YOUR radio receiving set will afford new possibilities and new thrills if equipped with Bradleystats and Bradleyleaks. Many radio dealers replace the wire rheostats of ready-built sets with Bradleystats, and they invariably recommend them to set builders who seek the best in radio.

The Bradleyleak has the endorsement of Amrad, Flewelling, Kennedy, Crosley, Clarkson, Cockaday, and other radio engineers. It is pronounced "the perfect grid leak" by all users.

Get the benefit of the graphite disc design by avoiding all substitutes. Carbon or metallic powder was abandoned, years ago, as impractical and unreliable. Insist that your dealer supply you with the genuine Bradleystat and Bradleyleak.

Send for the latest bulletins on closer tuning and perfect grid leak control

Allen-Bradley Co.
Electric Controlling Apparatus

288 Greenfield Ave.  Milwaukee, Wis.

There is no substitute for the scientifically-treated graphite discs

THE ALLEN-BRADLEY CO. HAS BUILT GRAPHITE DISC RHEOSTATS FOR OVER TWENTY YEARS

Radio (Apr. 1924), p. 96

Radio News (Sept. 1926), p. 313

NOW—A Lamp Socket Set $150

Model AC-5
Set $65.00
Power Unit $85.00

AMRAD MUSICONSOLE
Equipped with Model S-522 $92.00
With AC-5 (Lamp Socket Set) $182

THE Amrad power-driven Neutrodyne meets the demand for the best possible radio at a price that makes luxury economical.

Hear it demonstrated by your dealer. And for the first time hear full, rich symphonic character from bass to treble. Enjoy the thrill of perfect tone quality.

The Amrad Lamp Socket Set represents five years of organized research, development and a year of actual test in homes. It is selective, simple to operate, has full tone range, is cased in a handsome cabinet and has the exclusive lamp socket feature which does away with A, B and C Batteries. At $150 it is the most amazing Neutrodyne value in the world of radio.

Send for folder giving complete description of all Amrad Neutrodynes and name of a near-by dealer where they can be seen and heard.

Dealers
For full particulars regarding this newest Amrad development—and the Amrad line write to your jobber, or direct to us.

Address:
Department L-S
Amrad Corporation
Medford Hillside, Mass.

5-Tube Neutrodyne
Selective, powerful, simple in operation; genuine licensed Neutrodyne, only.....$60.00

AMRAD Power Driven Neutrodyne

The Amrad Corporation
NEUTRODYNE

S-522 May 1926 $60 AC-5 Aug. 1926 $150

AMRAD Neutrodyne

Radio Retailing (Mar. 1927), p. 101

Serve Your Community Completely with this Complete Line ~ priced*far below competitive quality

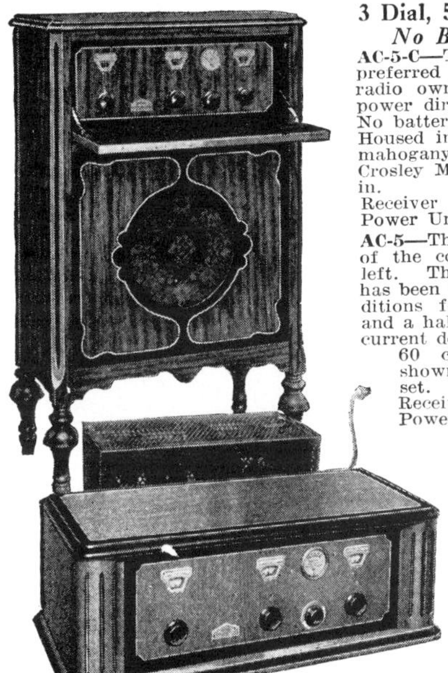

3 Dial, 5 Tube, Batteryless
No Batteries Needed

AC-5-C—This type of control is preferred by many experienced radio owners. All necessary power direct from light socket. No batteries, no trickle charger. Housed in a beautiful two-tone mahogany cabinet with genuine Crosley Musicone skilfully built in.

Receiver $115 **$175**
Power Unit $60

AC-5—This is the table model of the console pictured to the left. The Amrad power unit has been tested under home conditions for more than a year and a half. Operates from A.C. current delivering 100-120 volts, 60 cycle. Power unit is shown behind table model set.

Receiver $65 **$125**
Power Unit $60

3 Dial,
5 Tube Models
Battery Type

S-522-C—Beautiful in appearance, a wonder in selectivity and performance. Console model—two-toned mahogany — with Crosley Musicone skilfully concealed behind silk screen. Delicate vernier controls make tuning extremely easy and efficient.

$110

S-522—Amrad quality is again exemplified in this beautifully made and proportioned set. The simple, elegant lines of this set win the admiration of all. The first of a line that has kept production at peak in the Amrad factory ever since.

$60

CONE TABLE

A mahogany cabinet finished in two-tones with the genuine Crosley Musicone cleverly concealed — the Musicone that won universal praise. Cone table matches Amrad table model receivers in design and finish. **$32**

* Amrad low prices reflect the influence of the Crosley purchasing power and the Crosley resources. This, coupled with Amrad engineering skill and the technical achievements of Amrad laboratories, is a wonderful combination.

Write for sales franchise at once. Many desirable territories still open.

7 Tube S-733 Models
Battery Type—2 Dial

S-733-C—This exquisite model is a handsome piece of furniture to grace any home. Delivers the utmost in radio enjoyment at a very reasonable price. Crosley purchasing power and Amrad engineering skill is reflected in this set. With the simplified 2 dial control, it becomes a super value at **$127**

S-733—Owners of this Amrad Neutrodyne report complete satisfaction. High ratio vernier controls make tuning easy. Sockets are mounted on rubber cushioned base. Volume is controlled by a single adjustment. Two-toned mahogany cabinet, a wonderful value at **$77**

7 Tube, 2 Dial Batteryless
No Batteries Needed

AC-9-C—This radio is designed especially for A.C. power. It is designed for use with the Amrad power unit—a thoroughly tested batteryless power supply of great efficiency. The value of this set will be appreciated by many.

Receiver $132 **$192**
Power Unit $60

AC-9—This set is similar in construction to the console. Two dials simplify tuning. The elegant lines of this model—two-toned mahogany—will harmonize with the surroundings in any home. Receiver $82. Power Unit $60 **$142**

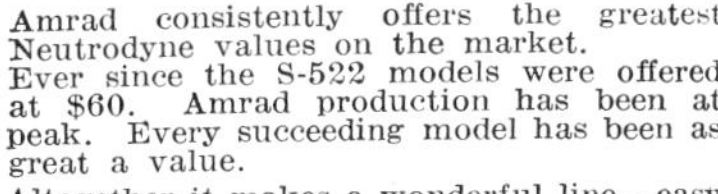

"B" Eliminator

The famous "Mershon" Condenser in the design of this "B" eliminator makes possible a source of "B" power supply with finest TONE QUALITY. The unit is housed in a metal cabinet and finished in black enamel. All parts are easily accessible, in plain view and not covered with a compound. No variable controls. Furnished with the Famous Amrad S-1 Tube, Voltages—22½ or 45; 67, 90, 135 or 180. Max. volts at 50 mils.

$35

Amrad consistently offers the greatest Neutrodyne values on the market.

Ever since the S-522 models were offered at $60. Amrad production has been at peak. Every succeeding model has been as great a value.

Altogether it makes a wonderful line—easy to sell and easy to service.

Write Dept. 8C7 for descriptive literature.

THE AMRAD CORPORATION

Medford Hillside, Mass.

Voltage Regulator

Will feed a steady current no matter how badly the line voltage wavers. Many homes can now operate Amrad Batteryless Lamp Socket Sets. Heretofore many could not enjoy the convenience of lamp socket power because of badly fluctuating electric currents, causing broadcasts to swing in and out, or fade. Requires 1 UX—876 Ballast Tube. For complete satisfaction a voltage regulator should be included when sets are sold.

$15

AMRAD *The Voice of the Air*

THE "INDUCTROLE"—A RADIO MARVEL

THIS newest AMRAD has the great long distance and "selective" features of the famous AMRAD 35 Models. In addition, it marks a radical improvement in simplified operation, beauty of appearance and all round efficiency.

No Radio at any price compares in all fundamentals with the remarkable AMRAD "Inductrole" which utilizes Inductance Tuning.

As an example of careful engineering, this Model may be operated with either One Tuning Dial or Two Tuning Dials. One Dial Operation is sufficient for receiving broadcast programs under ordinary conditions. This means there is only one Dial to rotate to tune in various stations—a distinct AMRAD feature. But when special selectivity is desired, *i.e.*, when one or more broadcasts "interfere" with one another and it is impossible to distinguish one from the other, then change to Two Dial Operation (two dials for tuning). By carefully following the directions, it is then possible to "tune-out" the broadcast not wanted.

INDUCTANCE TUNING

The secret why AMRAD is the "World's Finest Radio" lies in the scientific principle incorporated in each AMRAD—namely, Inductance Tuning. Generally speaking, distance range and volume, prime necessities in the Radio Set, depend on the amount of voltage furnished the vacuum tubes. By Inductance greater voltage reaches the tubes than in any other way.

To design a Set with Inductance Tuning requires organized research. It is a far more difficult achievement than the capacity principle included in the average set. In fact, it is practically impossible unless the tuning elements are very low in distributive capacity. The marvelous Basketball Variometers, developed exclusively for AMRAD—provide this low distributive capacity. Thus AMRAD — and AMRAD alone—is able to utilize the highly desirable engineering principle of Inductance Tuning.

SPECIFICATIONS

The "Inductrole" is encased in a handsome, hand-rubbed mahogany cabinet, 17" x 14" x 12", with special compartments for batteries. Use of storage battery is recommended to light the tube filaments. 4 "B" Batteries (large size) for plate supply.

Wavelength Range 220 to 560 meters, which covers all broadcasting. May be used on either indoor or outdoor antenna. Requires 4 vacuum tubes — Cunningham or Radiotron. (See page 4.)

The head telephone set is plugged into the front panel. Removing this plug, automatically transfers to Loud Speaker which is connected to "Horn" terminals of the Set.

3500-3 Inductrole Mar. 1924
$100, $180, $285, $425, $800, $1200
Technical article in *New York Evening World*, Aug. 16, 1924 p.3

Radio Retailing (July 1927), p. 144

The Windsor, $195
AC operated, $295

7-tube Compact, pure one-dial control, extremely selective, operated on loop or antenna, all parts completely copper shielded.

DC6	Warwick	July 1928	$138
AC6	Warwick	July 1928	$238
DC7	Windsor	July 1928	$195
AC7	Windsor	July 1928	$295

Short AC7 technical article in *Radio Broadcast*, June 1928, p. 91

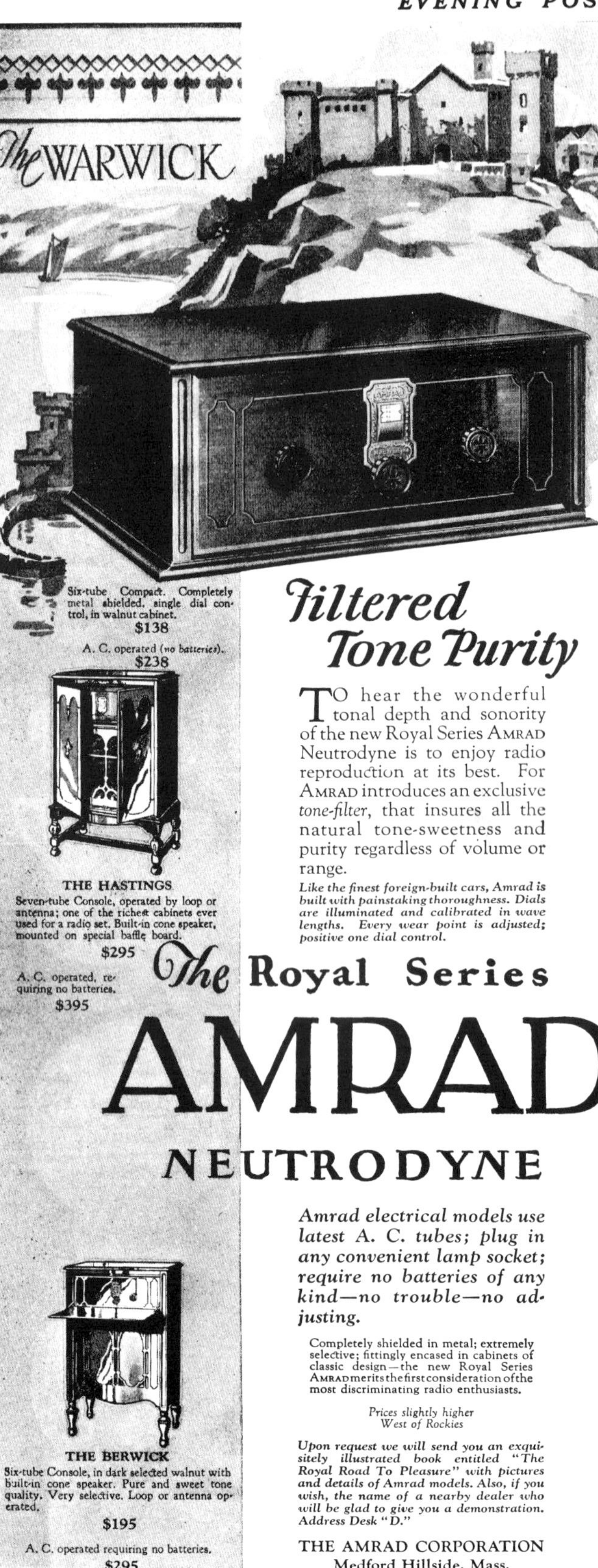

The SYMPHONIC Series

NEW DYNAMIC MODELS

Purely Electrical Operation!

Radio Retailing (June 1928)

The SONATA

Louis XVI, finished in light walnut with Butt Walnut doors and inside panel. Decorative panels of satinwood and zebra wood, add character as well as beauty. A fine piece of furniture for any home, in harmony with modern design.

The inside panel contains a handsome bronze enameled escutcheon plate with illuminated single dial control. The radio is the most modern purely electrical design, using power tube UX-250 or UX-210. Double shielded, extremely sensitive and designed for utmost selectivity, with a full rich tone that has never before been achieved in a Console radio, and is due to the use of the very finest type of Dynamic Loud Speaker (R.C.A. 105) giving the entire range of the musical scale with exquisite beauty and power. Dimensions 51½x34x17 in. Price **$475** (with out tubes).

Priced slightly higher West of the Rockies.

The CONCERTO

This beautifully proportioned cabinet reflects modernistic tendencies in furniture. Finest veneers are used with top and sides of American walnut and front of diamond matched oriental walnut. Exquisite satinwood border. Doors swing fully back. The decorations are of genuine solid brass in antique finish. Dimensions 49½x30½x17⅜ in.

The CONCERTO contains the purely electrical Amrad Chassis using power tube UX-250 or UX-210. The unique tone quality is achieved by an electric Dynamic Power Speaker built into the cabinet, with exclusive Amrad construction. It is double shielded and has extreme selectivity and sensitivity. Illuminated single dial control and bronze escutcheon plate enameled in color. Price **$320** (without tubes.)

Priced slightly higher West of the Rockies.

Special Features

The Chassis includes a tone control in the rear of the Chassis, enabling the user to adjust the tone of the receiver to suit his taste, emphasizing either the high or the low notes.

It also has an electrical phonograph pick-up attachment which becomes effective by throwing a small switch, employing the audio amplifying system and electric dynamic speaker for phonographs of any type —giving the full richness and volume of tone, even in the case of small portable phonographs.

One of the outstanding features is that which permits the use of the electrical connection for the receiver (through the lamp socket) as an antenna and ground, in which case neither outdoor nor indoor antenna is necessary. Best results, however, can be obtained by use of an outdoor antenna and ground wire. The principal use of the antenna plug-in is in demonstrating the receiver either in the home or in the display room.

The NOCTURNE

A beautiful Console model of walnut veneer, with doors of choicest Butt Walnut stock. Finished in the most modern trend. A built-in Dynamic Power Speaker gives a tone production of marvelous fidelity and rich beauty. The radio is purely electrical house current type, uses power tube UX-250 or UX-210. Specially designed to give utmost selectivity, double shielded and extremely sensitive. Illuminated single dial control, and bronze escutcheon plate enameled in color. Dimensions 50x30x17 in.

Price **$295** (without tubes).

Priced slightly higher West of the Rockies.

Also *Opera* $875 Model 70 Technical article in *Citizens Callbook* vol. 10 no.4, Nov. 1929, p.92.

THE BEL CANTO SERIES

Prices slightly higher West of the Rockies.

The ARIA

Doors of selected Butt Walnut Veneer, with African Walnut overlay top and bottom. The inside of this modified Art Moderne cabinet has a fine figured walnut face in an attractive Gothic design. New ultra-sensitive Amrad chassis using shielded grid tubes; equipped with nine-inch Peerless Dynamic Speaker. *List*........$198

The SYMPHONY

Beautiful cabinet of Art Moderne design. Front and sides veneered in highly figured East Indian Laurel Wood, with base rail of Macassa Ebony, decorated with inlays of ebony and holly Rounded and recessed top of Oriental Walnut. Inside panel of matched Oriental Walnut Veneer. The Amrad screen grid chassis is especially designed for utmost selectivity and sensitivity. The special audio system, in combination with the built-in R. C. A. 106 Dynamic Speaker, gives an unequalled rich tone production. The chassis uses 8 tubes. *List*........$295

The SERENATA

The simplicity of this modern sliding door cabinet sets off the rich beauty of diamond matched Oriental Walnut, and other fine woods. Uses standard Amrad shielded grid chassis with R. C. A. 106 Dynamic Speaker. Escutcheon plate and door pulls of old bronze finish, in harmony with console. *List*........$245

The DUET

A combination electrical radio and phonograph, inspired by the finest Art Moderne furniture. The beautiful veneers in this cabinet are of Oriental Laurel, appropriately decorated by inlays of ebony and maple with Macassa Ebony base rail.

The inside front panel is of finely matched Oriental Walnut with inlaid border. The grille is extremely beautiful and the escutcheon plate is of silver with a shield enamelled in scarlet and blue. The door pulls are of antique silver.

Shielded grid tubes are used in the chassis, which is extremely powerful and unusually sensitive and includes the R. C. A. 106 Dynamic Speaker built in cabinet. *List*........$495

Radio Retailing (June 1929) p. 211

Model 81 Technical article in *Citizens Callbook* vol. 11 no.2, Mar. 1930, p.81.

AMRAD S-TUBE

Amrad S-tube

Although the S-tube is not directly related to radio-receiver manufacture, it was one of Amrad's major products and has never been adequately covered elsewhere.

In 1919 Vannevar Bush hired for the Amrad laboratory Charles Grover Smith, who had studied mathematics and physics at Harvard after graduating from the University of Texas. Smith worked quietly in the lab with his rare gas mixtures and spectroscope, accumulating data in reports known among Amrad people as the "S" or "Smith" papers. In 1921 he was able to make a workable rectifier using the principles he had discovered, presenting a paper to the IRE (*Proc.*, Feb. 1922, pp. 41–51). This first rectifier required an external magnetic field, but in early 1922 he had further developed one that used no magnet, only electrodes of a certain shape (*QST*, Aug. 1922 pp.11–13; *Trans. AIEE*, June 1922 pp.402–11). This model, known as the S tube, went into production in May 1922 as the model 3000, for $8. Due to high voltages required, it was suitable only for transmitters, not receivers; it was far better than mechanical rectifiers or chemical "slop jars" but rather expensive for the average ham.

When Amrad folded, Laurence Marshall of the American Appliance Co. (later Raytheon) purchased Smith's pending patent applications from J.P. Morgan Jr. and hired Smith to keep developing his inventions, resulting in the Raytheon B rectifier in November 1925. While the B worked on the much older principle of rectification by unequal electrode areas, it did use the "short-path" principle of the S-tube for insulation around the electrodes, thus giving Raytheon full patent coverage over all practical gaseous rectifiers. The story of Raytheon has been written in detail in *The Creative Ordeal* (New York: Atheneum, 1974) but many of its statements about Amrad are wildly inaccurate.

In April 1927 Amrad advertised a B-eliminator using its own "S-1" gaseous rectifier but this lasted only about a month—no doubt it was squelched by Raytheon.

Aria

Flea Market

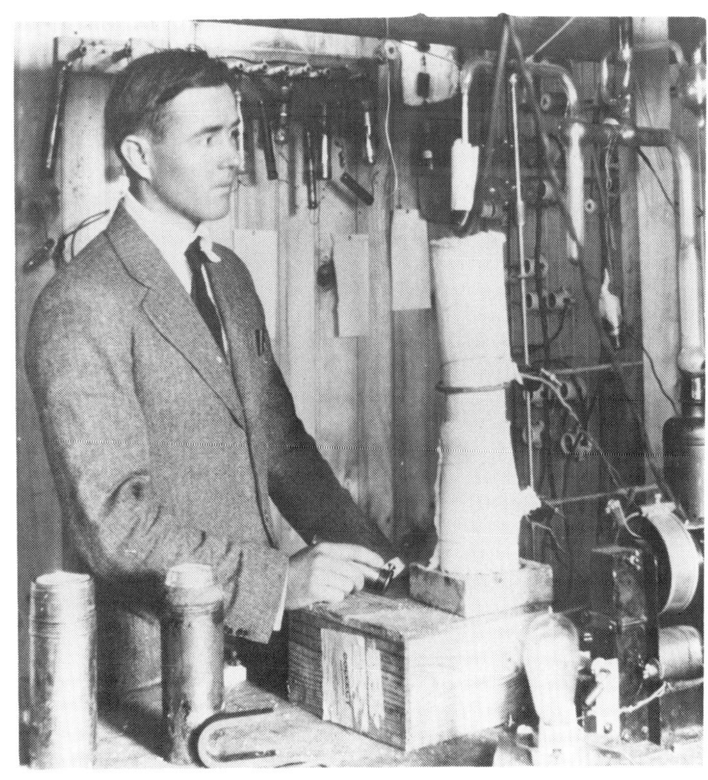

APEX, INDIANA (CASE)

Apex Electric Manufacturing Co. Indiana Electric & Manufacturing Co. (Case)

The Apex Electric Manufacturing Co., incorporated in 1911, made automobile windshield cleaners and pumps. It entered the radio field in 1925, a year too late, and got off on the wrong foot by overproducing. It made so many Super-Fives for the 1925–1926 season that they were still being offered in 1926–1927, finally being dumped to a New York surplus dealer by April 1927.

Too small to qualify for an RCA patent license ($100,000 yearly minimum royalty), Apex combined on August 20, 1927 with Case, Sentinel, Slagle, and Workrite to form the United States Electric Corporation. Case had begun in 1923 as the Indiana Electric & Mfg. Co. in Marion. However, this merger lasted only until January 1928 when one of the companies failed. Apex and Case (and two sales companies) subsequently re-merged on November 30, 1928 as the United States Radio & Television Corporation, dropping the Case tradename in favor of Apex.

U.S. Radio & TV continued to make money in the early 1930s, merging with General Household Utilities (Grunow refrigerators, formed by William C. Grunow in 1932 after being kicked out of Majestic) on July 12, 1933. Grunow, however, filed for bankruptcy in November 1935 and reorganized in 1936, discontinuing radio production in October 1937 and finally selling its Marion plant (where all radios were made) to Farnsworth in March 1939.

Consolidated Income Account: Apex, Case, and sales companies:

				(10 mos.)
year ending	Dec. 31, 1925	Dec. 31, 1926	Dec. 31, 1927	Oct. 31, 1928
net sales	$1,293,321	$2,365,216	$1,897,465	$2,572,840
profit(deficit)	$82,505	$167,058	($60,764)	$154,861

			(6 mos.)
year ending	Nov. 30, 1929	July 31, 1930	Jan. 31, 1931
net sales	$10,032,396	not avail.	not avail.
profit	$5,597	$365,467	$715,931

Radio Dealer (Aug. 1925), p. 180

Radio Retailing (Oct. 1927), p. 111

A. E. Case

500		July 1925	$62.50	503		Nov. 1925	$100
700	console	Oct. 1925	$175	701	console	Nov. 1925	$200
702	console	Nov. 1925	$175				

INDIANA HYPERDYNE RADIO

WILL HEREAFTER BE MARKETED UNDER THIS NEW NAME "CASE RADIO APPARATUS"

Nothing Changed but the Name!

MODEL 503

A 6-tube receiver using same parts and circuit as No. 500 It is the final refinement in radio and uses 3 vernier-controlled, silver-plated pointers. Works on inside or outside aerial. Retails $100.00. West of Rocky Mountains, $105.00.

Model 506 — Same as above only 5 tubes. Lists at $75.00. West of Rockies, $90.00.

If you haven't heard this set—do so. It will surprise you as it has others. Selectivity, distance, volume and tone—quality unequaled in any radio near the price. Every part and set complete, built in our own modern plant, keeping down price.

Buy radio merchandise that is built right—that stays sold, when sold. Jobbers all over the country are doing exceptionally well with our line. Do not delay, the time is opportune. We have a distributing proposition that will interest you. Write today for full descriptive literature and dealer's attractive helps.

Description No. 500—Five tubes using Tuned Radio Frequency—no howls or squeals. Solid mahogany cabinet 24 x 8 x 12; 15° sloping panel of black crystal lacquer on aluminum. Best material throughout. No exposed wires and many other good features........ $65 List

CONSOLE
MODEL 701

Same as Model 503, placed in this beautiful walnut console with built-in Utah speaker. Retails $200.00. West of Rocky Mountains, $210.00.

Offices

NEW YORK CITY
Geo. L. Holmes
1819 Broadway

SAN FRANCISCO, CAL
T. A. Mitchell Co.
Balboa Bldg.

FT. WORTH, TEX.
F. J. Keller Co.
905 W. Magnolia Ave.

MINNEAPOLIS, MINN.
J. C. Roper
310 Market Bank Bldg.

ATLANTA, GA.
Darling & Nevins
411 Ga. Sav. Bk. Bldg.

MODEL 500

Factory and General Offices

Indiana Manufacturing & Electric Co.

530 CASE BLOCK MARION, IND.

Radio Apparatus and Automotive Accessories

The Radio Dealer (Jan. 1926), p. 49

600	May 1926	$90
606	May 1926	$100
603	May 1926	$110
703	Apr. 1926	$170

6 Tubes Multi-Power $85.00

Think of it! Six tube reception with four degrees of power amplification for $85.00. Case Radio model 503, gives four degrees of power, a degree of power for every condition. This is only one of our nine models.

A full and complete line of radio apparatus to meet every demand. Built in five and six tube models, table type without speaker, table type with speaker and console models, listed as follows:

Model	Price	Model	Price	Model	Price
500	$65.00	600	$90.00	702	$160.00
506	75.00	606	100.00	703	170.00
503	85.00	603	110.00	701	180.00

The CASE line of Radio Receivers has been highly complimented from every section of the country. Strict economy in production, fabricating from raw materials and building the set complete in one plant keeps down the cost of CASE RADIO APPARATUS. They give positive performance at all times.

Jobbers and Dealers—A few splendid franchises remain. Our advertising will help you. Get our plans and general bulletin. Write today.

Quality Radio Reception At All Times

Indiana Manufacturing & Electric Co.
600 Case Block, Marion, Indiana

Radio News (May 1926), p. 1602

Radio Dealer (Sept. 1926), p. 166

New table type CASE set; 3/4-inch solid mahogany cabinet; dual vernier control; list $85

61A	Feb. 1927	$85
61C	Feb. 1927	$135

New type "C" CASE console; beautiful two-tone walnut cabinet with built-in speaker; standard CASE circuit of tuned radio frequency with six tubes and dual vernier control; list $135

Radio Retailing (Feb. 1927), p. 115

60A	Aug. 1926	$75
60B	Aug. 1926	$100
60C	Aug. 1926	$125
60D	Aug. 1926	$170

A June ad claimed 14 different models, including eight-tube.

Radio Retailing (June 1927), p. 154

The NEW "CASE 90"

Loop operation ℭ Single tuning control ℭ Self interstage shielded ℭ Self-shielded loop ℭ Compact ℭ Full-sustained quality of tone over complete scale ℭ A set for the man who wants a precision instrument ℭ Licensed under Technidyne patents

NINE TUBE SETS

MODEL 90A

Table model in beautiful walnut cabinet. Can be furnished for AC tubes if desired. Without accessories, list....

$225.00

The new CASE Line for 1927-28 features higher standards of precision and performance —and beautiful cabinets

MODEL 90C CONSOLE

A masterpiece of furniture craftsmanship plus *real radio*. Loop-enclosed and panel-operated. Full throated concert speaker. Can be supplied for AC tubes if desired. Without accessories, list.........

$350.00

THE 90 LINE

Model 90C as illustrated. Retails $350.00.

Model 92C same as 90C except wired for A C tubes and equipped with tubes ready for attaching to light socket. Retails $475.00.

Model 90A as illustrated. Retails $225.00.

Model 92A same as 90A except wired for A C tubes and equipped complete with tubes ready for attaching to light socket. Retails $350.00.

CHASSIS SHIPPED SEPARATE FROM CABINETS, ELIMINATING BREAKAGE DUE TO ROUGH HANDLING IN TRANSIT.

THE 60 LINE

Model 61A retails $85.00, less accessories. Six tubes. Table Cabinet. Two tuning controls and battery-operated.

Model 60A retails $65.00, less accessories, six tubes. Dial control.

Model 62B retails $185.00 complete. Six tubes. Table Cabinet. Equipped complete with 6 A C tubes ready to plug in light socket.

Model 61C retails $135.00, less accessories. Same as 61A except in High-Boy Console Cabinet.

Model 62C retails $235.00 complete. Same as 61C, except wired for A C tubes and equipped with tubes complete ready to plug in light socket.

Indiana Manufacturing & Electric Co., Marion, Ind.

RADIO PRODUCTS

Radio Receiving Sets ~ Automatic Chargers

90A		June 1927	$225
90C		June 1927	$350
92D	radio-phonograph	Oct. 1927	$775

Radio Retailing (May 1927), p. 122

Model 62 C

Case Console includes 6 AC Tubes, "B" eliminator, everything complete ready to plug in light socket..... $235^{00}

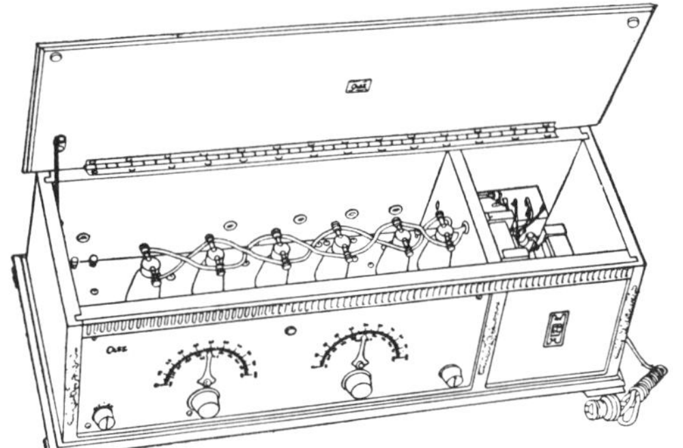

Model 62 B open, showing compact arrangement of AC Tubes, "B" eliminator, etc.

62B May 1927 $185

Radio Retailing (May 1927), p. 122

Battery Set

True single tuning control without sacrifice. One control operates sensitivity, dial light, and filament switch. Illuminated drum scale, calibrated in meters and kilo-cycles and with logging space.
Becoming to the furnishings of any home.

61A Oct. 1927 $88

Electrical Set

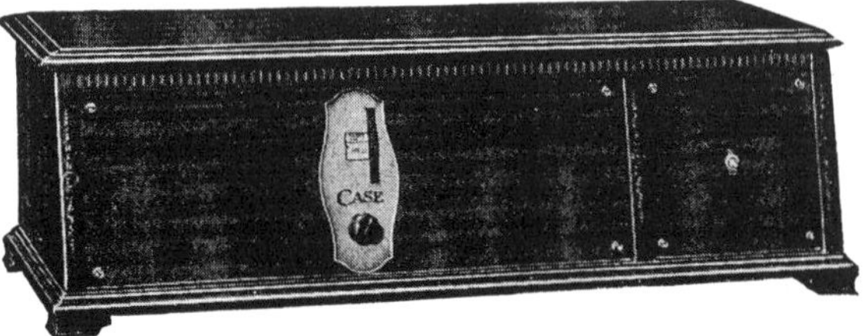

Same as model 61A except wired for AC tubes and equipped with tubes "B" and "C' eliminators, etc., complete ready for attaching to light socket. Same cabinet except 7" longer.

62B Oct. 1927 $200

MODEL 66A
All-Electric Table Model
6 tube. Neutrodyne with A.C. tubes. Illuminated drum, single control tuning. Wooden cabinet, with flush front adaptable to any console. Dimensions—20x 13x9-in. high. Price—less all tubes—**$98.00.**

66A		June 1928	$98
63C	console	Jan. 1928	$250
67A	(similar, 7 tubes)	about July 1928	$98

MODEL 73B

73B June 1928 $175

MODEL 73C
All-Electric Console

73C June 1928 $250

Radio Retailing (Oct. 1927), p. 111

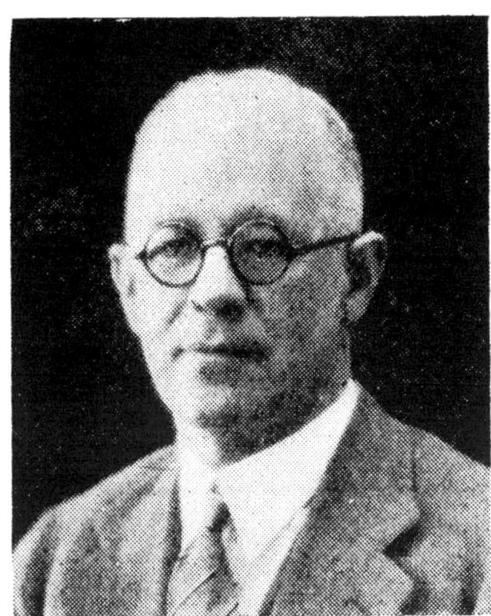

O. G. Nilson
President

APEX VERNIER DIAL
PAT. PEND.
MADE IN U.S.A.
It Brings 'Em In!
Get more stations—greater range—bigger volume—finer selectivity—less interference. Lasts forever. The one big advance yet made in tuning. Ratio 12 to 1. Quickly applied to any shaft. For sale by all good Radio Dealers. If unable to obtain from dealer, enclose $2.50 for nickel-silver finish, or $3.50 for De Luxe satin finished gold.
Microdyne SEVEN TUBE RECEIVER WHEN ASSEMBLED
7-Tube Super-Heterodyne for $97.50
Receive the parts complete to assemble your own set. Coast to coast on an 18-inch Loop. Assemble this 7-tube Microdyne Super-heterodyne on a 7x18 in. panel in three hours. Parts complete, including drilled and engraved panels, condensers, sockets, transformers, dials connecting plugs, cables, etc., with drawings, diagrams and instructions. Price of cabinet—to fit—on application. If your radio dealer cannot supply parts for complete Microdyne Radio Set, send check or money order for $97.50 and name of your dealer.
APEX ELECTRIC MFG. CO., Dept. 203
1410 W. 59th Street, CHICAGO

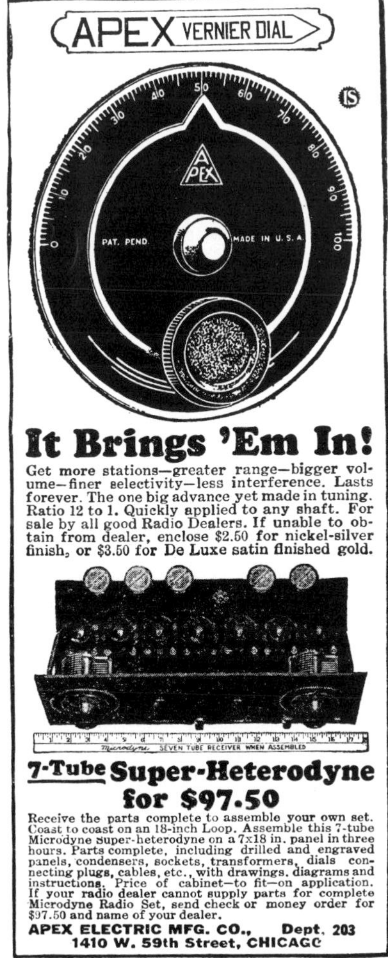

Popular Science Monthly (Jan. 1925)

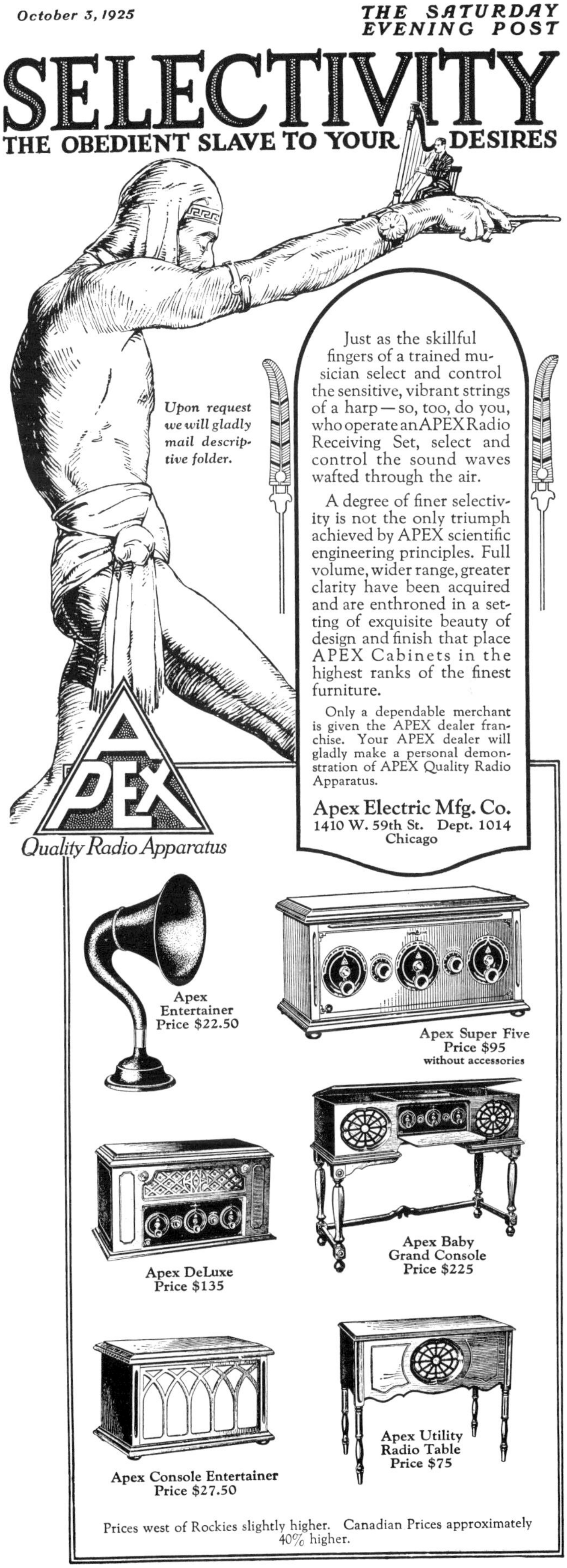

October 3, 1925
THE SATURDAY EVENING POST
SELECTIVITY
THE OBEDIENT SLAVE TO YOUR DESIRES
Upon request we will gladly mail descriptive folder.
Just as the skillful fingers of a trained musician select and control the sensitive, vibrant strings of a harp—so, too, do you, who operate an APEX Radio Receiving Set, select and control the sound waves wafted through the air.
A degree of finer selectivity is not the only triumph achieved by APEX scientific engineering principles. Full volume, wider range, greater clarity have been acquired and are enthroned in a setting of exquisite beauty of design and finish that place APEX Cabinets in the highest ranks of the finest furniture.
Only a dependable merchant is given the APEX dealer franchise. Your APEX dealer will gladly make a personal demonstration of APEX Quality Radio Apparatus.
Apex Electric Mfg. Co.
1410 W. 59th St. Dept. 1014
Chicago
APEX
Quality Radio Apparatus
Apex Entertainer
Price $22.50
Apex Super Five
Price $95
without accessories
Apex DeLuxe
Price $135
Apex Baby Grand Console
Price $225
Apex Console Entertainer
Price $27.50
Apex Utility Radio Table
Price $75
Prices west of Rockies slightly higher. Canadian Prices approximately 40% higher.

Super-Five first advertised in Jan. 1925, other models in Aug. Apex vernier dials were used by many other manufacturers. Technical article in *N. Y. Evening World*, Jan. 3, 1925, p.31.

THIRD RADIO SPACE BB - 11 WORLD'S FAIR

Patented Compensator

That the new Apex Six will be one of the most popular sets on the market this season and for many seasons to come—is a foregone conclusion. Consider these important improvements, Patented Compensator (patented in 1924 and withheld from the market until its merit had been established beyond all question)—one dial control—impedance coupled—the entire sphere of radio at the command of the turn of a single dial—all distortion eliminated — illuminated dial—automatic filament control.

This notable receiver is housed in furniture worthy of its quality and designed in combinations which provide a wide selection on small investment. For example—receiver No. 6 combined with cabinet No. 100 makes set No. 106.

Sixteen years of unimpeachable manufacturing activities vouch for the integrity and stability of the Apex organization and for the value of Apex products.

With but few exceptions Apex dealers of yesterday are Apex dealers of today. Conclusive evidence that Apex products give satisfaction and the Apex policy proves profitable to dealers. Apex sets have never been "junked" or price-slashed and never will be. Progressive dealers are invited to write for particulars regarding liberal Apex dealer franchise.

Apex Electric Mfg. Co.,

Dept. 916, 1410 W. 59th St., Chicago, Ill.

Apex Model No. 106
Without Accessories $175

Apex Model No. 5
Without Accessories
$85

Apex Model No. 116
Without Accessories $210

Apex

Radio Dealer (Sept. 1926), p. 74

5	Sept. 1926	$85	106	Sept. 1926	$175
6	Sept. 1926	$115	116	Sept. 1926	$210

Radio Retailing (June 1927), p. 162

Table models: left, Lyric, $80. Right, Corsair, $170 (Technidyne). Consoles: left, Minstrel, $225. Right, Troubadour, $295 (both Technidyne).

Radio Retailing (Jan. 1928), p. 107

Presents Most Extra-ordinary Quality at

Startling Low Prices!

Apex, consistently first in radio progress, is again first with the two new neutrodyne models illustrated on this page. Both of these splendid receivers are outstanding values in the world of radio. There is nothing to equal either the *Milan* A. C or the *Music Chest* at their prices.

The New
6-Tube Shielded
Neutrodyne
"MUSIC
CHEST"

Less Accessories, $65

Licensed Under Patents of

Radio Corporation of America and Associate Companies
Hazeltine Corporation
Latour Corporation

The
Milan Electric
6-Tube Neutrodyne
Complete $220
(Without Tubes, $192.50)

Apex Electric Mfg. Company
1410 West 59th Street, Chicago
Division of
UNITED STATES ELECTRIC CORPORATION

Everything In Radio — Everywhere In America

Radio Retailing (June 1929), p. 197

MODEL 89—A *genuine* Neutrodyne, nine - tube radio — single control, without compensators. A great value that will attract customers—$89.95.

$89.95

A GREAT organization — the United States Radio and Television Corporation — makes the APEX radio. Its commanding situation in the industry; its ample financial resources; its wealth of man power and engineering skill is sufficient guarantee of excellence in its product. Five great plants with manufacturing facilities for 6500 sets per day assures prompt shipments. It's wise to tie-up with such an organization. Past incidents prove it.

United States Radio and Television Corporation

CHICAGO, ILLINOIS

Apex Receivers

The 1930 line of Apex radio sets built by the U. S. Radio and Television Corporation, 1340 Michigan Ave., Chicago, Ill., is now ready.

Model 11 with the super-screen-grid chassis, which has three tuned circuits using two 224's in r. f. amplification, two stages of audio with two 245's in push-pull in the last stage, two 227's and a 280 rectifier, comes in an American walnut veneer cabinet, 40¼ in. x 27⅛ in. The speaker in this model, which is standard for all the new models, is a 12 in. electro-dynamic reproducer. The intended retail price is $115.

Model 115 with the "high-gain Neutrodyne" circuit, utilizing five 227 tubes, two 245's and a 280 rectifier in three tuned stages of r.f. and two audio stages, is inclosed in the same style cabinet. Price, $115.

Model 160

Model 14 may be had with either chassis. It comes in a taller cabinet (47¼ in. high) with sliding doors. Price, $140.

Model 24 with screen-grid chassis and remote control, in a cabinet similar to that used with Models 14 and 140, is $240.

Model 160, the super-high-gain Neutrodyne set in a high-boy cabinet is $160.

Model 60, a table set with the Neutrodyne circuit, in a metal cabinet, size 18½ in. x 13½ in., x 7½ in., is $60.

Model 45, a battery operated table set using five 201A's, one 171A is $45.—*Radio Retailing*, October, 1929.

Model 11

89 John Bayusik

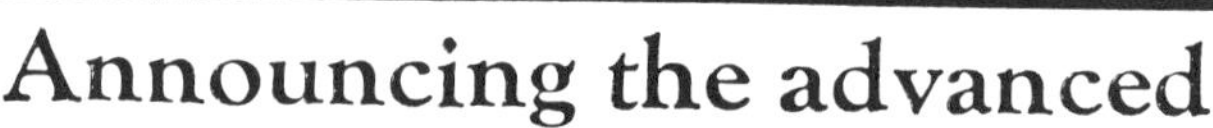

Announcing the advanced

APEX
ALL ELECTRIC
NEUTRODYNE

$85
without tubes. Add 5% West of Rocky Mountains

See it at the Chicago Show JUNE 11-15

Eighteen years of diligent manufacturing activities are back of the exceptional performance so apparent in the NEW APEX ALL-ELECTRIC NEUTRODYNE. This advanced set represents one of the greatest values in the RADIO field today—it means enormous sales and good profits for every dealer who is progressive enough to see the great possibilities of the APEX Receiver.

This genuine NEUTRODYNE has self-contained and highly-perfected power-pack. Employs six tubes, plus one rectifier. Only one tuning-knob. Illuminated dial. Metal cabinet of attractive walnut finish. Greater distance and selectivity than many high-priced sets. Has a deep, mellow and natural tone. Easiest set to operate.

See the APEX at the Radio Manufacturers' Association Trade Show—Booth B-87, Grand Ball Room, Stevens Hotel, Chicago, June 11th to 15th inclusive.

Apex Electric Mfg. Co.
Dept. A., 1410 W. 59th St., Chicago

Gentlemen:

I am interested in an APEX franchise and would like to receive complete details relative to same.

NAME..

STREET..

CITY..

Radio Retailing (June 1928), p. 107

ATWATER KENT

Atwater Kent Manufacturing Company

"Mr. Kent, of the Kent Electric Co., sojourned with us for the space of one term, during which time he held the purse of the class. Either the duties and cares of this office were too burdensome, or his outside electrical work too engrossing, for he failed to appear at recitations after the mid-year exams. More self-confident than ever in his ability to bluff, he entered the Class of 1900 in the following year, and, of course, his relations with us became more or less indirect. His bluffs worked well for a time (as might be expected in a class of bluffers) but they didn't 'score points' on the exams, and now Arthur devotes the most of his time to the affairs of his company. A good natured fellow with a pleasant smile. May be seen at his best Sunday evenings at Piedmont Church receiving the offering and (he fondly imagines) the admiration of the young ladies."

from the 1899 Aftermath,
Worcester Polytechnic Institute.

Arthur Atwater Kent's fellow students apparently understood him better than the professor who warned him that without a diploma from "Tech" he would never amount to anything. Because he did amount to something, after all. Not only did he invent the ignition system used in almost every automobile from the 1920s to the 1970s, but he did more: he also became the world's largest radio manufacturer.

Attending Worcester Tech was natural since he lived in that city, and had a strong mechanical inclination. However, as noted, he didn't continue his studies very long; about 1895 he founded the Kent Electric Manufacturing Co. in the back room of his father's machine shop, where he made small motors and fans. And not only made them, but advertised and sold them. He was not afraid to take out ads in the largest electrical trade papers, whatever their cost.

In 1900 he sold his design for a small but efficient motor to Kendrick & Davis of Lebanon, New Hampshire, makers of watch tools, electrical specialties, and toys, and he lived there for a time before leaving to sell electrical equipment. While on a business trip to Philadelphia in 1902, Kent decided it was the ideal place to start another company, founding the Atwater Kent Manufacturing Works in the loft of a rented building at 6th and Arch Sts. Here he made meters, intercommunicating telephones, and other small electrical items. Legend has it that he never had to sweep the floor at this location because of the wide cracks between the boards.

In 1905 Mr. Kent felt prosperous enough to purchase his first one-cylinder automobile, as he put it, "not being married and not having to conserve cash." The troubles he encountered with this automobile were the beginnings of his rise to fame and fortune. By the end of 1905 he was manufacturing automobile timers, trigger ignition systems, and switches. This necessitated a move to larger quarters on Arch St.

Within a few months, Kent hit upon his first real invention, the Unisparker, an improved ignition system which integrated the usual series of weak sparks into a single hot spark for ignition. The Unisparker combined contact points, condenser, centrifugal advance mechanism, and distributor into one compact unit to be used with an ignition coil. This was the same ignition system used in most cars until the recent adoption of electronic ignition. For this achievement, Kent was awarded the John Scott Legacy Medal and Premium by the Franklin Institute in 1914.

By 1912, the success of the Unisparker forced him to move again, this time to a much larger facility on Stenton Avenue. Soon, self-starters and lighting systems were added to the Atwater Kent line of automotive products. By World War I, the Atwater Kent Manufacturing Works was large enough to land government contracts for fuse setters, clinometers, and optical gun sights.

Employing about 125 workers, the Atwater Kent Manufacturing Co. was incorporated in 1919, but due to the economic slump, its post-war business outlook was not good. More and more autos were being sold with factory-installed ignition, so the market for Kent's equipment was shrinking. Always a versatile inventor (he had 25 U.S. patents by this time, and would eventually take out 93) he got hooked on radio in 1921 along with thousands of other fans. It didn't take him long to realize that radio was the perfect field for his company to enter. He had all the equipment and know-how for Bakelite molding, a relatively new process at the time, as well as metal-forming and electrical coil-winding. And he had a national reputation and a network of dealers well-suited to handling radio.

He tooled up (Bakelite molding requires expensive steel dies) and in June 1922 began advertising variometers and audio transformers. Other components were soon added to the line: variocouplers and tuners, detector and amplifier units, RF transformers. By December and January, Atwater Kent's ads showed five models of factory-wired receivers, using his coupled-circuit tuner and various detector or amplifier units, mounted on wooden "breadboards." All these models were regenerative when a variometer was added by the purchaser, but Kent was scrupulous in not mentioning this in his ads or literature, nor showing the variometer in position. He was not licensed by Armstrong (not having been in radio in 1920 when licenses were available) and while he could have done as several other companies did, and bought out a moribund licensee, patent-owner Westinghouse was already making trouble for them, and Kent had too much to lose, to take a chance. He preferred to improve his radios' performance by adding RF amplification ahead of the detector and avoiding the regenerative circuit. This was more expensive, but his receivers were not cheap in any event, and up to then the public had seemed willing to pay more for a quality product.

Two models embodying RF amplification, again assembled from standard parts, were described in Kent's literature in January and advertised in March 1923: the 4052 and 4066. These two sets, along with the five previously announced, made up Atwater Kent's 1922–23 line. A little late for the Christmas selling season, but a creditable showing nonetheless, for a first try.

Aside from a few bastions of low-priced radios such as Crosley, where the regenerative circuit was necessary to get maximum performance from few tubes, the industry trend was toward RF amplification. Beginning in May 1923 broadcast stations were shifted away from the two frequencies where they had been before (360 and 400 meters, 833 and 750 kc). Now radios needed more selectivity, particularly in urban areas where more and more stations crowded in, and tuned RF amplification was the most practical way to get it. Furthermore, the prices of tubes were coming down, and they used less current (the ¼-ampere 201A was announced in December 1922), making it practical to put more of them in a radio. The Neutrodyne makers started the trend, making the five-tube, three-dial radio commonplace; even at an average price of $150, roughly 100,000 such sets were sold in the 1923–1924 period. But Atwater Kent was quick to spot the trend, and react. On September 7, 1923 specifications were drawn for a five-tube, three-dial set, the model 10 or Radiodyne. Unfortunately, Western Coil in Racine, Wisconsin had registered this trade mark one day earlier, and had claimed prior use in August, so Kent soon had to discontinue its use, but whatever its name, the 10 became a runaway best seller, no surprise since it sold for one-third less than a Neutrodyne and worked just as well (as far as the public was concerned).

$100 was still a lot of money, though, and not everyone needed the Radiodyne's performance, so Kent also specified two other models at the same time. The model 5 was a repackaged version of last season's 4066; electrically identical, but with a minimum of separate components and wiring. It was intended for locations with only one or two broadcast stations, and cost just $55. An intermediate model, the 8, was a 5 with one tuned circuit added for improved selectivity, at $70. It was dropped quickly in favor of the 9, a four-tube tuned-RF set with variometer tuning that performed better but cost only $68. The 9 was electrically identical to the old 4052, except that a tuneable RF transformer (a coupled-circuit tuner) replaced the untuned one.

For the few buyers who wanted more power to drive a loudspeaker, Kent added an extra audio stage to the 10, following a circuit he had published the year before, creating the six-tube model 12. These four models, 5, 9, 10, and 12 were Atwater Kent's lineup for the 1923–1924 season.

Atwater Kent radios, more than most other brands, show a logical evolution from one model to another, and from one year to the next. The 10 illustrates this process, as it was continually modified to improve its performance

THE KENT ALTERNATING SEWING MACHINE MOTOR.

THE extended use of the alternating system of electric lighting has created a large demand for small power specialties for such circuits. The Kent Electric Manufacturing Company, of Worcester, Mass., have lately been bending their energies in this direction, and one of their results is a sewing machine motor illustrated in the accompanying engraving.

The great difficulties encountered in this work are the small starting torque, and the non-variable speed of these motors. Both of these obstacles are overcome in the device shown. The motor, which is of the induction type, is mounted on a base, to move with a lever, connected to the treadle of the machine, and the speed regulation is obtained by tightening and slackening the belt. By this method, any desired speed may

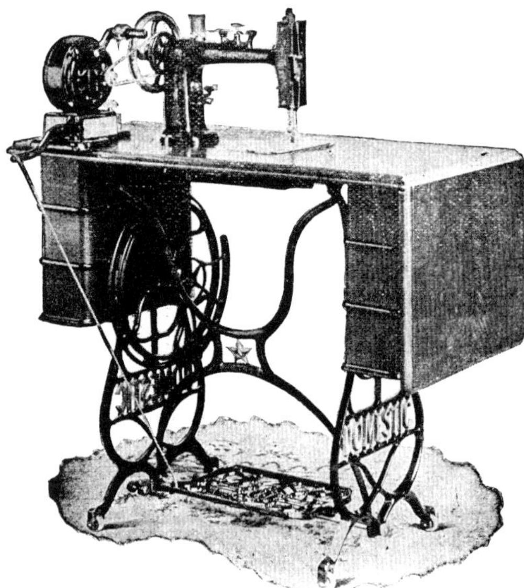

THE KENT ALTERNATING SEWING MACHINE MOTOR.

be maintained. In this way even more perfect regulation can be obtained than by direct current, as the speed can be changed in an instant, and the brake which touches on the circumference of the balance wheel stops the machine instantly.

A switch located on the base is turned on when the operator sits down to the machine. The motor is thus started without a load, and is running constantly while the machine is in use. It is compact, and neat in appearance.

The company also manufacture a variety of other apparatus such as battery fan outfits, alternating current fans, and small machines which can be run either as a dynamo or a motor, capable of lighting from one to four 8 candle-power lamps, or of running a 6-inch fan. The voltage varies with the speed from 1 to 18 volts.

Electrical Engineer (Dec. 9, 1896), p. 609

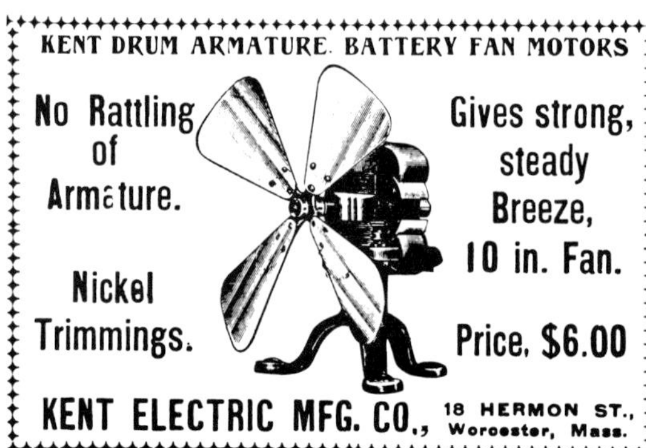

(1898)

or to reduce its manufacturing cost. The 5 had been dropped for inadequate performance; the inductively-tuned 9 gave way to the condenser-tuned type (really a 10 with one less RF stage); and the 12 had limited sales since its extra audio power was unnecessary; but the 10 was so good that it remained in production for two more years.

Yet there was a strong demand for an enclosed or "cabinet" set that Atwater Kent couldn't ignore, even though his factory was not geared to cabinetmaking. Most radio fans wanted a piece of furniture, not laboratory apparatus. For the 1924–1925 season, then, Kent took the circuitry of the 10, indeed many of the same parts, and enclosed them in a simple mahogany cabinet, reminiscent of his earlier ignition devices. Instead of copying the engraved Bakelite panel of most other makers, a production bottleneck, he chose crinkle-finished metal (a harbinger of things to come). The model 20 was even more successful than the 10; along with the Freshman Masterpiece introduced in the same month, it knocked the stuffing out of the Neutrodynes, for besides costing one-third less, it looked more attractive and up-to-date.

As the 10 had its counterpart 9 with four tubes, so the 20 had its 19. At only $10 less, its sales were low, but apparently enough to justify its production. The six-tube set, however, was made only as an open model, the 12, an updated version of the previous 12. Buyers who wanted a bit more than the 20 were offered the 20 Deluxe, or 24, in a fancier cabinet.

In 1924 Kent spent half a million dollars on advertising, $190,000 of it in national magazines, but large as this amount was in comparison to most other companies' budgets, it was nothing like his future expenditures. In October 1925 he began sponsoring the Atwater Kent Hour, featuring the top musical names of the day; this quickly became the most acclaimed, and most popular, regular radio program. When it resumed for the 1926–1927 season it was costing Kent $7000 per week, and a typical year's outlay for printed advertising was three to four million dollars. It was worth every dollar.

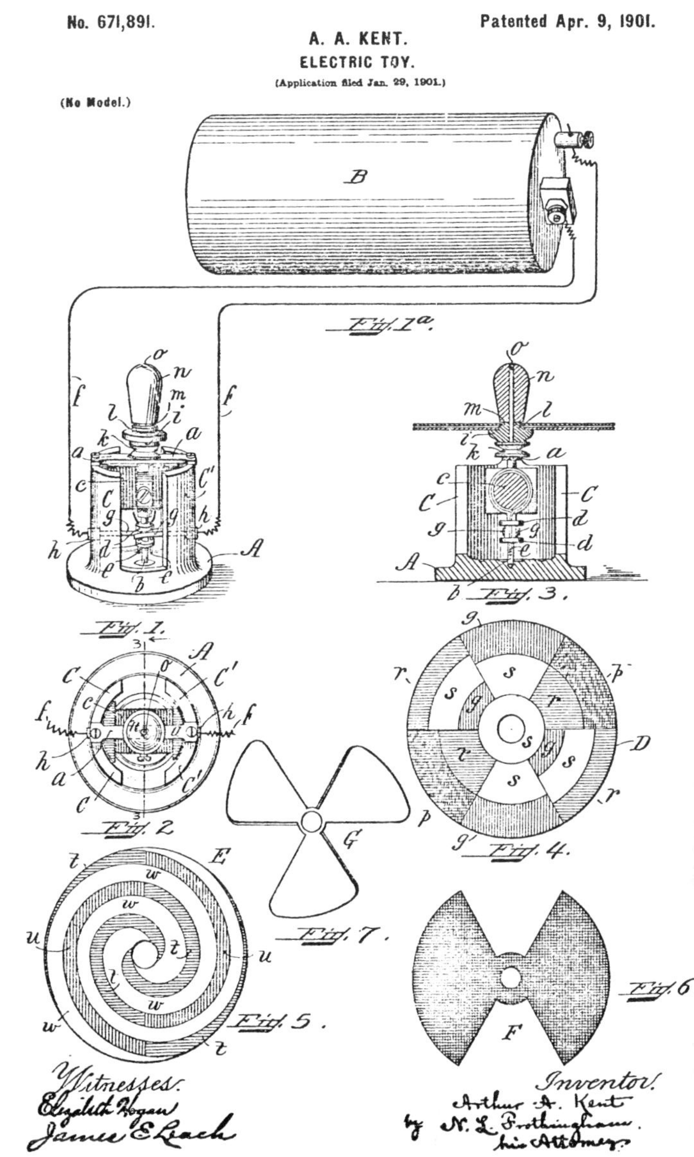

Kendrick & Davis motor, originally designed by Atwater Kent in 1896.

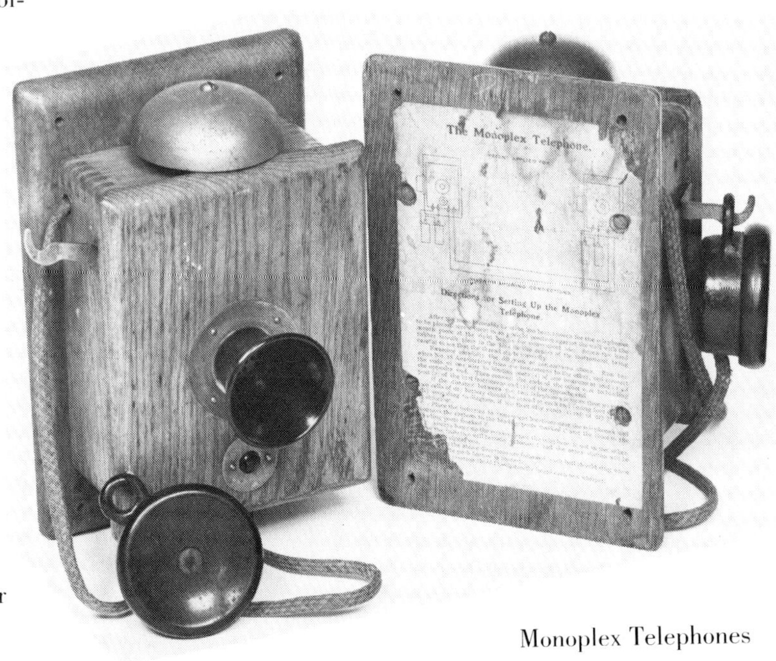

Monoplex Telephones

As Atwater Kent's production expanded, the company occupied a series of adjacent buildings on Stenton Avenue (which by the way are still standing, very much as they were then) but by 1924 there was no more room. Kent bought a large tract of land on Wissahickon Avenue and erected a two-million-dollar factory in mid-1924, covering five acres (with additions, it would eventually cover 32! It too still exists). His future was limitless.

1925–1926 was another season of evolution for Atwater Kent. On the theory that the less obtrusive a radio looked, the more people would buy it, he squeezed his model 20 into a cabinet of half the volume. The contrast with the old 20 and the 10C (which remained in his lineup, with the 24 and the 12) was striking, and Kent's judgment of the public's taste was, as usual, accurate.

Since one new model was hardly enough for a season, Kent also made a dry-cell version of the 20 Compact, the 21, using UV199 tubes, figuring to do away with the messy storage battery that was such a sore point with radio owners (RCA owed a good deal of its success to exclusive production of dry-cell models). He barely got the 21 into production when RCA obsoleted the UV199 in favor of the standard-base UX199, which would not fit the old sockets; factory records show that of 17,584 model 21s made in 1925, 7,208 of them were returned in 1926.

Atwater Kent's remaining dream for 1925–1926 was a one-dial radio. Several other companies had produced them, and there was no question that they were the wave of the future. But the obvious expedient of coupling all three dials together wouldn't work, not if the new model were to use many parts in common with the 20 Compact. While the second and third tuning dials tracked together, the antenna coupled to the first stage affected its tuning drastically.

Then Atwater Kent's chief theoretician John M. Miller (discoverer of the Miller Effect during his previous tenure at the Bureau of Standards) came to the rescue. He added a sixth tube between the antenna and the first tuned circuit, isolating the circuit and making it tune in unison with the others. Now the three dials could be belted together and controlled by one knob, and with only minor mechanical changes to the 20 chassis. Miller applied for a patent on December 4, 1925, and specifications for the new model 30 were drawn on November 22; all indications are that Atwater Kent decided on a crash program to produce his one-dial sets before the winter selling season melted away. It was too late to make the necessary parts (this usually took the first nine months of each year) but, fortunately, there was already a stockpile available: 100,000 20 Compacts, of the older 7570 pattern made for UV201A tubes. The 7960 Compact, for the new UX tubes, was already in production, and since Atwater Kent had overproduced by up to 300,000 sets, he had nothing to lose by scrapping 100,000 7570s to turn out new models.

Actually he scrapped very little, as nearly all the old parts were re-used in making the 30s. Making the extra tube socket was easy. Even the old cabinets were modified by plugging holes and drilling new ones, and hollowing out the wooden backs to make room for the new chassis. The one-dial model 30 was announced in February, and was in active distribution by late March, at a hefty increase in price too, to compensate Kent for his labors. It was a virtuoso performance.

Naturally the later production model 30 was made more conventionally, in a cabinet designed to fit; it was a mainstay of the 1926–1927 season.

1903 Monoplex Literature

Ralph & Elinor Williams

Test meters

The Literary Digest for November 22, 1919

ATWATER KENT

AUTOMOTIVE EQUIPMENT

Ignition, Starting and Lighting

¶ *Hundreds of thousands of discriminating users of motor cars today enjoy the satisfaction that goes with dependable, unfailing spark performance—the result of Atwater Kent equipment.*

¶ *Accuracy, precision and quality of workmanship—ceaseless vigilance in designing, engineering and inspection of the finished product—sixteen years of specialization and concentrated experience. These are reasons for the reputation of your Atwater Kent ignition system.*

ATWATER KENT MFG. COMPANY

Philadelphia

SEE YOUR DEALER OR WRITE TO 4939 STENTON AVENUE

69

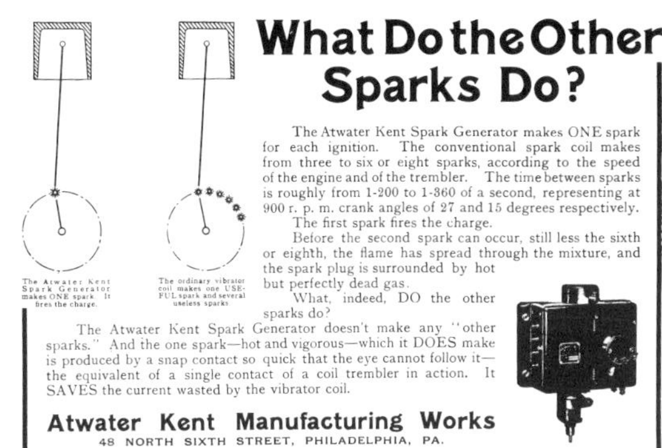

(1907)

As 1926 progressed, other manufacturers brought out one-dial models that both looked and worked better than Atwater Kent's 30, and Kent realized that he would either have to lower the price or improve the performance, to stay competitive. He reduced his manufacturing costs somewhat by changing to metal-frame variable condensers, but the big stumbling block was the wooden cabinet. He knew he could stamp out cabinets of sheet steel much more cheaply, and disguise the material with a two-color crinkle finish and gold trimmings, but would the public buy a metal radio? Kent decided that the public's taste could be changed, if not by advertising alone, then certainly by the $15 difference in price. And so he turned out 200,000 model 35s in 1926. His dealers would rather have sold the model 20, $10 cheaper since it was obsolete, but Kent would only give them twenty of those for the show windows to every thousand of the "tin cans." According to engineer John Dreyer Jr., the 35 cost Kent $12 to produce.

The "performance" model in 1926–1927 was the 32, a seven-tube set with one more RF stage than the 30 or 35. Never a big seller, it did give him a comprehensive line to match his competitors.

Atwater Kent's sales did not always follow the industry's overall trends (he had increased his sales in 1926, even though industry-wide numbers were down) but 1927 was a slow year for everyone. He had run out of tricks. He tried to push the 35 (but had to take back many unsold ones from the dealers) and he held over the 30, at $15 more than the 35, to cater to the die-hards who wanted a wooden cabinet. His major advance, early in the season, was the model 33, nearly equal in performance to the expensive 32 because he added an extra control to tune the antenna circuit, allowing the sixth tube to amplify as well as isolate. Significantly, he didn't try to force a "tin can" on his customers with this model; for the relatively smaller number built, a metal cabinet wouldn't have shown the cost savings that the 35's cabinet did.

When RCA began licensing other manufacturers in early 1927 to use its patents, at a royalty of 7½% of net sales price and $100,000 minimum, Kent was in no hurry to sign up. Aside from giving his competitor an effective 15% price advantage, he was theoretically liable for substantial back royalties. Furthermore, his legal talent had an excellent defense to counter Alexanderson's TRF patent, the basis of RCA's lawsuits against Kent and others: knowledge of prior work by B.F. Miessner, the same defense being employed by Splitdorf (recounted by Miessner in *On the Early History of Radio Guidance*, San Francisco Press, 1964, quoted in the Splitdorf chapter).

If Kent had pressed his case and (as seems likely) won, RCA's licensing structure would have disintegrated. RCA had to act quickly, to enroll industry-leader Kent, while it had momentum and a particularly favorable court decision on the Alexanderson patent that might be lost if it were appealed. While the exact license terms may never be known—Kent purposely destroyed his business records after 1936—it is unlikely that the Atwater Kent Company ever paid any back royalties, and quite possible that it paid less than 7½% later.

Kent was waiting to introduce an AC set. As with the one-dial sets, there was no question that AC radios were the wave of the future, but until 1927 there was no practical way to make one. "Practical," at any rate, by Kent's definition: a cheap, reliable set that would use most of the same internal parts as his older models. For that, he had to wait until RCA had developed and released the 26 and 27 tubes; none of the old cranky '99s or off-brand disasters from Kellogg or Arcturus for him. He couldn't tool up for half a million AC radios and *hope* they worked properly, or keep tinkering with small production runs until they did. Besides, his RCA license, once he had acquired it (August 11 or 18, 1927), forced him to equip his sets initially with RCA tubes.*

But the right tubes did finally arrive, and when they did, Atwater Kent was not all that far behind even RCA's own Radiola 17 in using them—only one month. It was not for nothing that most of his sets were alike, under the skin. His first AC set, the 36, was simply an electrified 33, with a separate power pack adapted directly from his old B-eliminator. No great engineering advance, to be sure, but it did keep his dealers up-to-date with something to match the competition, while Kent's engineers worked feverishly to produce the ideal set.

And, just in time for the Christmas season (the 24th, to be precise), they did. The model 37 was a bombshell in the industry. For $88 it offered most of the performance (and half the size) of RCA's $130 model 17. A big chunk of that difference was in the metal cabinet. This time, the cost and space savings were too compelling to ignore, whatever the public's taste, but customers did accept metal now; in fact, metal cabinets became somewhat of a fad by mid-year.

With the AC radio's arrival, battery models were all but forgotten; Kent could only update the 30 and the 33, paint their panels a different color, and call them the 48 and 49. His engineering efforts were all with the AC sets. For the 1928–1929 season, his peak of perfection was the model 40. Basically like the 37, but given another six months to get out all the bugs and to tool up properly, it sold nearly a million units and kept Atwater Kent at the top of the industry. It took RCA nearly a year to match it with the Radiola 33.

*This was the famous "clause nine" in the RCA patent license. Unable to maintain its monopoly in vacuum tubes after the de Forest grid-audion patent expired in 1925, RCA aimed to put independent tube makers out of business by requiring all licensed radios to be equipped initially with RCA tubes. Clause nine, never enforced because of a storm of protest and lawsuits, was struck down in 1931.

Gunsight

Atwater Kent, Jr. on loan to Ralph & Elinor Williams

Ignition Components

Ralph & Elinor Williams

In an almost-unbroken trend since 1922, the average price of a radio had been going up each year, to $122 in 1928 and $135 in 1929. This meant that a lot of consoles were being sold, but Atwater Kent was in a poor position to take advantage since he emphasized table models. The *dealer* created an Atwater Kent console by placing a table set in a cabinet made by someone else. For the 1929–

1930 season, the plan was to downplay table models and go for the higher-priced console market. To this end, Kent arranged with several furniture makers to produce console cabinets, while he would supply the chassis. This had been done before, but not on such a scale.

The new screen-grid models 55 and 60 came off the lines as scheduled in mid-1929 and into console cabinets, and indeed a large number were sold, but perhaps for the first time, Kent had seriously misjudged the future. Danger signals were up by mid-1929; with or without the stock-market crash, the Depression was inevitable. By 1930, the average price of a radio had plummeted to $78. Of course, Kent was not the only one to misjudge; Majestic did too, and paid the price of bankruptcy. But some could react and change strategy in time: Philco switched to midget sets and soared to dominance in the radio industry. Atwater Kent, who ironically had made his fortune selling compact radios, was left in the dust. He lost interest in piloting his company and let it slide downhill until 1936 when he closed the doors forever. With his personal fortune largely intact, he retired to Hollywood to live among high society and movie stars, an association he had always enjoyed. From all indications, he was fully as happy in retirement as he had been while inventing ignition systems or running the world's largest radio company. He died in 1949.

Ralph & Elinor Williams

3925 Jan. 1923 $32

Ralph & Elinor Williams

3945 Jan. 1923 $35.50

Ralph & Elinor Williams

3955 Jan. 1923 $37.50
3960 Jan. 1923 $23.50 (without 2-stage amp.)

3975 Feb. 1923

Ralph & Elinor Williams

ATWATER KENT

RADIO RECEIVING SETS

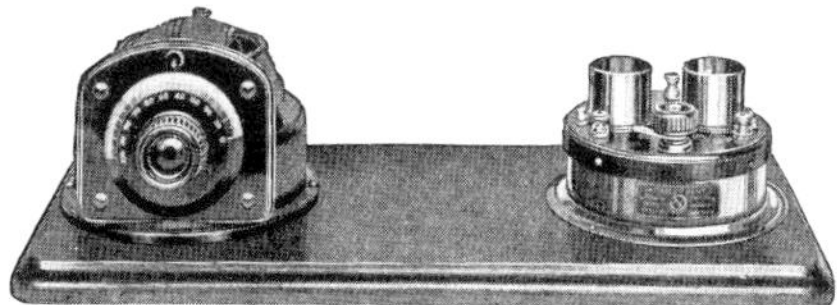

The two instruments shown above comprise an excellent and complete receiving set. The Coupled Circuit Tuner and Detector 1-Stage Amplifier on the mahogany mounting board, present a beautiful appearance.

Complete Outfit, as above, wired . . . $32.00

The above set, consisting of Coupled Circuit Tuner and Detector 2-Stage Amplifier, is an ideal set for either phone or loud speaker use. Note that this set includes two stages of audio frequency amplification.

Complete Outfit, as above, wired . . . $35.50

The Coupled Circuit Tuner and Detector Unit only, comprise a complete receiving set. Later, if desired, the 2-Stage Amplifier can be added for two stages of audio frequency amplification as shown above.

Complete Outfit, as above, wired . . . $37.50
Complete Outfit, as above (without Amplifier), wired, 23.50

The Mounted Variometer carries through the standard quality of Atwater Kent products. For an open set it supplies a finished instrument unsurpassed in appearance and performance.

Mounted Variometer . $10.00

An Excellent Merchandising Proposition

Atwater Kent Manufacturing Company
4943 Stenton Avenue Radio Dept. Philadelphia, Pa.

Radio News (Feb. 1923), p. 1505

PRODUCTION DATES
COPIED FROM AN OLD FACTORY LISTING, WITH CORRECTIONS.

Date	*Part No.*	*Model and description*
5-26-22	3590	Det. plus 2 stage amp. in wooden cabinet*
11-23-22	3925	Tuner, detector and amplifier
11-27-22	3945	Tuner, detector and 2 stage amplifier
12- 2-22	3955	Tuner, detector unit and 2 stage amplifier
12- 4-22	3960	Tuner, detector unit and 2 stage amplifier
12-23-22	3975	Vario-coupler, variometer, detector and 2 stage amplifier
1-14-23	4052	Tuner, pot., RF trans., tube unit, det. & 2-stage amplifier
1-24-23	4066	Single cir. tuner, 2 RF trans., 2 tube units, det. & 2 stage amplifier
3-15-23	4120	Single circuit tuner, RF trans., tube unit, det. unit
4-30-23	4205	Same as 4066 (4135 tube unit for use with ¼ amp. tubes)
5- 8-23	4207	Same as 4066 (Cabinet board)
5-15-23	4220	Model 15 cabinet receiver (4066)
6- 5-23	4275	Single cir. tuner, 2 RF trans., 2 tube units, det. unit
9- 7-23	4325	Model 8 Duplex
9- 7-23	4333	Single circuit tuner, det. and 2 stage amp.
9- 7-23	4340	Model 10 set (plain gray)
10- 9-23	4445	Model 9 (cable type) 2 tuners
11- 1-23	4480	Model 9 in 4217 console with cable
11- 9-23	4490	Model 10 in 4427 cabinet with cable
12-14-23	4535	Model 9 in cabinet with cable
12-14-23	4540	Model 10 in console with cable
12-14-23	4550	Model 10 in brown finish, cable
1-31-24	4560	Model 10 in black finish, cable
2-16-24	4590	Model 10 on Pooley board
2-22-24	4600	Model 10 brown finish
2-28-24	4610	Model 10-B in 4217 console
3- 3-24	4620	Model 12 six tube set
4-18-24	4660	Model 9 "C" set, 2 var. condensers and 2 RF trans.
4-30-24	4640	Model 20 set in cabinet (large)
5- 6-24	4650	Model 10-B brown set in 4427 cabinet
5-29-24	4700	Model 10-C set
7- 7-24	4880	Model 19 set
7-23-24	4910	Model 12 set (early type)
7-23-24	4920	Model 20 Deluxe
7-29-24	4930	Model 22 six tube cabinet set
8-25-24	4950	Model 10 Pooley set
3- 2-25	7570	Model 20 compact
6-17-25	7780	Model 21 dry cell type
7-30-25	7800	7780 set with 7790 dry cell container
11-18-25	7950	Model 30 set (one rheostat)
11-22-25	8000	Model 30 set (two rheostats)
1-15-26	8100	Model 35 set (metal cabinet)
5-15-26	8270	Model 32 set (seven tubes)
11-12-26	8000A	Model 30-A
11-12-26	8100A	Model 35-A
11-22-26	8450	Model 25 set
1-10-27	8500	Model 50 set
3-16-27	8820	Model 50 console set
5-11-27	8930	Model 33 set
5-24-27	9040	Model 30 console set
5-24-27	9050	Model 33 console set
6-29-27	9090	Model 51 set

*Probably a TA unit in a wooden case, like an early ignition coil, not a complete radio set. "TA" stands for "Table (-mounting) Audio."

1,719,014. RADIO APPARATUS. ARTHUR ATWATER KENT, Ardmore, Pa. Filed July 21, 1922. Serial No. 576,423. 16 Claims. (Cl. 250—14.)

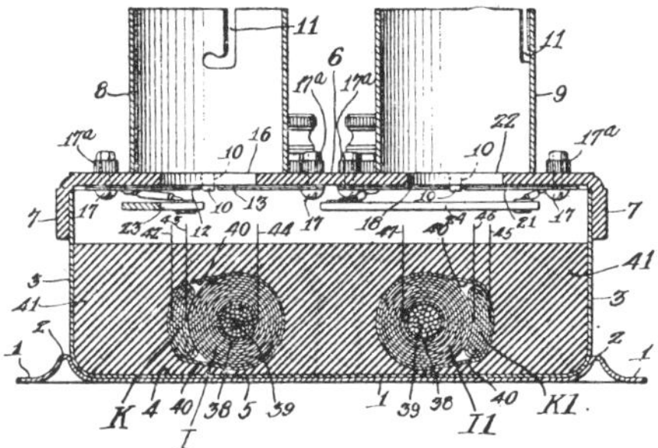

14. An audion coupling unit comprising an enclosing casing, insulating material in said casing, a plurality of audion coupling means embedded in said material and held thereby in predetermined position with respect to each other and to said casing, whereby substantial coupling between said means is prevented.

Ralph & Elinor Williams

4052 Mar. 1923 $60

4120 same, with 1-tube detector in place of 3-tube unit. $45

Ralph & Elinor Williams

4066 Mar. 1923 $72

4205 same, for ¼-amp tubes.

4275 1-tube detector in place of 3-tube unit.

1,679,310. DETECTOR APPARATUS. ARTHUR ATWATER KENT, Ardmore, Pa. Filed May 28, 1923. Serial No. 641,835. 16 Claims. (Cl. 250—31.)

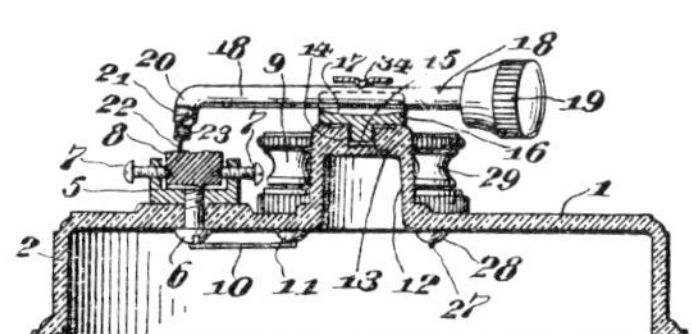

1. In a detector, a member rotatable in but one plane and having a groove in a face thereof, a rod disposed in said groove for longitudinal and rotative movement therein, and a detector element carried by said rod.

Ralph & Elinor Williams

While Atwater Kent never sold a complete crystal set, local entrepreneurs did, using his parts. The few that were made in the factory, were donated to hospital patients. Kent did make electrical and mechanical prototypes for two crystal sets, the size of a type 11 tuner base, one of them using a compression tuning condenser, but the project was abandoned.

Ralph & Elinor Williams

5 (4333) Oct. 1923 $55

The 8 was similar, with a condenser between the two units, on a longer board.

Ralph & Elinor Williams

9 (4445) Nov. 1923 $68

Entertainment, Education and Enjoyment Are Combined in This Christmas Gift

ATWATER KENT

Each of the three Atwater Kent Radio Receiving Sets shown below fulfils a particular demand.

Atwater Kent Model 10 (Radiodyne)

CAT. 4340 Grey PRICE $100.00
CAT. 4600 Brown PRICE $104.00

The Atwater Kent "Model 10" set is the De Luxe outfit with the maximum selectivity, tone quality and volume. It includes 2 stages of tuned radio frequency amplification, detector and 2 stages of audio frequency amplification. The instruments included are 3 Variable Condensers, 3 Radiodyne Transformers, 2 Tube Units, Potentiometer and Detector, 2 Stage Amplifier. The "Radiodyne" cuts out local interference and is easily operated.

Atwater Kent Model 9

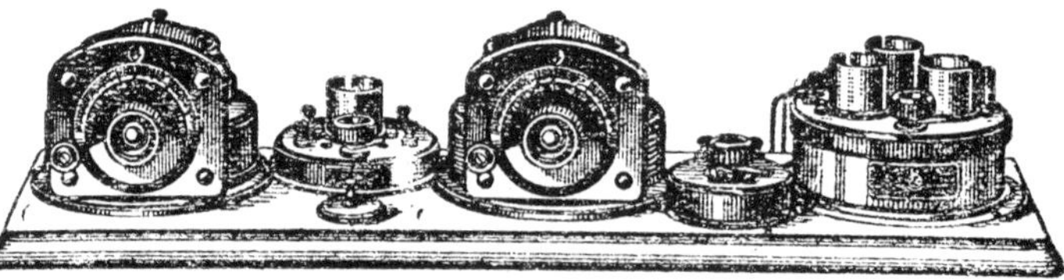

CAT. 4445 PRICE $68.00

"Model 9" Atwater Kent Receiver is a four tube set which gives exceptional results, clear quality reception without that annoying interference. It incorporates the ultra-selective tuned radio frequency principle used in the De Luxe set. This set is particularly recommended for use with an indoor or loop aerial. There is a world of value in this $68.00 set.

Atwater Kent Model 5

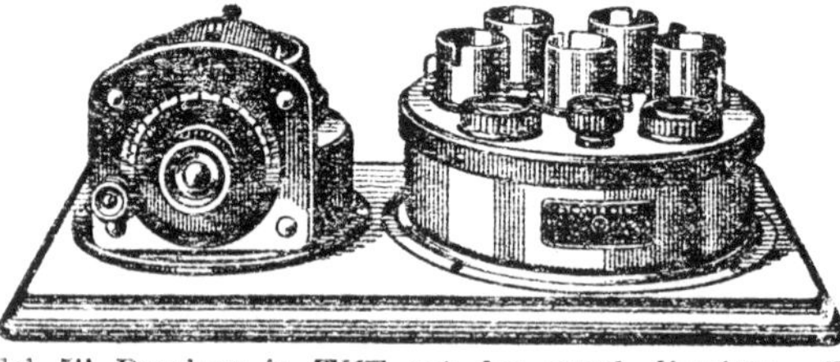

CAT. 4333 PRICE $55.00

The Atwater Kent "Model 5" Receiver is THE set for rural districts. The same excellent results secured with the No. 4066 outfit of last season are had with this $55.00 set. The tuning arrangement affords great ease of operation and the two stages of radio frequency, detector and two stages of audio frequency amplification give ample volume for Loud Speaker operation on both local and distant broadcast programs. No. 4333 complete set as above, wired, $55.00.

NEW YORK

Klein Radio & Electric Co., 34 Park place, 48 Fulton Street and Barclay Street and West Broadway.
S. Shearn, 1122 Madison Avenue.
Radiophone Equipment Co., 436 Seventh Avenue.
J. Haslinger, 77 West 23rd Street.
Marks Radio & Electric Co., 68 West 23rd Street.
R. H. Macy & Co., 34th Street and Broadway.
J. L. Lewis, 132 West 32nd Street.
Vim Electric Co., 66 Cortlandt Street.
Rush Radio & Appliance Co., 8 East Fordham Road, Bronx, N. Y. C.
Cortlandt Electric Co., 175 Greenwich Street.
Davega, Inc., 125 West 125th Street.
Davega, Inc., 111 East 42nd Street.
Sol. Langer, 1783 Southern Blvd., Bronx, New York.
Stephens Shop, 15 East 3rd Street, Mount Vernon, N. Y.
Terminal Cycle & Sporting Goods Co., 42 Cortlandt St.
Electric Service Engineering Co., 105 West 47th St.
John Wanamaker, Astor Place.
Morris Radio Co., Inc., 160 West 23rd Street.
R. W. Zundel, 47 Whitehall Street.
W. H. Straut, Haverstraw, N. Y.
Modells, 6 Church Street; 191 Fulton Street.
Gimbel Bros., 33rd Street & Sixth Avenue.
Mielke-Eberhart Radio Stores, Inc., 1263 Lexington Ave.

LONG ISLAND

M. M. Griffin, Rockville Center, L. I.
Mielke-Eberhart Radio Stores, Inc., 543 Steinway Ave., Astoria, L. I.
Bangert Electric Co., 297 Fulton St., Jamaica, L. I.
Morch Radio & Electric Co., 11201 Jamaica Ave., Richmond Hill, Long Island.
H. J. Wicks, Bayshore, L. I.

BROOKLYN

Tollner Electric Co., 523 Nostrand Ave., Bklyn.
Brooklyn Electric Lamp & Novelty Co., 278 Fulton St Bklyn.
Brooklyn Radio Center, 311 Broadway, Bklyn; 651 Manhattan Ave., Bklyn.
Morch Radio & Electric Co., 3135 Fulton St., Bklyn.
Geo. Moskowitz, 416 Central Ave., Bklyn.
Electrical Necessities Shop, 778 Rogers Ave., Bklyn
J. C. Hoose, 80 Livingston St., Bklyn.

NEW JERSEY

Amboy Lighting Co., Perth Amboy, N. J.
Heraco Exchange, 615 Bergenline Ave., West New York, N. J.
West New York Heraco, 593 Bergenline Ave., West New York, N. J.
National Electric Co., Passaic, N. J.
Independent Electric Construction Co., Passaic, N. J.
Joseph Czukor, New Brunswick, N. J.
R. B. Jolley, Morristown, N. J.
Eagan Radio Shop, 66 Hudson St., Hoboken.

Literature describing the entire line of Atwater Kent Radio Sets and Parts sent on Request

Wholesale Only **E. B. Latham & Co.** Wholesale Distributer

550 Pearl Street New York City

N. Y. Sun & Globe (Dec. 23, 1923)

12 (4910) Oct. 1924 $105 Ralph & Elinor Williams

10 (4340 Radiodyne) green Oct. 1923 $88 ($100 after Dec.) Ralph & Elinor Williams

10 (4340) black Ralph & Elinor Williams
Component binding posts removed.

10 (4600) Dec. 1923 $104 Ralph & Elinor Williams
Cans painted brown.

10A (4550) brown Feb. 1924 $104
10A (4560) black Feb. 1924 $100

Battery cable instead of binding posts.

10B (4550) brown $104
10B (4560) black $100

Orthogonal RF coils.

10C (4700) $85
Potentiometer removed, shorter board.

10C (4700) compact $85

Still shorter board with splined ends.

ATWATER KENT RADIO

Good Housekeeping (Dec. 1925), p. 182

In CHARLES DANA GIBSON'S *home there is an Atwater Kent Model 20 Compact. The price of this Receiving Set is eighty dollars.*

. . . only by hearing it

ONLY by hearing an Atwater Kent Radio can you understand exactly what we mean when we say "all-round" performance. Only by contrasting it with other sets you have heard can you appreciate an ideal combination of performance features: tone and volume, distance and selectivity.

If you are not entirely satisfied with your present set, or if you have not as yet had a radio—and want one that will please your friends as much as it pleases you—put your faith in Atwater Kent Receiving Sets and Radio Speakers.

Send for illustrated booklet telling the complete story of Atwater Kent Radio.

ATWATER KENT MFG. CO.
A. Atwater Kent, President
4745 WISSAHICKON AVENUE
PHILADELPHIA, PA.

MODEL 20, $80

MODEL 10, $80 (without tubes)

RADIO SPEAKERS priced from $12 to $28

Saturday Evening Post (Apr. 4, 1925)

New Model No. 12 (Six Tubes)

12 (4375,4620) Apr. 1924 $105

Ralph & Elinor Williams

9C (4660) May 1924 $70 (later $65)

1,655,372. RADIO APPARATUS. ARTHUR ATWATER KENT, Ardmore, Pa. Filed June 19, 1924. Serial No. 720,890. 15 Claims. (Cl. 250—16.)

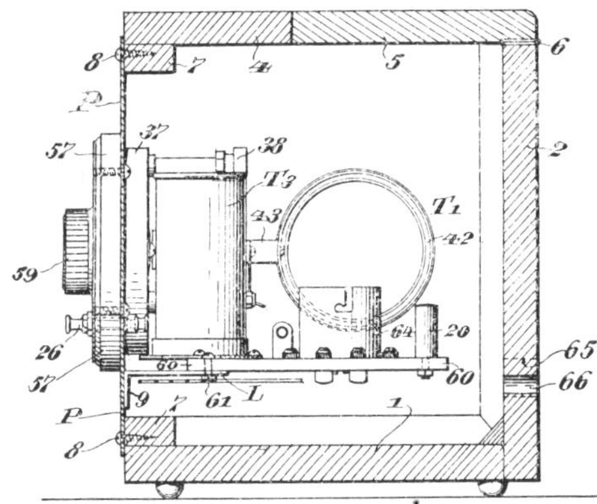

1. The combination with a housing, of a panel structure forming a detachable wall of said housing and comprising a sheet metal panel and a member secured thereto and having substantial extent normal thereto for stiffening said panel substantially throughout its entire extent, and radio receiving apparatus carried by said panel structure within said housing and removable as a unit with said panel structure.

Ralph & Elinor Williams

20 (4640) Oct. 1924 $100

24 (4920) or 20 Deluxe. same chassis

Feb. 1925 $120

Technical article in Popular Radio, June 1925, pp. 532–542.

Ralph & Elinor Williams

19 (4880) Oct. 1924 $90 (later $85)

Ralph & Elinor Williams

20C (7570) Sept. 1925 $80

20C (7960) $60 by July 1926

21 (7780)

32 (8270) July 1926 $140 Ralph & Elinor Williams

35 (8100) July 1926 $70 Ralph & Elinor Williams

Atwater Kent's one-millionth set was a 35, made on Dec. 3, 1926.

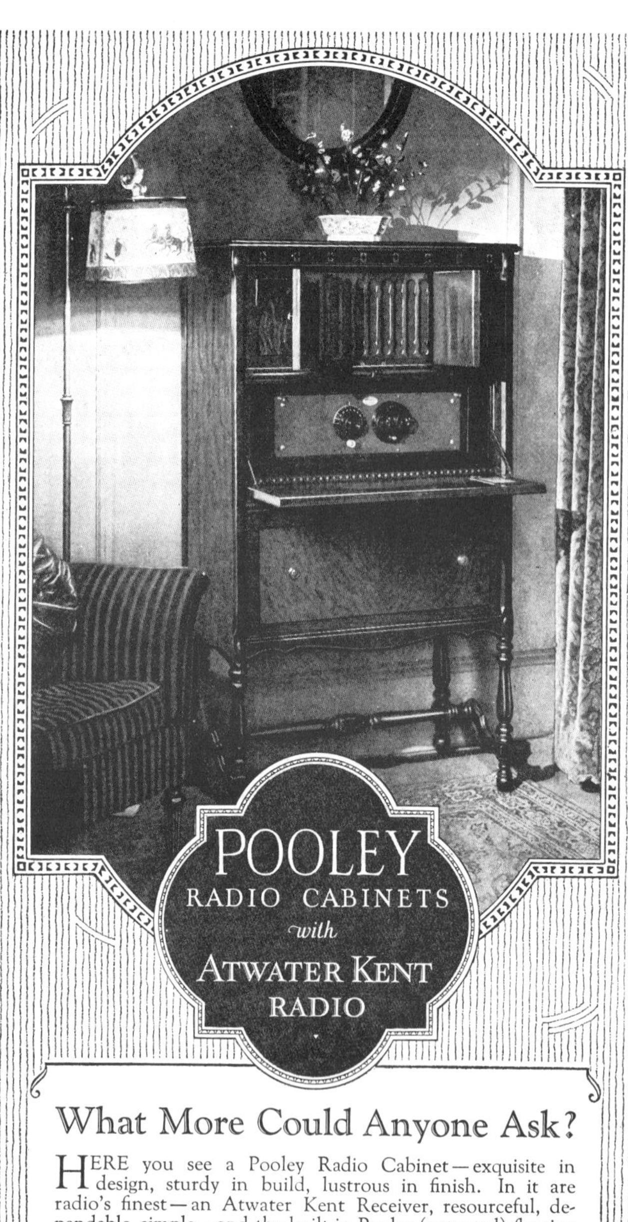

What More Could Anyone Ask?

HERE you see a Pooley Radio Cabinet—exquisite in design, sturdy in build, lustrous in finish. In it are radio's finest—an Atwater Kent Receiver, resourceful, dependable, simple—and the built-in Pooley (patented) floating horn, clear-spoken, rich and true.

And here is an opinion on radio cabinets worth knowing.

> "The Pooley Radio Cabinet is approved for Atwater Kent Radio because of the design and quality of Pooley cabinet work and because of the tone qualities of the Pooley built-in floating horn. Both meet the standards we set and maintain for Atwater Kent Receivers and Speakers." (*Signed*) A. ATWATER KENT

A gift that will make Christmas memorable—a lovely addition to any home. What more could anyone ask? Pooley Radio Cabinets bring you complete and lasting radio satisfaction. See them—hear them—there's a Pooley dealer near you.

Beware of imitations—look for the name "Pooley" before you buy

THE POOLEY COMPANY
1672 Indiana Avenue Philadelphia, U. S. A.

Model 1300-R-2D (*shown above*)
Equipped with built-in Pooley (patented) Floating Horn and Atwater Kent Receiving Set. Prices complete, without tubes and batteries, $225 to $305. Other Pooley Radio Cabinets, equipped with Atwater Kent Radio, from $135 to $390. Write for complete illustrated booklet—today.

Prices slightly higher west of the Rockies and in Canada. Canadian Pooley Radio Cabinets are manufactured by Malcolm and Hill, Ltd., Kitchener, Canada.

POOLEY
CABINET SPEAKERS
with
ATWATER KENT
UNIT

Model 2200 (*shown at left*)
Pooley Cabinet Speaker, with built-in Pooley (patented) Floating Horn and Atwater Kent Reproducing Unit. Ample battery space. Price, without set, tubes and batteries, $50.

Saturday Evening Post (Dec. 11, 1926)

RED LION CABINET—ATWATER KENT RADIO

Words and Music~

The usefulness and beauty~ of a personal writing desk and the luxury of a wonderful radio....

For the woman who is interested more in the sound of radio than the sight of it, we have designed this smart, *usable* desk which almost hides its world-famed Atwater Kent receiving set and speaker.

As an article of furniture, the Red Lion copyrighted Desk, with its soft toned, hand-rubbed finish of highly figured veneers, is a thing of beauty among even the most luxurious surroundings. And as for its musical qualities—it is equipped with Atwater Kent radio. Need we say more?

Expensive? Quite the opposite. Visit an Atwater Kent dealer today and ask to see the downright value Red Lion Desk Cabinets offer you in these exclusive models at $40.00 to $50.00.

With Atwater Kent Receiving Sets $94.00 to $133.00

A little higher west of the Rocky Mountains and in Canada

AN ATWATER KENT RADIO IN A Red Lion Cabinet

RED LION CABINET COMPANY, RED LION, PA.

Makers of the famous Red Lion Furniture

Saturday Evening Post (Nov. 19, 1927)

Radio Broadcast (Feb. 1926), p. 489

The Atwater Kent plant in early 1926. A 3½-acre addition was due to be completed in May, bringing the total floor area up to nearly 15 acres.

30 (8000) Apr. 1926 $105 ($85 after June 1)
48 July 1928 $49

Ralph & Elinor Williams

Ralph & Elinor Williams

33 (8930) July 1927 $90
49 July 1928 $68

THE FACTORY BEHIND THE PRODUCT

Service Manual Cover

ATWATER KENT RADIO

ELECTRIC! modern!

A·C

MODEL 37

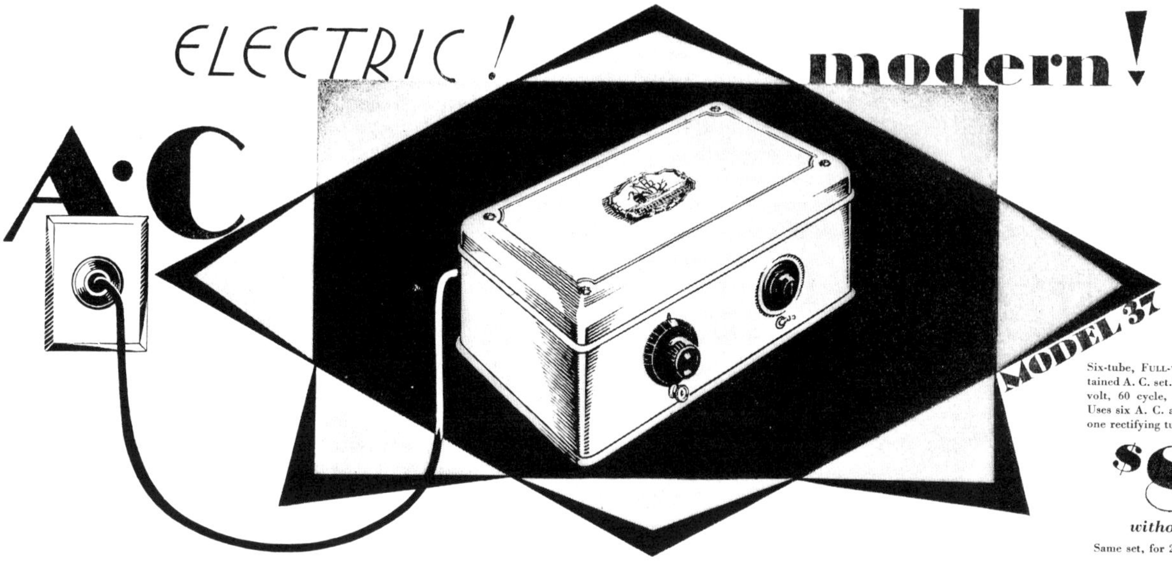

Six-tube, FULL-VISION Dial, self-contained A. C. set. For use with 110-115 volt, 60 cycle, Alternating Current. Uses six A. C. amplifying tubes, and one rectifying tube.

$88

without tubes

Same set, for 25 cycle, A. C., $98

200,000 in four months...

ONLY four months ago the compact, self-contained Atwater Kent A. C. set was introduced. *Only four months ago!* And already more than 200,000 families are enjoying this simpler and better radio—making a total of more than 1,600,000 Atwater Kent receivers in American homes.

A record! But more than a sales record. A record of satisfaction—of jubilant owners who write us:

"You didn't tell us half the story. This set is much better than we expected. We didn't know radio *could* be so good."

Radio *can* be so good. Let the Atwater Kent A. C. set tell you with its own clear voice. Radio's most thrilling summer is just ahead. The Atwater Kent A. C. set will give you the last tingling pulsation of every one of those thrills.

Get yours now and be ready for the political conventions with the set that's *always* ready. This is going to be a radio campaign. Travel to Kansas City and Houston at the speed of light and at a cost of only a fraction of a cent an hour. Sit up on the platform with the orators. Get down among the delegates and hear the voting. Listen to the bands, the singing. Have at your side a man who knows what's going on every minute and will tell you what it means.

Hear the candidates nominated—and hear them later as they knock at your door to make their plea—by radio. Take in the Tunney-Heeney fight, too. Sit at the ringside with Graham McNamee.

Let the experience of more than 200,000 owners guide you to this simple, modern, beautiful, reliable, wholly satisfying set. Why not see, right away—in time for *tonight's* fine program—just *how* good and inexpensive an electric receiver can be? The nearest Atwater Kent dealer will be glad to help you.

Model 38, without tubes, $125

An extra powerful A. C. set of particular value where distance getting, or maximum daylight reception is essential or an inside antenna is necessary. Beautiful tone quality. Unusual selectivity. Uses seven A. C. tubes and one rectifying tube.

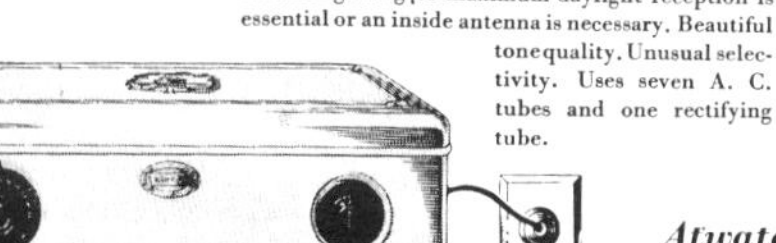

ONE Dial Receivers licensed under U. S. Patent 1,014,002

Prices slightly higher West of the Rockies

Write for illustrated booklet of Atwater Kent Radio

ATWATER KENT MANUFACTURING CO.

A. Atwater Kent, President

4703 Wissahickon Ave. Philadelphia, Pa.

Model E
Radio Speaker, $24

Radio's truest voice. All parts protected against moisture. Comes in a variety of beautiful color combinations.

Atwater Kent Radio Hour every Sunday night on 19 associated stations

Saturday Evening Post (May 26, 1928), p. 85

A Mar. ad stated that by the middle of Feb., Atwater Kent had made more than 100,000 model 37s, and was turning out 3000 per day. 12,240 were made during 1927.

The model 50 used untuned RF amplifiers, avoiding patent claims of Hazeltine and Alexanderson (relative impedances in RF interstage coupling, and cascaded TRF stages). It was designed by Dr. John M. Miller (perhaps why Miller was hired from NRL in 1925) and the prototype built by his assistant Sarkes Tarzian. John Dreyer related that one of his first jobs after arriving full-time in May 1926 was to improve the 50 with individually-shielded RF coils, although the set as finally sold in 1927 was the earlier version.

The 50 replaced the top-of-the-line model 32 for a few months, perhaps only to use up stocks of parts, as Kent by that time had signed with RCA and had no need of this unusual circuit.

Miller left Atwater Kent in 1936 and was hired by RCA at Dreyer's suggestion (as was Tarzian, Dreyer having joined RCA in 1932). He returned to the Naval Research Laboratory in 1940.

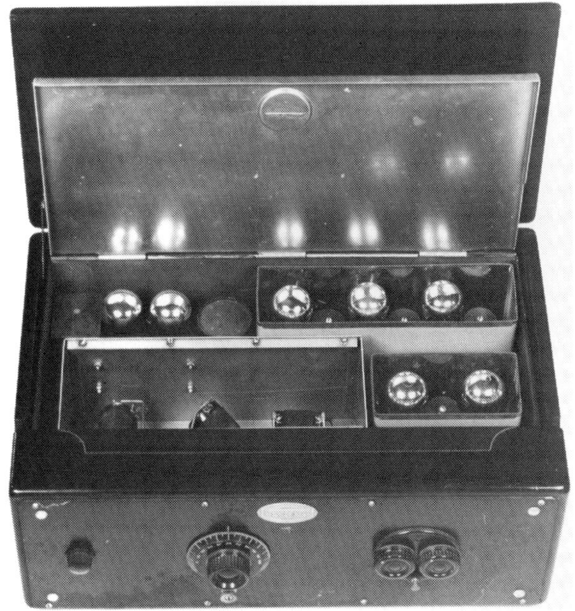

Ralph & Elinor Williams

50 July 1927 $150

Rich Elskamp

36 Nov. 1927 $125 (in a table cabinet with separate power pack)

Ralph & Elinor Williams

37 Dec. 1927 $88

According to John Dreyer, the ship is the Mayflower.

Ralph & Elinor Williams

46 (green & black)

38 Apr. 1928 $125 Technical article in *Citizens Callbook* vol. 11 no.1, Jan. 1930, p.72.

42 Ralph & Elinor Williams

78,030. CABINET FOR RADIO RECEIVING APPARATUS. ARTHUR ATWATER KENT, Ardmore, Pa. Filed Aug. 23, 1928. Serial No. 27,942. Term of patent 14 years.

The ornamental design for a cabinet for radio receiving apparatus, as shown.

52 Aug. 1928 $117 Ralph & Elinor Williams

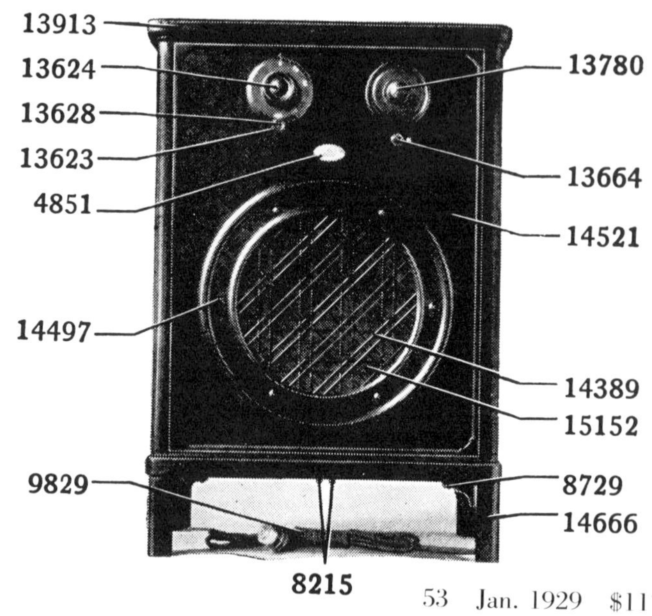

53 Jan. 1929 $117

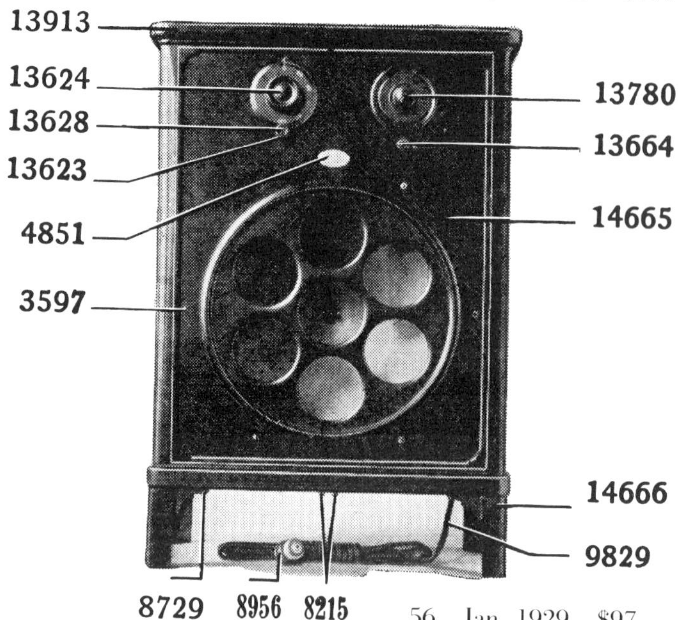

56 Jan. 1929 $97

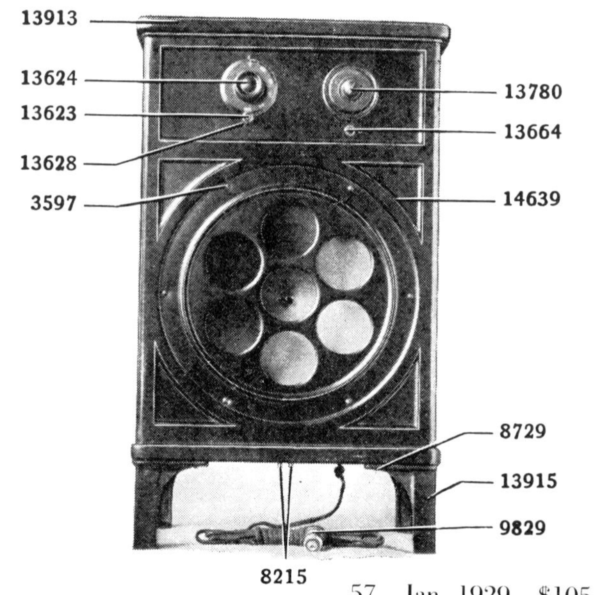

57 Jan. 1929 $105

ATWATER KENT RADIO

Saturday Evening Post (Dec. 8, 1928), p. 87

The Christmas Gift *that keeps on giving*

All that a big set could offer you, now yours in a little one

The slimmest, tidiest, daintiest, friendliest little radio companion you could ever wish for—Model 52, the new all-together set for 1929. Yet it has the tone, power, range, all-round efficiency of a big set.

It presents the famous Atwater Kent compactness in a new all-electric form—your receiver and speaker *combined* in a shielding cabinet only 30 inches tall, 11 inches deep, 18 inches wide. Nowhere near as high as your waist!

No brow need be wrinkled over the placing of this convenient radio. Any little corner—any little niche—is just right. Let it place itself and efface itself. That's the companionable little thing it is. Lay your book or magazine on the golden top panel, and it's a reading table. Fine place for a bowl of flowers, too.

Unobtrusive Model 52 in the home of Ellis Parker Butler

MODEL 52 A. C. Combining electric receiver and speaker in satin-finished compact cabinet. FULL-VISION Dial. Uses 6 A. C. tubes and 1 rectifying tube, with automatic line voltage control. Without tubes, $117.

All four sides have a rippling satin finish. There are two speaker grilles—front and back. Cords for antenna and ground connections are twenty feet long. Place your radio anywhere—out in the room, if you like, and hear the music clearly and in full volume from any position.

The tone is even lovelier because of the blending of all that is best in Atwater Kent Radio in this complete modern instrument. Everything you could hope for in a big set at a big price is now offered in a little one at a little price. . . . And *have* you tingled to the thrill of easy, instantaneous program-selection with the Atwater Kent FULL-VISION Dial?

On the air—every Sunday night—Atwater Kent Radio Hour—listen in! Write for illustrated booklet of Atwater Kent Radio

ATWATER KENT MANUFACTURING COMPANY *A. Atwater Kent, President* 4703 Wissahickon Avenue, Philadelphia, Pa.

MODEL 40 A. C. A powerful, compact, all-electric receiver in a satin-finished shielding cabinet. FULL-VISION Dial. Uses 6 A. C. tubes and 1 rectifying tube. Without tubes, $77.

MODEL 41 D. C. Without tubes, $87.

MODEL 42 A. C. Crowned lid, panelled corners, ball feet. FULL-VISION Dial. Uses 6 A. C. tubes and 1 rectifying tube, with automatic line voltage control. For 110-120 volt, 50-60 cycle alternating current. Without tubes, $86.

Prices slightly higher west of the Rockies

"RADIO'S TRUEST VOICE"

Atwater Kent Radio Speakers: Models E, E-2, E-3, same quality, different in size. Each $20.

MODEL 44 A. C. Extra-powerful, extra-sensitive, extra-selective. Local-distance switch, FULL-VISION Dial. Uses 7 A. C. tubes and 1 rectifying tube, with automatic line voltage control. For 110-120 volt, 50-60 cycle alternating current. Without tubes, $106.

Model	Date	Price
40	July 1928	$77
41	Aug. 1928	$87
42	July 1928	$86
43	about Dec. 1928	$83
46	Jan. 1929	$83
44	July 1928	$106
45	Jan. 1929	$94
47	about Jan. 1929	$100

55 June 1929 $88
Technical articles: *Citizens Callbook*, vol.11 no.2, Mar. 1930, p.85; vol.11 no.3, Sept. 1930, p.82.

A. ATWATER KENT
PRESIDENT AND FOUNDER OF THE
ATWATER KENT MFG COMPANY
Philadelphia

Norman Rockwell, the artist, has placed his Atwater Kent One Dial Receiver in his studio

ATWATER KENT RADIO

SCREEN-GRID

TRUE voices—the truth of music and of human speech! Power that hurdles horizons—fetches entertainment from the blue distance! Two—three—four—five years from now, you'll still be saying proudly, "Listen to *that!* It's the Atwater Kent that came to us at Christmas in 1929!"

What could possibly mean more to the whole family? Imagine the enduring pleasure of the gift that goes on and on, pouring forth its music, giving companionship through the years!

YES, AND FOR MANY A CHRISTMAS TO COME!

JUST a radio will never do. Only an Atwater Kent can give you Atwater Kent reality of tone—with the vastly more powerful Screen-Grid tubes used as Atwater Kent uses them.

You are not restricted to one or two cabinets. Choose your radio furniture just as you choose other beautiful things for your home. In any cabinet, you get Atwater Kent Screen-Grid performance—and that *never* varies!

On the Air—*Atwater Kent Radio Hour, Sunday evenings, 9:15 (Eastern Time), WEAF network of N.B.C. Atwater Kent Mid-Week Program, Thursday evenings, 10:00 (Eastern Time), WJZ network of N.B.C.*

ATWATER KENT MANUFACTURING COMPANY *A. Atwater Kent, Pres.* 4822 Wissahickon Ave., Philadelphia, Pa.

Rich Elskamp

60 Aug. 1929 $100
61 Aug. 1929 $88 (110VDC)
67 Aug. 1929 $77 (battery)
Available in metal cabinets, or chassis for installation in consoles such as this Kiel table (introduced in August 1929)

Technical article in *Citizens Callbook*, vol.12 no.2, Mar. 1931, p.69. Short technical article on 67, in *Radio*, May 1930, pp. 46–47.
Not shown: 66 (chassis only)
Nov. 1929 $135.

Radio Mechanics (Apr. 1927), p. 320

EDITORIAL

SUMMERTIME is the radio manufacturer's opportunity to take stock of himself and his organization. This, of all summers, is not a mere dull period to be passed away by trips abroad—it's an opportunity to groom for next winter's battle. And battle it will be, too, for we shall see whether or not some of the companies so heavily financed last fall, now closed down tight, will be able to come back and hold on this fall.

The greatest weakness today is in the mechanical design of the sets and the manufacturing methods used to produce them. Last season the general idea seemed to be: slap'em together and throw'em out. The result is proof conclusive that that method doesn't work.

Circuits? There are plenty of good ones. The manufacturer's problem is to use them. Radio engineers? There are too many. The manufacturer needs mechanical designers and competent production and factory superintendents.

Last year's cheap sets were costly to the producer and consumer because they were failures. If they were good, they were not good enough. If you don't believe this, give yourself a little education. Buy a Weston phone plug and break it apart. You will find out more from that plug at less expense than in anything else you can do.

Count the number of parts, look at their design, examine the quality of the workmanship. Compare it with any other plug at any price. Any B. C. L. can tell you that it wasn't made by a radio manufacturer. Yet the Weston Company makes a profit on it at a list price of seventy-five cents.

What's the answer? Factory brains and factory equipment. If you want a post-graduate course, buy a Weston 301 voltmeter and take it apart. What would a radio manufacturer charge for such an instrument? Of course no radio company could produce it. Who controls limits to one-millionth of an inch, puts on knurling so fine it can't be seen, cuts threads two hundred to the inch, or makes parts today, accurate to half a thousandth, which are interchangeable in instruments ten years old? Weston does. Weston does all this and dozens of other things as remarkable in the 301 instrument which you buy for seven dollars and a half—and makes money at it.

Sixty or seventy-five dollars for a cheap, thrown-together radio set? It's absurd. The cost isn't in materials. It's in unnecessary labor, the expense of defective sets, service and sales effort required for inferior merchandise, and the losses on over-production.

The Atwater-Kent Company is the most successful radio manufacturer in this country because there is manufacturing intelligence and built-in honesty and stability in Atwater-Kent apparatus. Others are trying, some don't know how to try, and some don't try at all. The president of a company which put out a large stock issue last winter told me rather proudly that their returned, defective sets were only six per cent. I know they were more than that, and I know that they have on hand parts which will never be used for one-fifth as many sets as they sold.

With such competition it is easy to understand the success of the Atwater-Kent Company. If the public's money had gone into factory equipment rather than into the pockets of the executives, radio would have been saved two or three years of growing pains.

M. B. SLEEPER,
Editor.

Radio Engineering (May 1925)

PRODUCTION FIGURES

PART	# Model	1923	1924	1925	1926	1927	Total
4325	8	17	(2)				15
4340	10	12270	2689				14959
4445	9	2449	3920				6369
4480	9	8	7				15
4490	10	8	3				11
4550	10	122	11831				11953
4560	10	80	18517			(1)	18596
4590	10		2363				2363
4600	10	1235	1614		1		2850
4610	10 B		49				49
4620	12		1520			(2)	1518
4640	20		78321	109027	2422		189770
4660	9 C		4180	1796	(23)	(6)	5947
4670	20 Panel			7218	1		7219
4700	10 C		23324	18151	(412)	2	41065
4880	19		2298	3418	(18)		5698
4910	12		5377	3193	(148)		8422
4920	24		4257	2943	1786		8986
4950	10 Pooley		7043	5790	29		12862
7570	20 Compact			163376	(119)		163257
7885	20 Panel			18648	1516		20164
7780	21 Dry Cell			17584	(7208)	(10)	10366
7817	21 Dry Cell Panel			15	6		21
7960	20 Compact				76150	6446	82596
8000	30				99297	21525	120822
8041	30 Panel				642		642
8100	35				199944	122075	322019
8186	30 Panel				28406	214	28620
8260	20 Panel				9693	(716)	8977
8270	32				35958	788	36746
8280	32 Panel				1817	13	1830
8432	20 Panel					44	44
8500	50					1540	1540
8930	33					70370	70370
8970	33 Panel					28	28
9043	30 Panel					19841	19841
9050	33 Panel					38725	38725
9390	36					36209	36209
9500	37					12240	12240
9520	36 Console					4033	4033
Special Sets			9				9
TOTALS		16189	167320	351159	449740	333358	1317766

Figures reconstructed from factory records, copyright 1987 by Ralph Williams. Used by permission.

BREMER-TULLY

BREMER-TULLY MFG. CO.
Chicago, Ill.

Radio Broadcast (Dec. 1926), p. 187

Bremer-Tully Manufacturing Company

Harry A. Bremer's interest in radio dated from 1905 when he took "High Frequency Currents" as his sophomore thesis at Armour Institute in Chicago, and built a coherer and spark coil. He graduated with an EE degree but went into the manufacture of auto radiators and the machinery to make them.

John C. Tully was also a graduate electrical engineer, though he held at one time or another such positions as newspaper reporter, advertising manager, professor of college mathematics, real estate sales manager, and assistant bank cashier. He was developing a domestic oil burner when he met Bremer; their first joint venture was a semiautomatic machine for tinning radiator tubes.

Their backgrounds and interests meshed perfectly, Bremer doing the engineering work (he had 7 U.S. patents) while Tully attended to the business end. Tully was active in other businesses and trade organizations simultaneously. The company was incorporated October 17, 1922.

Bremer-Tully's first product is said to have been a one-plate vernier condenser in 1921, followed by a three-circuit tuner. From the making of parts they progressed to pictorial wiring diagrams and, by July 1924, their first kit, the "Nameless." This was supposed to be a temporary title while they held a contest to choose a permanent one, but, dissatisfied with the entries, they awarded a prize and kept "Nameless" anyway.

In October 1925 they began advertising complete receivers under the name "Counterphase" with five and six tubes; the following July they switched to shielded six- and eight-tube sets.

By mid-1928 Bremer-Tully had moved to a new factory and was a substantial company, licensed by RCA and Hazeltine and employing about 200 well-paid men. Quality was stressed over quantity and about 50 sets per day were made. Bremer and Tully sold out to Brunswick at a high price around December 1928, the announcement coming in March 1929. Brunswick shortly closed the plant and moved the machinery to its own factory. John W. Million, formerly with King-Hinners, was chief engineer. When Brunswick in turn sold its radio division to Warner Bros. in April 1930, and Million joined Utah, the Bremer-Tully corporate name remained as a subsidiary of Brunswick Radio Corp., but Warners discontinued production within a few years and sold off the machinery.

Here is a corner of the experimental laboratory where measurements on condensers and inductances are made. It is also here where preliminary tests are run on all apparatus before the company commits itself to the manufacture of a new article

General offices of the firm are at 532 South Canal St., Chicago, Ill. In this picture is shown a portion of the office

Radio Age (June 1926), p. 5

John C. Tully, president of the Bremer-Tully Manufacturing Co.

In the foreground is one of the machines used in winding primaries and interiors of torostyle coils. Finished primaries are seen under the spool of wire, while to the right are a few of the finished coils. Special machinery had to be developed for winding this type of coil, as explained in the article

Harry A. Bremer, vice-president and secretary of the Bremer-Tully Manufacturing Co.

A section of the plant where condensers are assembled

Radio Age (June 1926), p. 5

Nameless Feb. 1925 $26.50 kit

Construction article (different mechanical layout) in *Radio Engineering*, Apr. 1925, pp. 203–210. Also in *N.Y. Evening World*, Mar. 21, 1925, p. 17.

Ralph & Elinor Williams

B-T Counterphase, Complete Receiver

Counterphase Six Oct. 1925 $165
Technical articles in *Radio Broadcast*, Jan. 1926, pp.350–352,4,6,8, and in *Radio News*, Nov. 1925, pp.616–617. Counterphase Power Six construction articles in *Radio*, Dec. 1926, pp.35–36,94, and in *Radio Listeners Guide* and *Call Book* and *Radio Review*, vol.1 no.12, Mar. 1927, pp. 88–92.

Radio Dealer (Sept. 1926), p. 60

Counterphase Eight Aug. 1926 $225
Counterphase Six Sept. 1926 $155

A technical article on a 7-tube version is in *Radio News*, Mar. 1927, pp. 1101–1102.

Vernon Oehlke

6-40 R Mar. 1928 $130

6-40 S Apr. 1928 $130
6-41 console $190

Radio Retailing (June 1929), p. 213

Model 81—DE LUXE STANDARD CONSOLE
Chassis identical with Model 82.
$164, less tubes; slightly higher west of Rockies

80 June 1929 $89.50
81 $164 82 (Illus. on p 95) $195

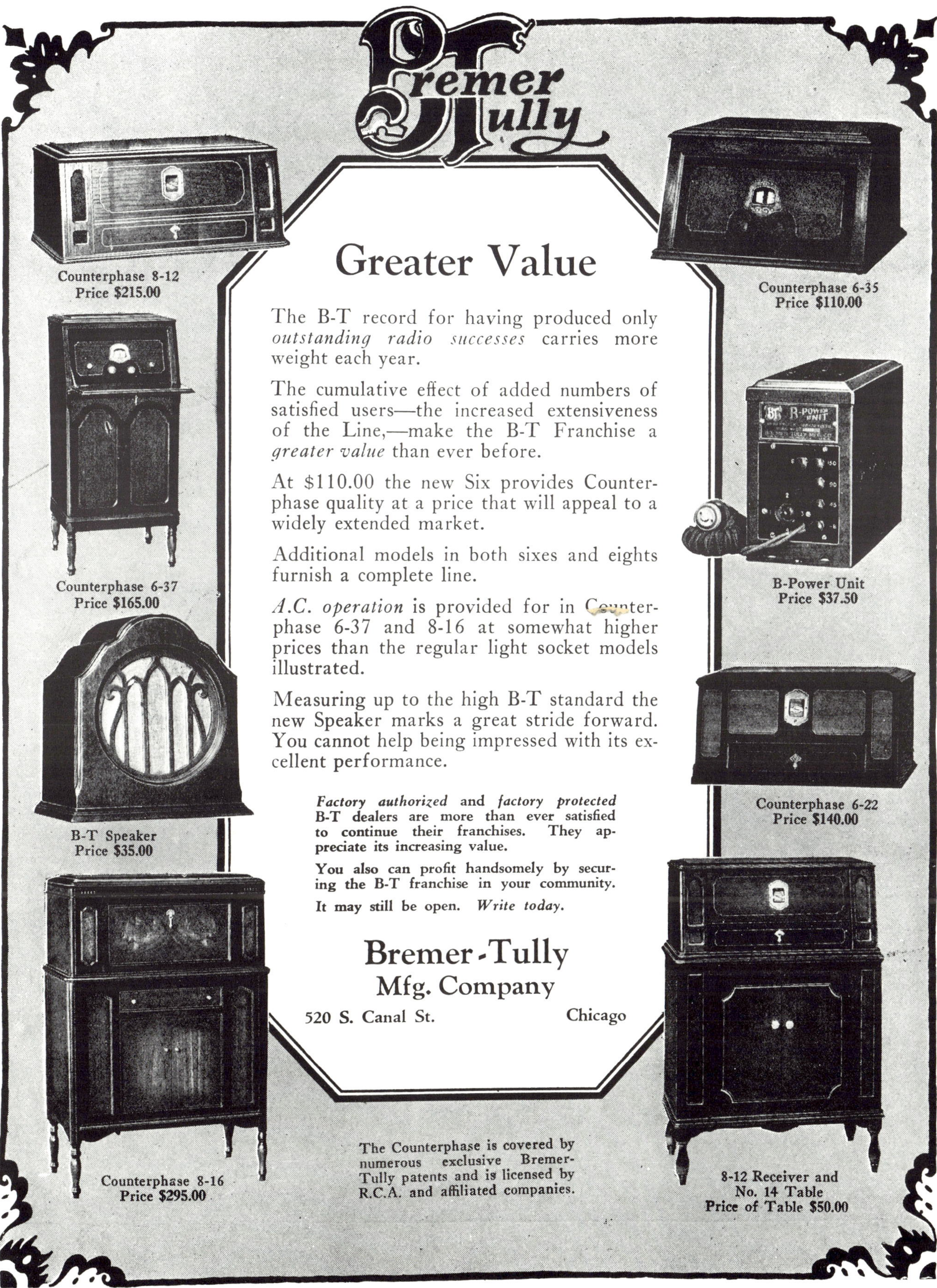

Radio Retailing (Oct. 1927), p. 51

These models were introduced in July 1927. In Sept. 1927 came two AC models: 6-38 (identical with the 6-37) and 8-17 (identical with 8-16). In Nov. the 8-13 (identical with 8-12) was added, along with a $175 model not pictured that may have been the 6-36.

Radio Retailing (Nov. 1929)

Brunswick
Highboy Console
Model No. 21
Price $174 Tubes Extra

Radio — Panatrope with Radio — Records

THE BRUNSWICK-BALKE-COLLENDER COMPANY

New York, Chicago, Toronto—Branches in All Principal Cities

Radio Retailing. November, 1929. Vol. 10, No. 5. Published monthly, McGraw-Hill Publishing Company, Inc., Tenth Avenue at Thirty-sixth Street, New York, N. Y. $2 per year. 25 cents per copy. Entered as second-class matter, April 10, 1925, at the Post Office at New York, N. Y., under the Act of March 3, 1879. Printed in U. S. A.

Nov. 1929	models:	14	$148	21	$174	31	$272.
Dec. 1929	models:	14S	$129	21S	$154	31S	$249 (screen-grid)

Same chassis as Bremer-Tully 80 and S80 series, which were still being advertised. Until 1928, Brunswick radios used RCA chassis.

Bremer-Tully Announces New and Improved Radio Models

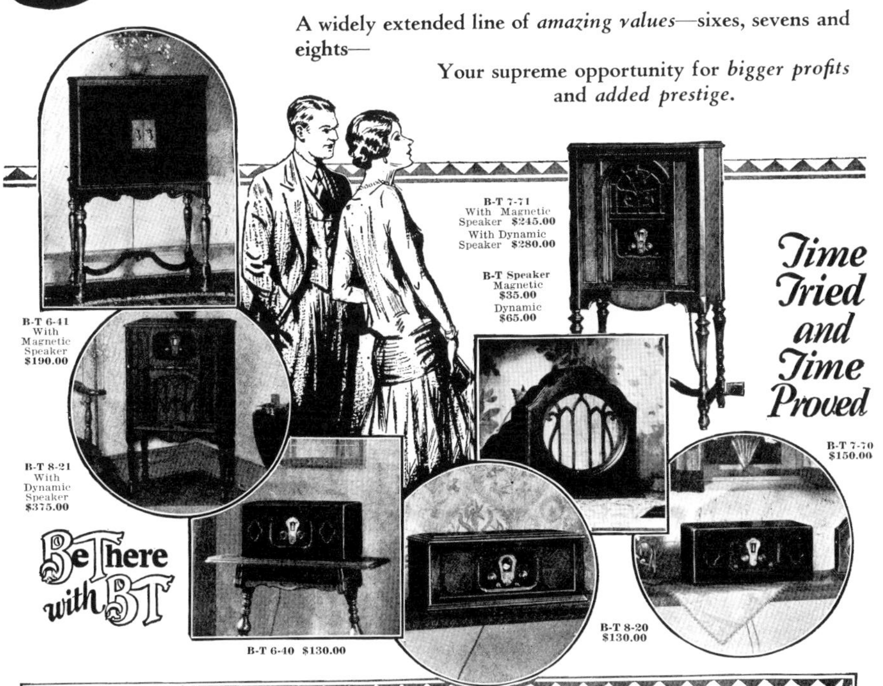

Every feature of this bigger and better B-T line reflects the extensive radio experience of the manufacturer.

Seven years of outstanding achievement and success is behind it.

Furniture of surpassing beauty.

Performance beyond anything at equal prices.

Selling policies that consistently protect the dealer.

No essential to radio merchandising success has been overlooked.

Get the B-T line now!

The coupon will bring further details.

COUPON

Name ....

Address ....

City .... State ....

Bremer-Tully Mfg. Company
656-662 Washington Blvd., Chicago

Model 82—DE LUXE FRENCH DOOR CONSOLE
Genuine butt walnut veneer with Carpathian Maple overlay.
$195, less tubes; slightly higher west of Rockies

Short technical article on the 8-20 in *Radio Broadcast*, Jan. 1929, p.177.

Radio Retailing (July 1928), p. 111

BROWNING-DRAKE

Browning-Drake Corp.

Although Browning-Drake was in business for more than ten years, the name is remembered for just one circuit. In August 1923 Frederick H. Drake, in his senior year at Harvard, made a mathematical study of tuned-RF amplification at the suggestion of the radio editor of the *Christian Science Monitor*, Volney D. Hurd. Drake approached Glenn H. Browning, four years older, a Research Fellow at Harvard's Cruft Laboratory after a stellar undergraduate career at Cornell, with the idea of making experimental measurements to confirm the math. In the course of a year, they found that the usual tuned-RF transformer had far too much capacitance between primary and secondary windings, which lowered the amplification, but that by making their primary of small wire wound in a thin slot, their transformer gave 90% of the theoretical voltage step-up.

Almost immediately the *Monitor* and other Boston papers picked up the "Browning-Drake" circuit. Since Browning and Drake had used National Co. condensers and vernier dials (National being located in Cambridge, with Harvard), president William A. Ready collaborated on the mechanical design, and by December 1924 was advertising "National Regenaformer" kits with the proper condensers and coils, attractively packaged with instructions for building the set. Arthur Lynch publicized the circuit in his magazine *Radio Broadcast*, *QST* printed an article, Browning and Drake read a paper before the IRE, and the race was on.

The radio magazines and newspapers were full of trick circuits, every one backed by some manufacturer with parts to hawk, but the Browning-Drake really was as good as it claimed to be. It continued to sell, with mechanical but few electrical changes, for several years.

By December 1925 Browning had formed the Browning-Drake Corp. and was advertising complete receivers, while National continued to sell the kits. In late 1928 one piece of advertising stated "over a million people listen in on Browning-Drake receivers" and that may not have exaggerated by much. Browning kept up a constant barrage of articles for any radio magazines which would print them. However, once the kit craze died down, and the company had to find something new to sell, it didn't do so well, being operated by a creditors' committee by March 1930.

Browning-Drake did continue until 1937 when Browning formed Browning Laboratories, making a diverse array of electronic devices over the years. Drake, meanwhile, having remained at Harvard to get his MA and PhD, joined Radio Frequency Laboratories at Boonton, New Jersey, then formed Aircraft Radio Corp. in Boonton in 1929 where he did outstanding pioneer work in airborne radio.

NATIONAL COMPANY, INC.

ESTABLISHED 1914

Engineers and Manufacturers

110 **BROOKLINE ST., CAMBRIDGE MASS.**

BULLETIN No. 105

THE NATIONAL REGENAFORMER

(PATENT PENDING)

FIG. 1

THE BROWNING-DRAKE RECEIVER USING THE NATIONAL REGENAFORMER

WHAT IT IS

To Glen H. Browning and Frederick H. Drake, of Harvard University, belongs the honor of designing the remarkable new type of Radio Frequency Transformer, known as the National Regenaformer, which has proved so efficient that over 90% of the value of amplification calculated by mathematics was actually produced when the transformer was subjected to laboratory tests. The National Regenaformer is incorporated in a hook up known as the *Browning-Drake Circuit* which has essentially two tubes—a radio frequency amplifier and detector (Fig. 1) to which two stages of audio amplification may be added, making a four-tube set for loud-speaker reception.

WHAT IT HAS DONE

Mr. L. D. Yont, of Brighton, Mass., received 35 stations in one evening on this set, among which were: KGO Oakland, California; KSD St. Louis, Missouri, and PWX Cuba. Dr. D. B. Cheetham, located in Worcester, Mass., says: "This set is a wonder—had Dallas, Atlanta and Kansas City the first night, before 9 P M. on the loud speaker. Marvelous clarity and wonderful volume. Can't use push-pull on stations under 1500 miles, as too much volume" The reason for these results is clearly shown by referring to Fig. 1. Curve "A" shows the amplification with National DX Condenser and Regenaformer type Radio Frequency Transformer, while "B" is the amplification of a Neutrodyne type of Radio Frequency Transformer.

FIG 2

Curve "A" shows the amplification with National DX Condenser and Regenaformer type Radio Frequency Transformer, while "B" is the amplification of a Neutrodyne type of Radio Frequency Transformer.

WHAT THE PRESS THINKS OF IT

The set was first presented before the American Institute of Electrical Engineers. It was then published by The Christian Science Monitor, which gave complete data for constructing the set in a series of articles which started June 7, 1924. So many favorable reports of the circuit came in that a second write-up appeared in the Radio World's Fair Supplement, Sept. 22. The Boston Post featured the set in its Sunday edition of Oct. 6, reprinting it again on Oct. 10 because of its great popularity. The Boston Globe also published it Oct. 10, speaking of it as "a new circuit of great possibilities" The Boston American published it on Oct. 16 to 20 in connection with the new American Super. Published in Radio Broadcast, December issue, and in Radio Engineering, December issue.

THE NATIONAL REGENAFORMER

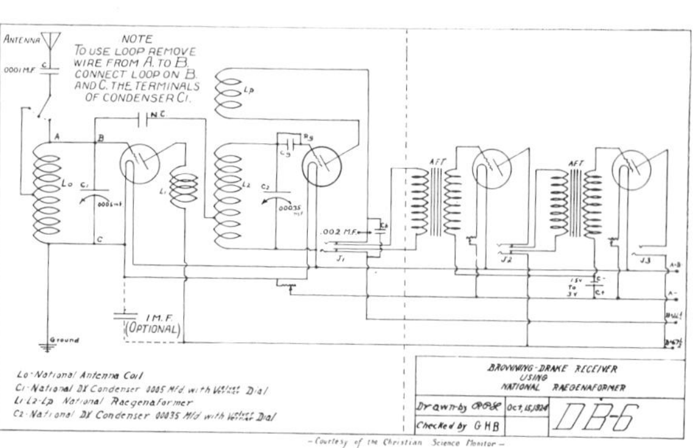

FIG. 193

PROPERTIES OF THE REGENAFORMER RECEIVER

1—**It is extremely sensitive, and therefore excellent for distant reception.**
2—**It is selective enough to cut out local radiocasting.**
3—**There is no radiation when properly constructed, and consequently no disturbance occurs in neighboring sets.**
4—**Stations may be tuned in with or without obtaining beat notes. On DX reception the beat note is a great help.**
5—**Stations come in at definite points on the second Dial, and, therefore, it may be logged for future reference.**

CONSTRUCTION

Fig. 1 shows a two-tube "Bread Board" model of the Browning-Drake receiver, while Fig. 3 shows the wiring diagram with two stages of audio amplification added (*Right of dotted line*). The principal parts of this set are embodied in a kit known as the National Regenaformer Kit, consisting of:

1 National Antenna Coil mounted on
1 .0005 National DX Condenser with
1 4" Velvet Vernier Dial
1 National Regenaformer, mounted on
1 .00035 National DX Condenser with
1 4" Velvet Vernier Dial
1 Blueprint of Wiring Diagram
1 Set of Hardware for Mounting.

PRICE OF KIT: **$22.00**

THE NATIONAL REGENAFORMER AND COIL, ONLY . . **$7.50**

Dealers' Standard Package 20

MANUFACTURED BY

NATIONAL CO., Inc.

110 Brookline Street, Cambridge, Mass.

The Browning-Drake set was described in *Radio Engineering*, Nov. 1924, pp. 327–335 (4-tube panel version), *Radio Broadcast*, Dec. 1924, pp. 282–287 (4-tube breadboard & panel versions), *QST*, Apr. 1925, pp. 21–23 (3-tube breadboard).

Introducing
the
BROWNING-DRAKE
RECEIVER

(Dec. 1925)

IN the quest for a set that would give greater distance and selectivity and still be economical and simple in design, Frederick H. Drake and Glenn H. Browning, both doing research work at Harvard University, set to work mathematically to see just how much could be developed by a tuned radio frequency transformer.

It was found that with the aid of a special primary to minimize capacity coupling, radio frequency amplification amounting to over 90% of the maximum could be obtained.

The resultant circuit, known as the Browning-Drake Regenaformer Circuit, has been so successful in getting extreme distance with exacting selectivity that it is now presented to the public in three styles of receivers.

Three stages of resistance coupled amplification give the Browning-Drake Receiver on unsurpassed quality of tone.

During the Trans-Atlantic tests made earlier in the year the Browning-Drake Receiver received Madrid, Spain. Owners of Browning-Drake Receivers living in the East have verified reception from KGO, Oakland, California; KFI, Los Angeles, California; Mexico City and Calgary, Canada.

Ask your radio dealer to show you the new Browning-Drake Receivers.

The B-D Standard
Five tubes with standard sockets. Provides for power tube. Mahogany cabinet with battery compartments at either end. List price, without tubes and batteries, **$130.**

Distributors and Dealers—Write us for details

BROWNING-DRAKE CORPORATION 

353 Washington Street **Brighton, Mass.**

BROWNING-DRAKE

μ

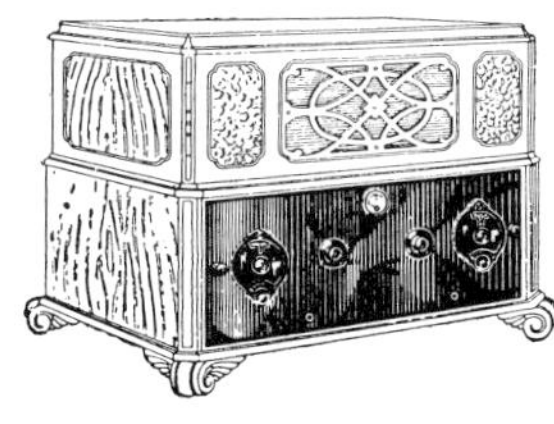

The B-D Senior
Six tubes with new type sockets. Built-in-loud-speaker. Battery compartments. List price, without tubes or batteries, **$185.**

The B-D Junior
Five tubes with new type sockets. Power tube may be used. Mahogany finish cabinet. List price, without tubes and batteries, **$95.**

★ Tested and approved by RADIO BROADCAST ★

(SPACE—CC-1)

Radio Dealer (Sept. 1927)

Presenting the
BROWNING-DRAKE
Line of Receivers

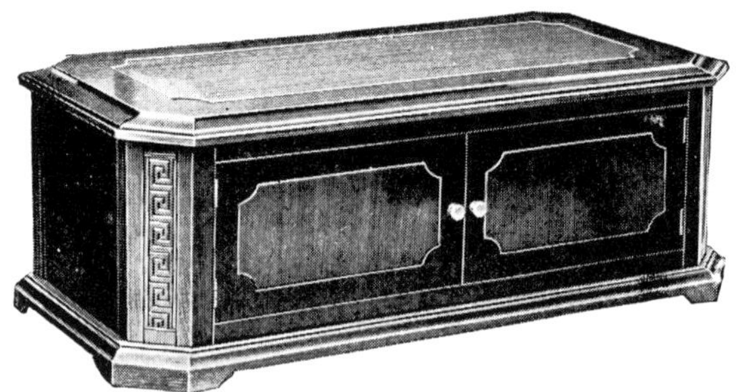

Model 6-A
SINGLE DIAL SIX TUBES
ILLUMINATED DRUM CONTROL
COMPLETELY SHIELDED
SUPER-SELECTIVITY

AFTER many years of extensive research work by Prof. Glenn H. Browning and Dr. Frederick H. Drake, the laboratories of the Browning-Drake Corporation have prepared for public presentation, an entirely new conception of the world-famous Browning-Drake Receiver.

Browning-Drake has long been known for its natural tone quality and only a few months ago made a record of transcontinental reception, Los Angeles from Boston, on seven consecutive nights. The ability of the new receivers to demonstrate even more remarkable distance performance, to give a finer tone, and to cut through the strongest local interference, gives them an unmistakable appeal.

The two new models are the first Browning-Drakes to use more than five tubes. An unusual single control drum dial is used which is exceptionally powerful in its action and operates without the slightest trace of backlash.

These receivers are the result of nearly five years of exhaustive research and steady progress. Backed by the reputation of the Browning-Drake Corporation, we believe they will occupy a paramount position in radio this year. Investigate the Browning-Drake opportunity TODAY!

Look for this

Trade Mark

Specifications:
Browning-Drake, Model 6-A: (illustrated above), uses conventional Browning-Drake circuit with slight modifications. Four audio tubes give natural tone and great volume when desired. Small auxilliary condenser is provided to bring signals of distant stations to maximum intensity. Beautiful two tone Duco walnut cabinet harmonizes with all home furnishings. Length 27 inches; depth, 15 inches, height 11 inches. List without tubes and batteries, $105.

You are invited to see our display in BOOTH 1; SECTION—CC, at the Radio World's Fair to be held in Madison Square Garden, New York City from September 19th to 24th.

BROWNING-DRAKE CORPORATION
CAMBRIDGE : : MASS.

MODEL 5-R
Five tubes. Uses three stages of resistance-coupled amplification. Provision for power tube in last audio stage. Cabinet is two tone Duco mahogany finish. List without tubes and batteries $95.

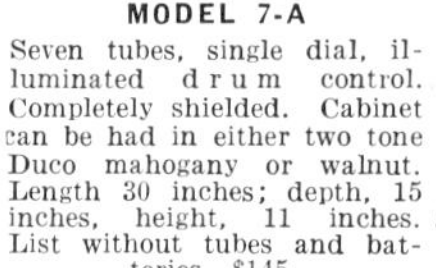

MODEL 7-A
Seven tubes, single dial, illuminated drum control. Completely shielded. Cabinet can be had in either two tone Duco mahogany or walnut. Length 30 inches; depth, 15 inches, height, 11 inches. List without tubes and batteries, $145.

BROWNING DRAKE

CABINETS
PARTS

RECEIVERS
KITS

7A first advertised in Aug. 1927

The National Regenaformer

30 July 1928 $215

34 (Eight in Line) Oct. 1928 $135
36 with speaker, 38 highboy. Short technical article in *Radio Broadcast*, Feb. 1929, p.266.

Model 83—
Battery Operated. Table Model, (less tubes)
$75.00

Model 53—
A-C Screen-Grid. Table Model, (less tubes)
$102.50

Model 63—
A-C Heater Type. Table Model, (less tubes)
$98.00

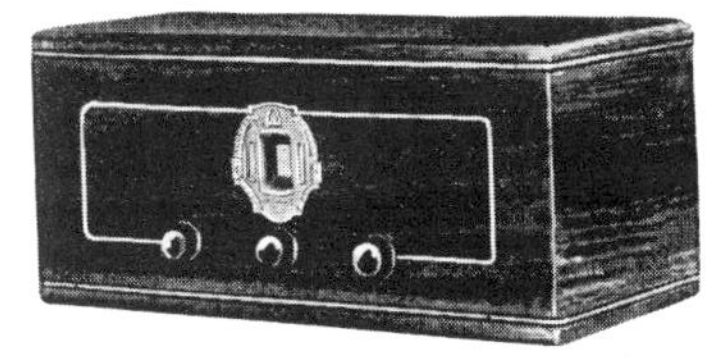

6A Sept. 1927 $105 Herb Parsons

5R changed from Jr. in June 1926 Warren Falls

32 July 1928 $295

Glenn Browning
Browning Labs.

Model 84
Small console (38x26x14) (less tubes) **$119.50**

Model 54
Small console (38x26x14)
Screen-grid
Less Tubes **$142.50**

Model 64
Small console (38x26x14)
Heater-type
Less tubes **$137.50**

Radio Retailing (Nov. 1929), pp. 14-15

Models			
	53, 54, 55	($172.50)	June 1929
	63, 64, 65	($167.50)	June 1929
	83, 84		Nov. 1929
	56, 66		Oct. 1929

Model 57 (illustrated above)
Screen-grid.
Less tubes **$188.50**

Model 67 (illustrated above)
Heater type.
Less tubes **$183.50**

Model 56 (42x25x15)
Screen-grid.
Less tubes **$154.50**

Model 66 (42x25x15)
Heater-type.
Less tubes **$149.50**

Radio Kit Reviews

ADAPTING THE BROWNING-DRAKE TWO-TUBE KIT TO THE SHIELDED GRID TUBE

The new double grid tube can be used in the r.f. stage of the Browning-Drake two-tube kit described in January RADIO by a few simple changes. These include (1) shorting out the primary of the r.f. transformer and substituting a ½ mfd. condenser and r.f. choke, (2) shielding the r.f. stage, (3) inserting a ½ mfd. condenser between the tube's plate and a connection to ground, and (4) putting a 10 or 15 ohm resistance in the filament circuit to reduce from 6 to 3.3 volts. A neutralizing condenser is used to balance the very small capacity between the tube's plate and control grid, thus giving greatest efficiency for the regeneration on the r.f. transformer. These changes are shown in Fig. 1.

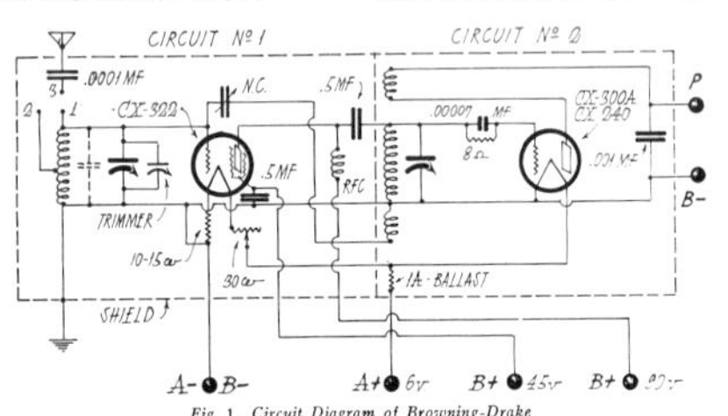

Fig. 1. Circuit Diagram of Browning-Drake

The primary of the r.f. transformer is shorted out because of the tube's very high impedance. This direct coupling (really an auto-transformer) requires a parallel feed system consisting of the ½ mfd. condenser and choke coil, thus keeping r.f. current out of the *B* supply. The tube's plate is connected directly to the stator of the second tuned circuit.

As this puts the plate-screen grid capacity across the second tuning condenser it may be necessary to put a 15 mmfd. condenser across the first tuned circuit, as shown by dotted lines in Fig. 1. In many cases, however, especially if the .0001 mfd. series antenna condenser adds enough capacity, the trimmer condenser will take care of any difference between the two singly-controlled tuning condenser settings throughout the waveband. A few experiments will determine whether it is needed in a given installation.

Suitable shields and instructions for their use may be secured from the Browning-Drake Corporation. The ½ mfd. condenser between the shield grid and ground is essential when using this tube.

The 10-15 ohm resistance in the filament circuit not only cuts the 5 volts used with the 300-A or 240 detector to 3.3 volts for the screened grid tube, but also supplies biasing current. If a 3 volt supply and —99 tube is used as a detector, this resistance is not necessary.

When the receiver is completed and connected up it may be neutralized by setting the dial at about 20 on the scale and turning the tickler in either direction until a distinct click is heard in the loud speaker or telephones. Adjust the tickler coil until this circuit is not oscillating. A test to determine whether or not the set is oscillating is to place the finger on the terminal of the ½ mfd. blocking condenser, which, if connected to the grid of the second tuning circuit (to get at this condenser the top of the shield on the r.f. compartment will have to be removed), a distinct click will be heard if this circuit is oscillating. Now turn the tickler back to where oscillating just ceases. Turning the trimmer condenser will then throw this circuit into oscillation if the neutralizing condenser is not properly set. The neutralizing condenser should be then set until the above test is satisfactory and the trimmer condenser has no effect on oscillations produced in the second circuit. It will be found that this trimmer condenser is almost at a minimum value.

Too much cannot be said in favor of using the screened grid tube as a radio frequency amplifier. The amplification obtainable is tremendous and the operator can easily get down to the noise level with little or no difficulty.

BOOK REVIEWS

"Practical Radio Telegraphy," by Arthur R. Nilson and J. L. Heornung; 380 pages, 5½ x 8 in., published by McGraw-Hill Book Co., New York City. Price $3.00.

This is an instructive test written primarily for those preparing to pass the examination given to applicants for license as a commercial operator. The authors have had long experience and great success in teaching students at the Radio Institute of the West Side Y. M. C. A., New York City. The book starts with an elementary explanation of electricity and gradually builds an understanding of the principles of radio transmitting and receiving equipment. As this is done without the use of mathematics it is easily comprehended by one who lacks such training.

Much of the material has not been previously published in book form. For instance, new R. C. A. vacuum tube transmitters are described and illustrated, including a complete ship's diagram for tracing trouble from the d.c. mains to the antenna insulator. Arc and spark apparatus, including that of independent manufacture, is completely covered. The operation of radio direction finders, and of remedying their troubles, are described in detail.

On the whole, it is a reliable, adequate and well-balanced text which admirably fulfills its primary purpose and in addition will be valuable to commercial operators who wish to brush up on the details of sets they may be called upon to handle.

Two-Tube Kit as Modified for Screened Grid Tube.

POPULAR RADIO WORK SHEET
The 1928 Browning-Drake

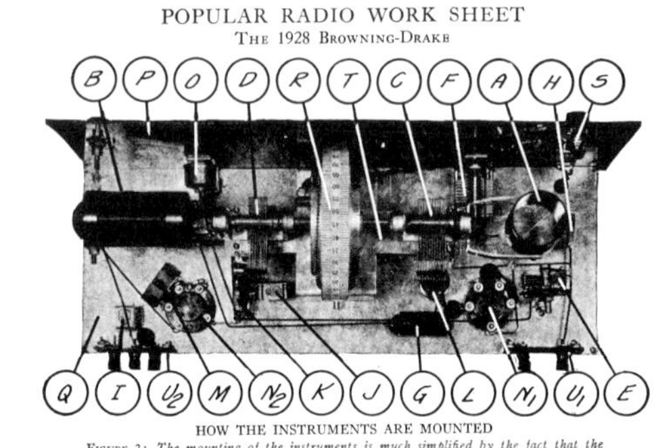

HOW THE INSTRUMENTS ARE MOUNTED

Figure 2: *The mounting of the instruments is much simplified by the fact that the sub-panel comes ready drilled with mounting hardware.*

List of Parts for Building This Receiver

Cost of Parts—*Not over $55.00*

Browning-Drake single-mount kit, containing:
- A—Browning-Drake antenna coil;
- B—Browning-Drake high-frequency transformer;
- C—Hammarlund midline variable condenser, .00045 mfd.;
- D—Hammarlund midline variable condenser, .0003 mfd.;
- R—Browning-Drake illuminated drum dial;
- T—Aluminum bracket, mounting brackets and hardware;

E—Browning-Drake special neutralizing condenser;
F—Browning-Drake variable midget condenser, .000135 mfd.;
G—Aerovox or Tobe .5 mfd. special blocking condenser;
H—Aerovox fixed condenser, .0001 mfd.;
I—Tinytobe fixed condenser, .001 mfd.;
J—Tinytobe fixed condenser, .00007 mfd.;
K—Tobe Veritas grid-leak, 6 megohms, equipped with a standard Lynch mounting;
L—Browning-Drake high-frequency choke coil;
M—General Radio center-tapped resistor, 100 ohms;
N1 and N2—Benjamin Cle-ra-tone 5-prong sockets;
O—Clarostat variable resistor;

Browning-Drake foundation unit, consisting of:
- P—Built and engraved front panel;
- Q—Drilled aluminum sub-panel complete with mounting hardware;

S—Yaxley offset filament switch;
U1 and U2—Micarta binding-post strips;
V1, V2, V3, V4 and V5—Eby binding posts;
Wire, solder, etc.

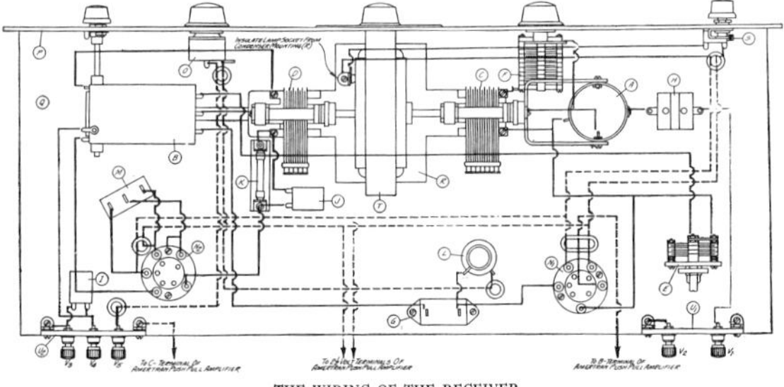

THE WIRING OF THE RECEIVER

Figure 3: *The instruments are in* BLACK. *The wiring above the sub-panel is indicated in solid* RED *lines, and that below the sub-panel in dotted* RED *lines.*

CLAPP-EASTHAM

Clapp-Eastham Co.

The Clapp-Eastham story begins with Melville Eastham in Oregon, where his father was instrumental in setting up the first electric generating system. In 1905, at the age of 20, he came East to work for the Ovington X-Ray Co., whose chief engineer was William O. Eddy, with J. Emory Clapp as sales engineer.

In 1906 the three decided to go into business for themselves, manufacturing X-ray equipment. As Clapp financed the venture, they moved to his hometown of Boston where they started the Clapp, Eddy, and Eastham Co. at 100 Boylston St. Eddy's coils were particularly good for X-ray and wireless work, where substantial current capacity was necessary.

When in February 1908 Eddy left, the company became the Clapp-Eastham Co., with $10,000 capital. They added other wireless components to their line, concentrating more and more in this area, to the exclusion of their X-ray business. Clapp, whose primary interest was in X-rays, sold out in 1910 to O. Kerro Luscomb, but the company remained Clapp-Eastham, moving to Kendall Square in Cambridge.

By 1915 Eastham was restless too. Envisioning a large future market for radio measuring instruments, he and Luscomb formed the General Radio Co. with three other investors. Starting with $9,000 in capital and a payroll of two, General Radio a year later had 30 employees, and was as substantial a concern as Clapp-Eastham. In 1917 Eastham and Luscomb each went his own way, exchanging his interest in the other's company. Eastham now controlled General Radio, and the Clapp-Eastham Co. employed neither Clapp nor Eastham!

Clapp-Eastham issued a series of 32 to 48-page catalogs during the teens, largely devoted to components, both transmitting and receiving. While it made a few complete receivers, these were a small part of its business. On April 18, 1920 the company obtained the first post-war Armstrong regenerative license; its subsequent quarterly royalty payments of $12.60 and $5.06 would indicate sales of just $252 and $101.20. George Eltz, Jr. was chief engineer after 1919.

Clapp-Eastham hit the jackpot during the radio boom of late 1921–early 1922, introducing the HR in December. But it was all down hill afterward, as the company couldn't match its competition. Clapp-Eastham models looked nearly as cheap as Crosleys, for example, but cost twice as much. A joint venture with United Cigar Stores, making its featured "Unico Special" model, might have given Clapp-Eastham the sales outlets it needed, as United planned to sell radio in all of its 1200 retail stores, after its first trials in August 1923. But when Westinghouse began harassing its Armstrong licensees, United by the end of the year had switched to other brands.

The factory in 1922. It still looks much the same.

Around April 1926 the company, or what was left of it, moved to Long Island City, New York and made some agreement with Westinghouse to produce regenerative receiver kits, to be marketed by Bruno Radio Corp. Whatever the deal was, it didn't work out; Clapp-Eastham was last heard from in 1929, collecting a debt from Bruno's financial backer.

G-R History, p. 1

Melville Eastham

Wireless Apparatus

Ceco Transformer—Closed Core Type

A 250 watt closed core transformer as powerful as the ordinary 6 in spark induction coil for wireless work, and costs but $30. It is especially adapted to tuned transmitting circuits, and can be run from any lamp socket giving alternating current. We also build larger sizes up to 10 kw. capacity at moderate prices. Complete parts for a 50 mile tuned transmitting set consisting of 250 watt transformer, large capacity glass plate condenser, oscillation transformer with adjustable primary and secondary inductances, and adjustable spark-gap making an accurate model of a modern commercial station complete for $60.

Send for our descriptive folder of experimental apparatus for receiving and transmitting, including

Transformers Telephone Head Receivers
Tuners Condensers Oscillation Transformers
Detectors Spark-Gaps Potentiometers

CLAPP-EASTHAM CO.

730 Boylston Street BOSTON, MASS.

When writing please mention "Modern Electrics."

Modern Electrics (Dec. 1908)

Right: D tuner 1914 (cat. S) $25
Last advertised in 1922. This marble base is a replica.
Left: Blitzen tuner Mar. 1912 $15
Last advertised in 1915. A similar "short-wave" type H cost $16 in 1919–1920.

Rich Wolven

Blitzen receiver Apr. 1912 $33 with phones (later $24 without phones). Last advertised in 1915. The right-hand switch, for a loading coil, was a $4 option.

The Blitzen Tuner

Constructed entirely of hard rubber on a mahogany base. Thirty primary and twelve secondary contact points. The price is **Fifteen Dollars.** Our new literature is full of surprises; sent for 4c stamps.

CLAPP-EASTHAM COMPANY

143 Main St., Cambridge, Mass.

Aylsworth Agencies Co.	**J. J. Duck**
143 Second Street	430 St. Clair Street
San Francisco, Cal.	Toledo, Ohio
Western Sales Agents	Central Sales Agent

Modern Electrics (Mar. 1912)

New England Wireless & Steam Museum

D receiver Dec. 1913 $90
Last advertised in 1915

F.A.R. (Z) 1920 $150

ZRFD Apr. 1921 $85

John Terrey

ZRF Nov. 1920 $38

ZR Double 1921 $290

ZRFDA Apr. 1921 $140

Ralph Muchow

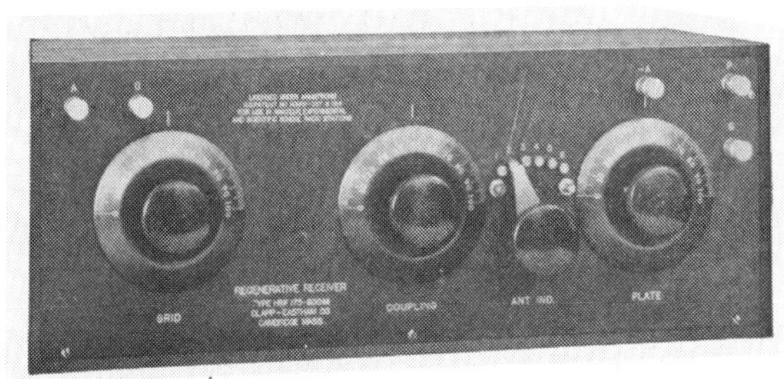

HRF 1922 $38

HR Dec. 1921 $35 After Apr. 1922, the oak cabinet was $35, the mahogany $40. A matching HZ amplifier appeared in Feb. 1922 for $35.

Radio News for February, 1922

CLAPP-EASTHAM QUALITY

Clapp-Eastham Type HR Regenerative Receiver. Wave length 180 to 825 meters.

$35 Buys this Set

Perfect Regeneration at all wave-lengths. Wonderfully simple control.

See this Set in CLEARWATER, FLA.

at our headquarters or order direct by mail. **Full details of this set and complete line of other radio equipment free on request.**

SOUTHERN RADIO SUPPLY COMPANY
CLEARWATER, FLORIDA

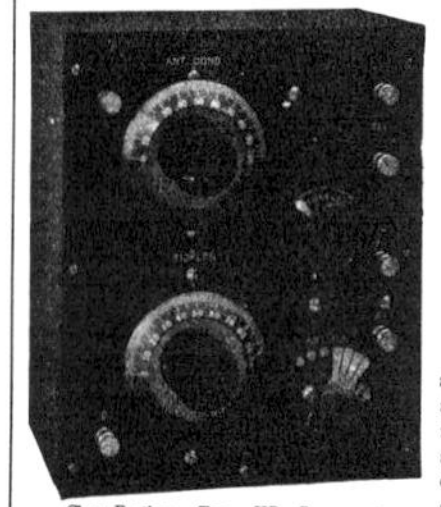

Clapp-Eastham Type HR Regenerative Receiver. Wave length 180 to 825 meters.

Low Cost But Big Results

Only $35, but absolutely guaranteed to give results equal or superior to any on the market regardless of price.

See this Set in NEW KENSINGTON, PA.

at our headquarters or order direct by mail. **Full details of this set and complete line of other radio equipment free on request.**

VALLEY ELECTRIC COMPANY
Copeland Building NEW KENSINGTON, PA.

Clapp-Eastham Type HR Regenerative Receiver. Wave length 180 to 825 meters.

A Set You Will be Proud of

Costs only $35, but will give results that will astonish your friends who own more expensive sets.

See this Set in DAYTON, OHIO

at our headquarters or order direct by mail. **Full details of this set and complete line of other radio equipment free on request.**

WILLIAM HALL ELECTRIC COMPANY
DAYTON, OHIO

Clapp-Eastham Type HR Regenerative Receiver. Wave length 180 to 825 meters.

The Latest Thing in Receiving Sets

You will be delighted at the distance, clearness and loudness of music, voices and signals received. Only $35.

See this Set in WESTERLY, R. I.

at our headquarters or order direct by mail. **Full details of this set and complete line of other radio equipment free on request.**

WHITALL ELECTRIC COMPANY
WESTERLY, R. I.

CLAPP-EASTHAM SERVICE

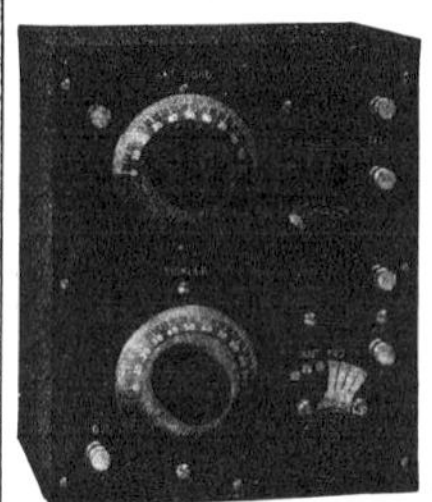

Clapp-Eastham Type HR Regenerative Receiver. Wave length 180 to 825 meters.

The Set with the Easy Control

This $35 set is guaranteed by the makers to give results equal or superior to any on the market, *regardless of price.*

See this Set in KANSAS CITY, MO.

at our headquarters or order direct by mail. Full details of this set and complete line of other radio equipment free on request.

CENTRAL RADIO COMPANY
575 Grand Avenue KANSAS CITY, MO.

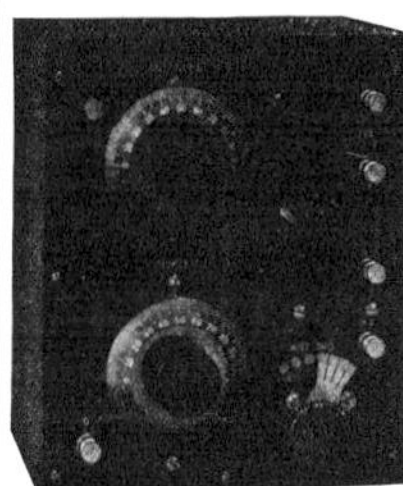

Clapp-Eastham Type HR Regenerative Receiver. Wave length 180 to 825 meters.

Wired Ready for Use

This set is complete when you get it. Includes tube socket and rheostat. Wired ready for use.

See this Set in NEW YORK

at our headquarters or order direct by mail. Full details of this set and complete line of other radio equipment free on request.

DREYFUS SALES CORPORATION
179 Greenwich Street NEW YORK CITY

Clapp-Eastham Type HR Regenerative Receiver. Wave length 180 to 825 meters.

For Distance Clearness Loudness

This $35 set will surprise your friends who own more expensive equipment. Remember that this set is fully guaranteed by the maker.

See this Set in DETROIT

at our headquarters or order direct by mail. Full details of this set and complete line of other radio equipment free on request.

DETROIT ELECTRIC COMPANY
434 Shelby Street DETROIT, MICH.

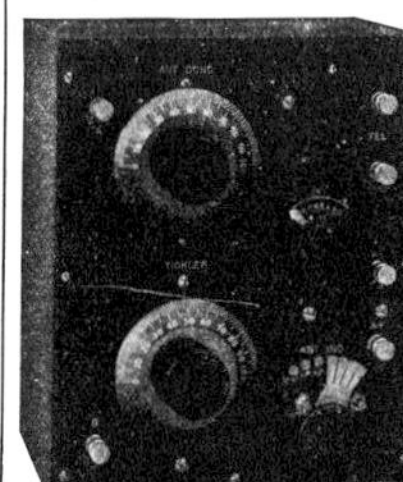

Clapp-Eastham Type HR Regenerative Receiver. Wave length 180 to 825 meters.

$35 Buys this Set

Perfect Regeneration at all wave-lengths. No capacity effect from hand or body.

See this Set in UNIONTOWN, PA.

at our headquarters or order direct by mail. Full details of this set and complete line of other radio equipment free on request.

REED AUTOMOTIVE SERVICE
97 West Main Street UNIONTOWN, PA.

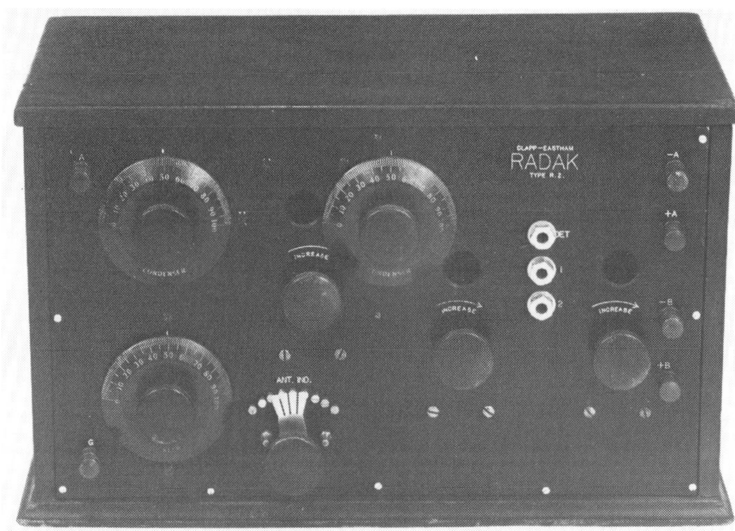

Ralph Thorn

RZ Aug. 1922 $100

A version with knife switches instead of jacks was advertised in June.

C3 Apr. 1923 $100

Ralph Thorn

C23 July 1923 $125

Robert Enemark

Country Gentleman (Oct. 14, 1922)

R3 June 1923 $45
R3 Amplifier $40

C63 ca. Aug. 1923 $290
C64 (not shown) Nov. 1923 $220

The Wireless Age (Mar. 1923)

MODELS R 23 AND A 23 RADAK
ONLY SET COMBINING REGENERATION WITH RADIO FREQUENCY AMPLIFICATION

Regenerative receiver with one stage each of **radio** frequency and **audio** frequency amplification. Model R 23 contains tuning elements only. Model A 23 contains detecting and amplifying elements. Binding posts inside cabinets, all connections being made from behind. Controls simplified to final degree—three simple tuning dials only, with two filament rheostats to control brilliancy. New Radak vernier dials. Receives 175-550-meter wave lengths. Handsomely made cabinets, 9 in. wide, 10¾ high, 7 deep. (Licensed under Armstrong U. S. Patent 1113149).

Radio Frequency Amplification Feature of this New Radak Set

AFTER a long period of experimental work we are now ready to offer the public a two-unit Radak set embodying the much-discussed feature of radio frequency amplification.

Model R 23 contains tuning elements only, Model A 23 embodies both detecting and amplifying elements. Model A 23, however, may be used in connection with any regenerative receiving set in which the detector is **not already mounted** and which has a **tuned plate circuit** (regeneration controlled by variometer).

The two units retail together for $100. Singly, Model R 23, $40; Model A 23 $60. If your electrical or radio dealer is not yet displaying these two new sets write us for further information. Send for new RADAK BOOK free.

CLAPP-EASTHAM COMPANY

America's oldest, largest makers of radio equipment exclusively

101 Main Street, Cambridge, Mass.

New York Sales Office, 295 Broadway Pittsburgh, 332 Third Ave. Chicago, 33 South Clinton St.

Ralph Thorn

R43 Aug. 1923 $125

R4 Oct. 1923 $25

Radio News (Nov. 1923)

RADAK R4

Complete Regenerative Receiver

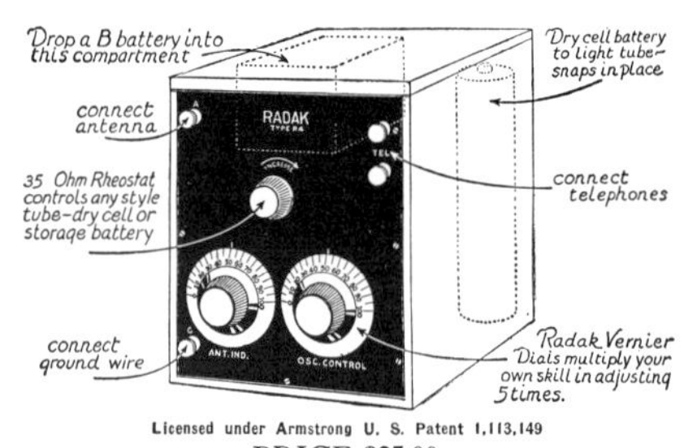

Licensed under Armstrong U. S. Patent 1,113,149

PRICE $25.00

A newly designed and thoroughly tested circuit of superior capability, solid mahogany cabinet, genuine Formica panel, remarkable Radak Vernier dials, all batteries inside the cabinet with overall size of but 6 x 8 x 10 inches.

Where else will you find these earmarks of quality in a set selling for $25.00? The new Radak "Governing Capacity" controls regeneration with surprising ease. Radak R4 is a self-contained set designed for use on dry cells and operating over a range of wave-lengths of 225 to 550 meters. Wherever you are, or wherever you go, you can take this set with you. Merely slip in a flashlight battery, a small "B" battery and a 3-volt vacuum tube, connect to a wire hung out the window, thrown over the limb of a tree or even laid on the roof if no antenna is available, and programs from considerable and often surprising distances may be received in a few minutes from the time you start. While easily carried to your summer home, camp, or on your vacation, the R4 is in no sense a portable or makeshift outfit, but its high quality of finish and workmanship will grace the most refined surroundings.

R4 Set complete, as illustrated and described, without accessories $25.00

A4 RADAK 2 STAGE AMPLIFIER
EXACTLY MATCHES THE ABOVE SET **$25.00**

From the R4 at $25.00 to the C64 five tube radio frequency set at $220.00 THE BASIS OF RADAK SUPREMACY lies in the fact that Radak sets are an engineered entity not a mere assembly of parts. Complete bulletin of all models sent on request.

Manufactured by

CLAPP-EASTHAM COMPANY

107 MAIN STREET, CAMBRIDGE, MASS.

New York, N. Y.	Cleveland, Ohio	San Francisco, Cal.
395 Broadway	Caxton Bldg.	709 Mission St.

The Radio Sun and Globe (Dec. 15, 1923)

Make this a Radio Christmas

Santa Claus Starts From Here!

ACKERMAN LOUD SPEAKER

"The small Horn with the BIG Voice"

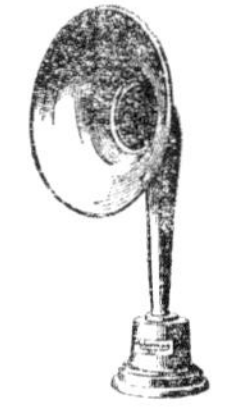

Standing 21" high, with 11" bell and made of heavy metal, eliminating vibration, together with its special loud speaking unit, this speaker reproduces voice and music far beyond expectations. Finished in plain black or brown; also special alligator grain in black and green or black crystalline.

$9.50

THE UNICO-FORD

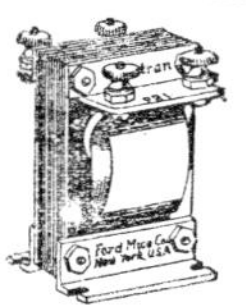

AUDIO FREQUENCY TRANSFORMER

Low organ notes and *base notes* as well as the *high tones* of the violin and piano are brought out with perfect clarity $4.50

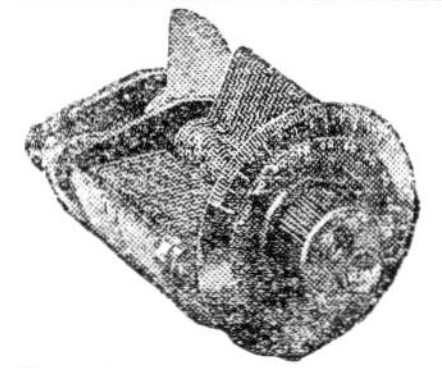

The Genuine

U.S. TOOL CONDENSER

A CHAIN is no stronger than its weakest link—and the most important link in a radio set is the condenser.

U. S. Tool Vernier Condensers are the last word in efficiency, and stand on their own reputation.

RADIO, the vague and wonderful mystery of yesterday, is Mankind's faithful servant and cheerful companion today.

Make this a Radio Christmas! Make your gifts Radio gifts! The sets and supplies illustrated here are only a small part of a vast assortment to be seen in our stores—*all products of manufacturers of highest reputation.* Select your Radio gifts at United—our salesmen will gladly give you expert advice, and your purchases will be covered by the unconditional

United Guarantee:

MERCHANDISE, QUALITY, SERVICE, SATISFACTION OR MONEY REFUNDED

"The Perpetual Present"

THE UNICO REGENERATIVE

"Unico Means Radio"

The "joy of receiving" becomes a daily experience when the Christmas gift is the Unico Regenerative Receiver! Although moderately priced, the Unico is beautifully made and handsomely finished—simple, sturdy, and easy to operate. You can be justly proud of such a gift—a source of never-ending pleasure to its recipients.

COMPLETE READY TO OPERATE NO EXTRAS $37.25

SPECIAL! Until New Years, above price includes an Extra Set of Headphones.

"EVERYTONE"

Guaranteed

Radio Headset

$3.50 2200 OHM

3000 Ohms $3.75

Reg. U. S. Pat. Off.

A TIMELY GIFT

YOUR radio friend can always use an additional headset. A Pacent "Everytone" is the logical choice. He will recognize it as a headset of high quality, and yet it can easily be included in your Christmas budget.

HEAR your favorite Christmas carols with all their clear, melodious beauty—exactly as the artists play and sing them!

The Radiotive Loud Speaker is an instrument made for people who appreciate a perfect reproduction of music and voice.

RADIOTIVE LOUD SPEAKER

Type "E" $45.00

UNICO-TESTED SETS *and* MATERIALS *of* QUALITY

Unico Two-Step Amplifiers

(Matching the Unico Receiver, or to be used with any other set)

- De Forest Four Tube Reflex Sets
- Freed-Eisemann Neutrodyne Sets and Parts
- Cunningham Vacuum Tubes
- Phonograph Attachments
- Frost Jacks, Plugs and Parts
- Burgess and all standard Batteries
- Eisemann Ready-Cut Panels and Combination Units
- Dubilier Tested Mica Condensers
- Exide Storage Batteries
- Voltmeters of standard makes
- Dictograph headsets (new type)
- Bakelite Panels—all sizes

and all other tested standard Radio parts

At Prices to Demonstrate that the Best Radio Values are at the UNITED

RADIO DEPARTMENTS OF UNITED CIGAR STORES

The first 6 of our Radio Departments are now operating at the addresses listed below:

23rd Street and 5th Ave., Flatiron Bldg.
33rd Street and Broadway
42nd Street and Park Avenue
Broadway and Chambers Street
*103 East 125th Street near Park Ave.
Brooklyn: 1369 Broadway at Gates Ave.

Mail orders filled. Address United Cigar Stores Co., Flatiron Bldg., New York

*New store

John Drew

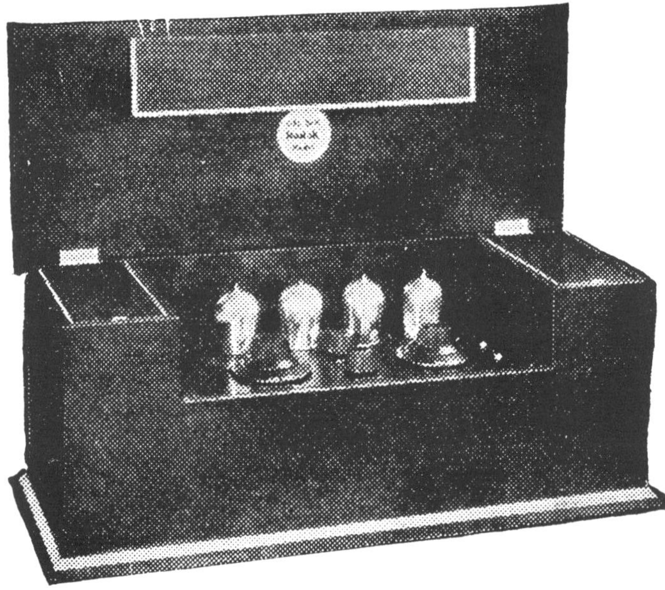

Gold Seal   Dec. 1924 $75

RHM   Jan. 1924 $100 (actually sold for $69.75 complete). Made for RH Macy

Electrical Merchandising (Dec. 1924), p. 4806

Unico Special   Nov. 1923 $37.25

R.H. Macy & Co.
34th ST. & BROADWAY *Inc.* NEW YORK CITY

Selectivity!

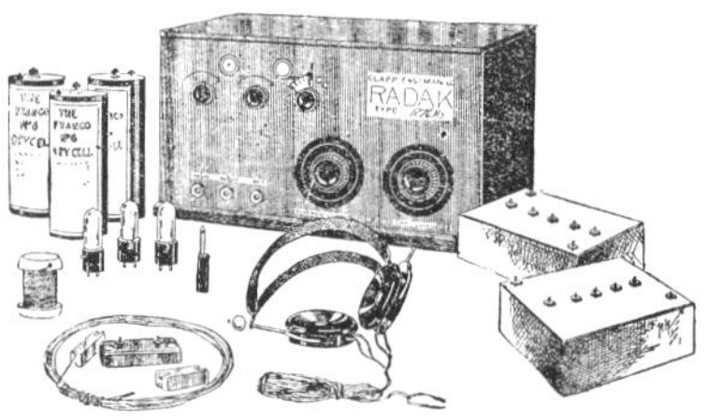

A special purchase and sale of the new improved model

THREE-TUBE

Regenerative, Clapp-Eastham

RECEIVING SETS

Complete with Tubes, Phones, Batteries and all Antenna Equipment

$69.75

Current Retail Catalog Price, complete, $133.80

IN our last sale of Clapp-Eastham receiving sets we were utterly unable to satisfy the demand, although we sold several thousands of them.

The New Improved Model will receive even greater distance than the original Clapp-Eastham. Broadcast reception will also be appreciably clearer.

With this set you can tune out undesired stations no matter how powerful

Points of Superiority:

1. Extreme Selectivity.
2. Set is regenerative—licensed under Armstrong patents.
3. Clapp-Eastham patented vernier dials, 5 to 1 ratio.
4. Eby binding posts.
5. Formica panel.
6. All connections and wiring inside cabinet.
7. Battery compartments inside cabinet, making set self-inclosed and portable.
8. Detector and two stages of audio frequency amplification.
9. Polished walnut cabinet.

We have received many letters telling us of clear reception, with the original model, of from 500 to 2500 miles. Lists often show 40 or more stations. Perhaps you can do as well as with one of these.

—Fourth Floor, New West Building

What You Get For $69.75

At current list prices the retail cost of this outfit would ordinarily be $133.80.

	Current Retail Catalog Price
1 Three-tube Radak set,	$100.00
3 UV 199 or C 299 tubes,	15.00
3 Adapters,	1.95
4 22½-volt "B" batteries,	7.00
3 "A" dry batteries,	1.05
1 phone plug (double),	.60
100 ft. aerial wire,	.75
25 ft. lead-in wire,	.50
2 insulators,	.30
1 lightning arrester,	.50
1 pair Frost phones,	6.00
1 ground clamp,	.15
	$133.80

Macy's special price means a saving, therefore, of $64.05.

N. Y. Evening Mail (Jan. 19, 1924)

Radio World (Aug. 30, 1924), p. 26

DARIOLA ONE

Original—Compact—Efficient

A unique receiver, embodying the 3 circuit regenerative principle. Very simple to tune. Built of the finest quality materials throughout.

Distance Reception Guaranteed

One thousand miles or more plainly heard under favorable conditions.

DARIOLA ONE is equipped with MYERS Famous Dry Battery Tube. Only extras needed are two Dry Cells, one 22½ V or 45 V "B" Battery, headset and wire for attaching.

Fully covered by manufacturers' guarantee. There is no better one tube outfit.

$16.50 Complete

At your dealer or sent postpaid on receipt of price.

DARWAL CORP.

Dept. A, 1261 Broadway New York City

Ralph Thorn

DD Mar. 1925 $38

Richard Hostler

Baby Emerson 1927–1928

Named after Col. Alfred E. Emerson, tube manufacturer, and used the Emerson Multivalve.

Billow - Armstrong Regenerative Set

A Compact Portable Set Complete with Genuine Myers Tube

The "Billow" is a newly designed receiving set. It has been built to meet the demand of the public for an inexpensive, yet efficient receiver, which is simple to operate, light in weight and compact. Consists of bakelite tube, wound with an aperodic primary and a periodic secondary, which is shunted with a .0005 MFD variable condenser, controlled by a dial at the top of the set. Knob on the side controls the regeneration. Comes equipped with a Myers tube. This tube is silent in operation, there being no tube hiss or mechanical noise. Signals, therefore, coming in loud and clear. Binding posts are in front, and are plainly marked. The operator will find the "Billow" to be very sensitive and sharp in tuning. By connecting any good amplifier unit of one or two stages from the telephone terminals to the input of the amplifier, greater volume can be obtained.

The circuit used is of the regenerative type, and this receiver is licensed under the Armstrong Regenerative patents. Mfg. by Clapp Eastham Company.

Our Price Including Myers Tube $12.50

CHICAGO SALVAGE STOCK STORE

The Crosley Radio Corp.

Powel Crosley was born in Cincinnati, Ohio in 1886, son of a prominent attorney. Fascinated with automobiles at an early age, and equally fascinated by the dream of becoming a millionaire, Crosley tried venture after venture. He studied engineering, then law, then left college to become a chauffeur. In 1907 he designed an inexpensive six-cylinder car and borrowed $10,000 to finance a company to make it, but, underestimating the capital needed, failed to survive the Panic of 1907. He drifted to Indianapolis and took a job as a driver for the Carl Fisher Co., scheduled to run one of their entries at the newly-opened Motor Speedway, but broke an arm cranking a car a few days before the race. Back in Cincinnati, another auto-manufacturing venture ended the same way as his first, followed in 1913 by a fiasco with cycle cars.

Crosley said "It was then that I woke up. I thought that I could finance million-dollar corporations on small amounts of capital that did not even belong to me. I promised myself then and there not to attempt more than I could safely manage, not to run my business on other people's money, and above all, to be strictly independent in my financial dealings. I made up my mind that I would finance myself even though I had to run a popcorn stand."

He sold advertising copy for a client who ran an automobile accessories business, and, with a $500 advance, formed the American Automobile Accessories Co. to sell these products by mail. His success allowed Crosley to buy out his client, and to add items to his line such as a radiator cap with flag holder, or the Litl Shofur which steered a model T back on the road after its wheels had hit a pothole. He bought out a printing company to produce his own advertising material, and, to take up seasonal slack in the auto business, a woodworking plant that made phonograph cabinets.

During this time, another company was active in Cincinnati, in the radio field. Thomas E. New founded the Precision Engineering Association in late 1918, incorporating on June 30, 1919 as the Precision Equipment Co. (trade name "Ace"). As a manufacturer and advertiser of regenerative receivers, Precision was signed up in mid-1920 by Armstrong's patent attorneys for a royalty license.

On Washington's Birthday, 1921, Crosley set out to buy a promised radio outfit for his nine-year-old son Powel III, and it was to Precision that he went. He was shocked to find that the "toy" he had expected to buy, cost $130; he and his son came home with a 25¢ book and a practice buzzer. Studying the book, Crosley learned how to make his own set, and shortly he and his son had built a crystal set and added an Audion detector, for a total outlay of $35, that to their amazement allowed them to hear broadcasts from Precision's 8XB, a whole seven miles away. "I unconsciously joined the class of radio bores. I told everyone I met about the distance our home-made set had covered."

Crosley was quick to spot the potential of low-priced radio apparatus. He designed a tube socket of porcelain, and a simple wooden variable condenser (patent applied for on May 23, 1921) which he advertised along with the cabinets turned out by his woodworking plant. Later that year he sold cabinets to other manufacturers, notably Grebe. He also hired Dorman Israel, a pre-junior in a five-year co-operative program at the University of Cincinnati, who designed the "Harko" series of sets and who worked for him part-time until 1923 (Israel returned in 1929, eventually went to Emerson).

For the low-priced market Crosley was aiming for, the regenerative circuit was absolutely necessary, to get the utmost from each tube, but he was not licensed under the Armstrong patent. He tried an arrangement with Tri-City (Tresco) who made regenerative Harko Seniors around December 1922, but it must have been unsatisfactory. The next month, he found the ideal solution: he bought Precision Equipment ("I worked out the details of the transaction at my sister's wedding, and bought the company the next morning").

The Armstrong license was not transferable, but all Crosley had to do was maintain the Precision (Ace) name and advertise its models separately. For the next year, Crosley and Ace typically advertised on facing pages, sometimes with the same address, sometimes two different streets on which the factory fronted. By the end of 1923, Crosley's production reached 1000 sets per day.

Still it was awkward having two different names, so in January 1924 Crosley went one better. He simply changed the name of the Precision Equipment Co. (who owned the license) to the Crosley Radio Corp., which then absorbed the old Crosley Manufacturing Co. Doubtless there was some behind-the-scenes dealing here, as Westinghouse (the patent owner) had made considerable trouble for most all of its other licensees, over a lot less. It's probably not coincidental that the two licensees who seemed to lead charmed lives, Crosley and Gene McDonald of Zenith, were old drinking and hunting buddies (Jim Millen tells of the time they stationed themselves on the top floor of the Illinois Athletic Club with a high-powered rifle and proceeded to shoot out street lights up and down Michigan Avenue). Political pull seems likely.

The merger of Precision and Crosley opened the way for two new models to replace the old line that, frankly, never sold very well except for the V. And one-tube sets were out-of-date now, at least as family radios. So Crosley took the old V and added an audio stage: V + 1, or 5 + 1, or 51. This was probably intended to compete with the Radiola III, which had just come out (advertised February 2, though Crosley must have known about it for weeks if not months) at $35 including two tubes listing at $5 each. According to a March 8 ad, Powel Crosley dreamed up the 51

Type ATC Detector Cabinet.
Price $12.00

A

Type A-2 Two-Stage Amplifier
Price $40.00

A ACE E

ACE C.W. Inductance
Price $6.50

ACE Vacuum Tube Socket
Price $1.50

DURING THE WORLD WAR

The word "ACE" was used to designate Aviators who had distinguished themselves by their exceptionally efficient performance of duty.

The word "ACE" engraved on Radio Equipment is your bona fide guarantee that the apparatus bearing this trade-mark is unsurpassed as to quality, workmanship and efficiency.

Equip Your Station With "ACE" Apparatus

Type AVA Regenerative Tuner
175-475 Meters. Price $45.00

Type AVC Short Wave Regenerative receiver, 175-475 Meters, Price, $56.00

Type AVB Regenerative Tuner
150-475 Meters. Price $32.00

ACE 6-Volt, 80 Amp. Hr. Storage Battery. Price $18.00

Type TTR Regenerative Receiver—150-3000 Meters. Price $75.00

ACE Hard Rubber Dial
Price $1.50

We carry in stock at all times a complete line of reliable Radio Supplies for the Amateur and Experimenter.

SEND 5c IN STAMPS FOR CATALOG

AMATEURS!
If your dealer cannot supply you write us direct
Radio 8XB

THE PRECISION EQUIPMENT CO., INC.

Manufacturers and Distributors

PEEBLES CORNER, CINCINNATI, OHIO

DEALERS!
Write us for trade Proposition

CABLE ADDRESS,
"ACE" Cincinnati

First advertising dates:
TTR Sept. 1920 AVA July 1921
AVB July 1921 AVC Nov. 1921

Radio News (Nov. 1921), p. 435

N.Y. Evening World (Sept. 1925)

1,611,000. VACUUM-TUBE RECEPTACLE. POWEL CROSLEY, Jr., Cincinnati, Ohio. Filed Dec. 7, 1921. Serial No. 520,644. 8 Claims. (Cl. 173—328.)

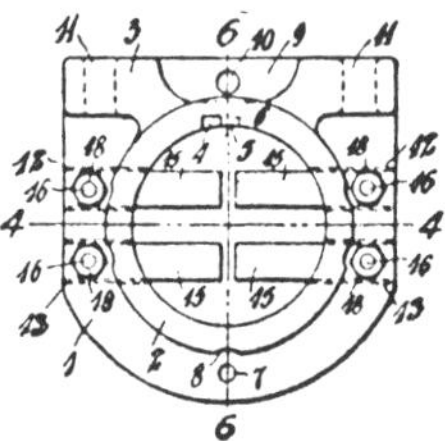

8. A receptacle for audions comprising a tubular socket member having an integral rib projecting from its outer surface and extending longitudinally thereof, said rib having a support engaging surface extending substantially parallel to the axis of the socket, and means providing additional support engaging surface disposed to one side of said rib.

Art Albion

TRU March 1922 $50
Matching two-step amplifier AV-2 advertised April 1923

on a two-week hunting trip, told his engineers about it on Monday February 4, had a model made by Tuesday, notified his dealers that night and put it into production with the first shipments on the 13th. By the 28th, production had reached 500 sets per day and was still climbing. And why not—who wouldn't buy a two-tube, hot-performing set for only $18.50? In fact it was so popular that several months later it had spawned the one-tube 50, three-tube 52, and add-on amplifiers to make a 52 out of the 50 or 51.

Once the 51 was under way, Crosley turned his attention to the top of the new line (but note that the "top" of his line was less than the lowliest Atwater Kent. For several years, Crosley made the most sets, but Kent made the profits). A three-tube model using both regeneration and reflexing to equal the average five-tube set, the $65 Trirdyn was not the wild success the 51 had been, but it was reasonably popular. 1924 was the best year Crosley had ever had, or would have for several years to come.

1924 also saw Crosley branch out again. He acquired a large interest in the De Forest Radio Corp. Ltd. of Canada who began advertising the 50, 51, 52, amplifiers, and Trirdyn in August, as well as the De Forest D12, all assembled in its Canadian plant. The following July De Forest-Crosley, as the new company was known, went its own way with a unique line of models. In December 1925 Crosley bought the remnants of Amrad for $39,000. Besides the Neutrodyne license, he also got a manufacturing plant, and a well-respected tradename that he could use for a higher-priced line.

(Wireless Age (Feb. 1920), p. 29)

Dorman Israel, prominent wireless amateur in Cincinnati

RRS $80

RT Nov. 1919 $55
renamed "TT" in June 1920

The New CROSLEY Variable Condenser

"Better — Costs Less"

MODEL "C"

The principle of this instrument needs no introduction. Thousands of the Crosley Model "A" Condensers have been sold with uniform satisfaction. The Model "A" is conservatively rated at .0005 Mf. The new Model "C" is conservatively rated at .0012 Mf.

While the Model "A" was made with wood frame and laminated wood plates the new Model "C" has ground porcelain plates with die cast frame. It is as efficient a piece of apparatus as you could desire. For tuning C.W. and for power transmission it cannot be equalled. It is tested on one thousand volts before shipment. No body or hand capacity. Low resistance due to absence of spring contacts, copper plates, brass binding posts, etc. We call it the "sensation" of radio—"Better—Costs Less".

Every CROSLEY VARIABLE CONDENSER is GUARANTEED to give absolute satisfaction or money refunded.

The CROSLEY VARIABLE CONDENSER is now made in three styles:

Model "A" with wood frame and laminated wood plates.

Model "B" with wood plates and die cast frame.
(Both models have the same capacity—.0005.)

Model "C" as illustrated—capacity .001—porcelain plates, die cast frame, etc.

Prices as follows:

	Without knob and dial	With knob and dial	With knob and dial mounted in cabinet
Model "A"	$1.25	$1.75	$2.50
Model "B"	1.75	2.25	3.00
Model "C"	2.25	2.75	3.50

Crosley Cabinets

We build a full line of cabinets in gum, mahogany or quartered oak and furnish them with genuine formica panels. Prices on cabinets range from .... $2.50 to $10.50

Crosley Two Step Amplifier

Complete with amplifying transformers, sockets, rheostats, switch, binding posts, etc., mounted on formica panel in mahogany finished cabinet. Price complete as shown in illustration $25.00

Crosley Vario-Couplers

Consists of formica tube, rotor and brass hardware. Price, complete as shown in illustration not wound or assembled, $1.50. Stator only 40c

Crosley Variometer Parts

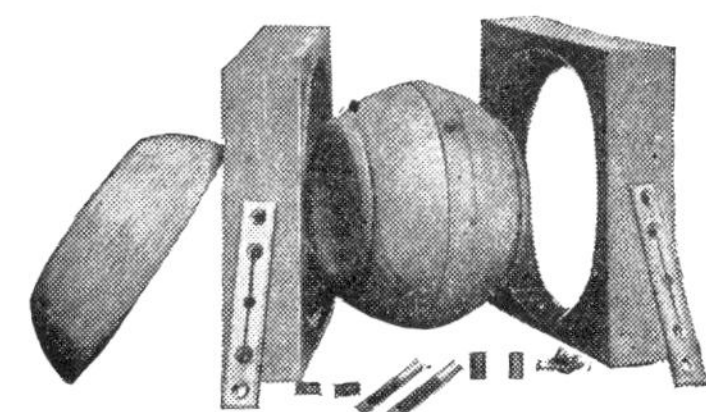

Consists of two stators, one rotor, the necessary hardware as shown in illustration. Shaft for knob and dial is 3/16" diameter. The wood parts are furnished either in poplar or mahogany. Price of Variometer parts, using poplar wood, $1.50. If wood parts are made of mahogany $1.75

Crosley Rheostats

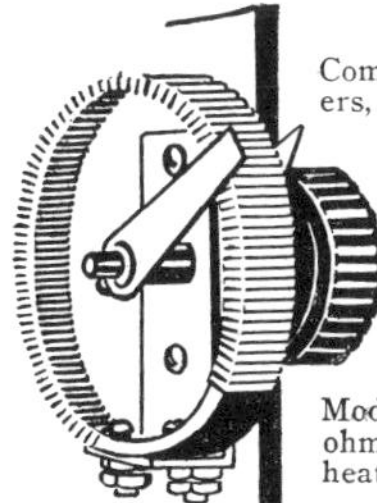

Complete with knob, pointers, etc., as shown in illustration.

Model "A"—overall diameter 1⅝". Resistance 7 ohms, one ampere without heating. Price...... 60c

Model "B"—Resistance 4 ohms, 3 amperes without heating. Price...... $1.25

CROSLEY V-T Socket -- 60c.

"Better -- Costs Less"

The biggest selling socket on the market. Practically unbreakable. For either base or panel mounting. Made of one piece porcelain; no metal shell, hence no ground hum. Better—and costs only 60c

Harko Radio Receiver

The most compact and complete efficent crystal receiving outfit on the market. Will tune from 200 to 600 meters bringing in spark, voice and music with amateur antenna. A wonderful little instrument. Price complete with battery, etc., $9.00. One thousand ohm single head set, 125 feet antenna wire regulators, etc., $6.00 extra. Complete outfit $15.00

Harko Senior Radio Receiver

Complete tuner and audion detector assembled on a formica or other high grade dielectric panel, mounted complete in mahogany finish cabinet. Range, 150 to 600 meters, non-regenerative hook-up. Price without battery, tube or phones $16.00

Crosley Detector Units

Furnished completely wired and mounted as shown in illustration or in knocked down form. Price completely assembled $7.50. Price of all parts including formica or other panel of high grade dielectric composition, not drilled $6.00

Every article guaranteed to give absolute satisfaction or money refunded. If your dealer can't supply you, send us his name and order direct.

Dealers and Distributors: Every item shown above should be in your stock. Write for proposition.

CROSLEY MFG. CO. Radio Dept. R-7 CINCINNATI, O.

In July 1926 he branched out still further, introducing his "Icyball" refrigerator, first in a long, successful line.

January 1927 began an association with the De Forest Radio Co. in Jersey City, New Jersey which had been operating under receivership since the previous July. Crosley offered to take over the management, and to supply $300,000 of operating capital, in return for a large block of stock when the company was again making money. Although he was elected president, he stalled on the rest of the deal, was ordered by the court to fulfill his contract in May, and in September was sued by the De Forest receiver (a year later, the suit was settled amicably and Crosley rejoined the Board of Directors). The timing of all these maneuvers coincides rather well with his negotiations with RCA for a patent license (signed May 22 after a month-long standoff) and it appears that he was using his association with De Forest as leverage for favorable license terms. The De Forest Co. controlled a substantial patent portfolio, and was also engaged in profitable tube manufacture.

1928 was Crosley's best year of all, with record sales and profits. He emulated Atwater Kent with a line of single-dial metal-cabinet sets, matching features and prices (and a million-dollar advertising budget) and customers couldn't get enough of them. Also like Kent, he saw a big future ahead, built a larage factory addition in 1929 and geared up for console production where the profits were. But his crystal ball was no clearer than Atwater Kent's. While 1929 sales were down only slightly from 1928, profits dipped by two-thirds, and in 1930 Crosley registered his first loss. He let Philco beat him at his own game of low-priced radios.

The Crosley Corp. survived the Depression with four years of deficits, and was sold to Avco in 1945. Crosley appliances, radio and TV were discontinued in 1956.

ANNOUNCING

The Crosley Variable Condenser

(Pat. Pend.)

"BETTER—COSTS LESS"

This Condenser works on an entirely new principle. The two plates are hinged and are opened and closed like a book by means of a specially designed cam. The plates are surfaced with copper. One copper sheet is covered with mica so that when the two plates are clamped tightly together the maximum capacity is obtained. The maximum capacity of this Condenser will average about .0008. We rate it conservatively, however, at .0005.

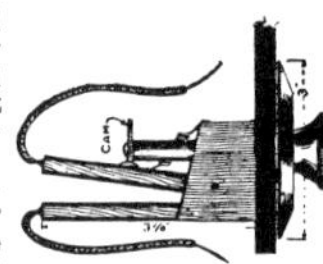

This Condenser has several advantages over the ordinary type of air condenser. Will stand 1000 volts without breaking down. It can therefore be used for C.W. work. Has no body or hand capacity effect. Has much greater signal strength due to the fact that mica is a much more efficient dielectric than air. The calibration curve of this Condenser is almost a straight line. Has unusually low zero capacity—.00006.

Price without knob and dial................$1.25
With knob and dial........................ 1.75
Mounted in cabinet with knob and dial..... 2.50

Sold on a GUARANTEE of absolute satisfaction or money refunded.

CROSLEY CRYSTAL RECEIVER

No batteries, tubes, etc., required. Hook it to your aerial and phones. It will tune from two hundred to six hundred meters, bringing in spark, voice, and music, with an average amateur aerial.

Complete with battery and interrupter for crystal testing, crystal, etc. Price $7.00. Phones extra.

DEALERS: This will help you get 'em started.

Radio News (Oct. 1921), p. 314

	1924	1925	1926	1927	1928	1929	1930(15mos)	1930 (Apr-Sep)	1930 (Oct-Dec)
net sales	5.2M	4.4M	5.5M	8.0M	17.5M	15.5M	9.0M	-	-
surplus	1.7M	.3M	.6M	.5M	3.4M	1.1M	(.9M)loss	(.4M)loss	.08M

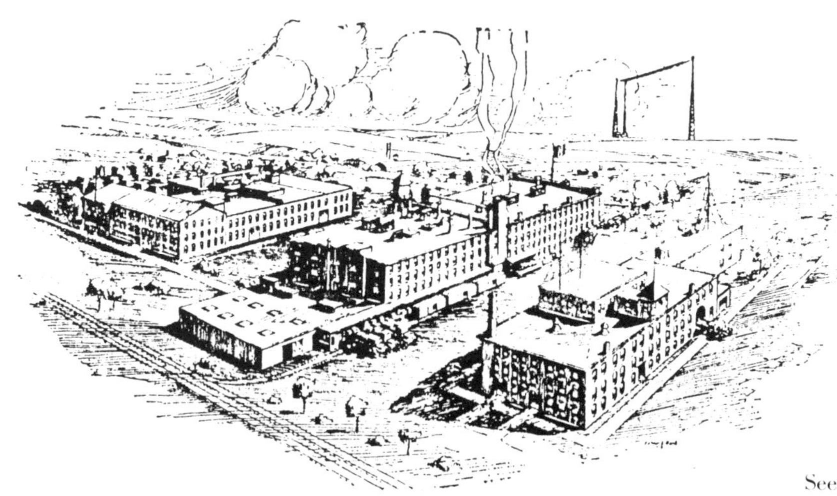

See p. 137

Crosley's most popular early model was the Harko Senior, originally advertised in February 1922 for $16. By June it was known as the "Harko Senior no. V" (five) costing $20. In February 1923, just after the arrangement with Tri-City ended, and Crosley had bought Precision Equipment, he offered a whole range of model V variations: three cabinet and panel styles, each in two types, regenerative or non-regenerative, at two different prices.

VA (overhanging lid) $16 or $14
VB (unmarked panel) $17 or $15
VC (engraved panel) $20 or $16

In April 1923 the VD replaced the VA, at $19. In June the VC began to be advertised under the Ace name, for $20. Finally, by early 1924, the V was once again sold under the Crosley name. Matching two-step amplifiers were available for all of these models.

The tendency in the radio field today is to put apparatus in cabinets not only for appearance's sake, but as a protection from dust, dirt, atmospheric conditions, etc. Realizing the demand for attractive stock cabinets of various sizes, we are building them in quantities in our large wood working plant. These cabinets are all uniform in style. The panels are rabbated in to the front. As the outside dimensions and inside dimensions are either larger or smaller than the panel itself, we show panel size and also inside dimensions. Prices quoted do not include the panels.

All cabinets are waxed antique mahogany finish. Wood used is either gum, genuine solid mahogany or quartered oak. Lids or tops are hinged. Sizes and prices are shown below:

CABINETS

For Panel Size	Inside Dimensions High	Wide	Deep	Gum	Mahogany or Quartered Oak
6x 7	5½"	6½"	7"	$2.50	$3.85
6x10½	5½"	10"	7"	2.75	4.40
6x14	5½"	13½"	7"	3.30	5.55
6x21	5½"	20½"	7"	3.90	7.30
9x14	8½"	13½"	10"	3.70	6.80
12x14	11½"	13½"	10"	4.40	6.80
12x21	11½"	20½"	10"	5.25	10.60

Cash must accompany order. No C.O.D's. We pay transportation charges.

We can furnish genuine formica panels 3/16" thick, cut to the following dimensions: 6x7; 6x10½; 7x9; 6x14; 7x12; 6x21; 7x18; 9x14; 12x14; 14x18; 18x21. Price of panels—2½c. per square inch. For odd sizes order the next largest size; we will trim. We pay postage.

Every article bearing the name "CROSLEY" is GUARANTEED to give absolute satisfaction or money will be refunded.

We shall be pleased to send literature describing the above mentioned and other radio apparatus to any one free of charge upon request. Get your name on our mailing list to receive latest Bulletins of other new Crosley products. If your dealer does not handle our goods, order direct and send us his name.

DEALERS—It will pay you to handle our line. Write for full particulars.

CROSLEY MANUFACTURING CO.

Radio Dept No. R-4B CINCINNATI, OHIO

Radio News (Oct. 1921) p. 314

John Wolkonowicz

Harko Sr. July 1922 $20

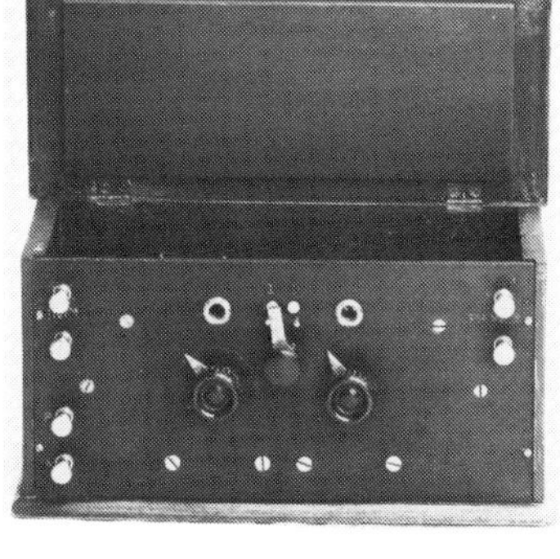

Dave Crocker

Two-step amplifier $25

Dave Crocker

VD Apr. 1923 $19

Dave Crocker

VC June 1923 $20

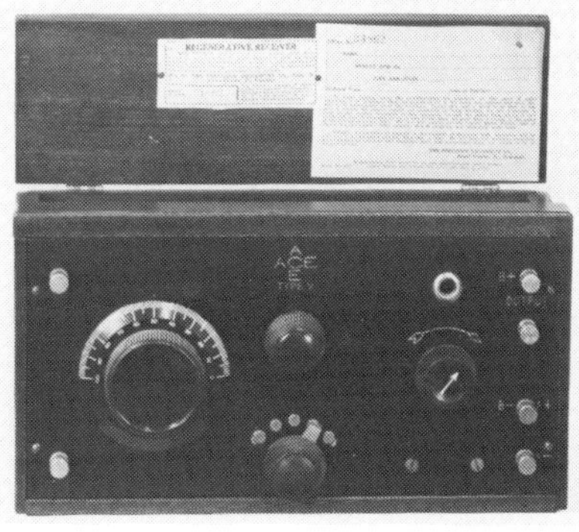

Dave Crocker

Late 1923

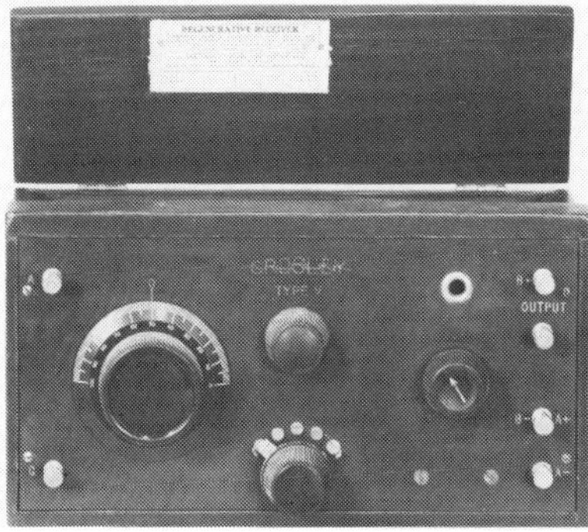

Dave Crocker

Early 1924

QST (Nov. 1922), p. 81

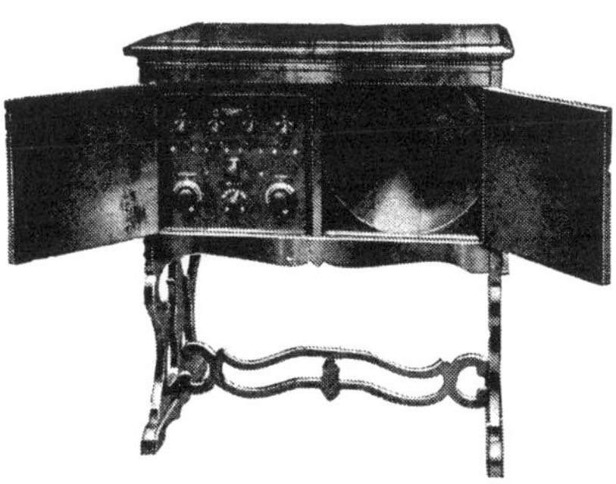

CROSLEY MODEL XXV

This beautiful mahogany cabinet is equipped with four tube panel incorporating the same units as the Model X, but the panel is of a different shape, as will be noticed from the illustration. This cabinet is arranged to take the Model R-3 Magnavox that can be quickly is not furnished at the price. Cabinet also contains installed and hooked up to the set, but the Magnavox space for "A" Battery, "B" Battery and battery charger if desired. A throw-over switch is provided to change from head phones to loud speaker. It is guaranteed to bring in broadcasting stations up to one thousand miles or more, loud enough to be heard all over the room. This beautiful instrument, without tubes, batteries or phones, sells for **$150.00**.

XXV Nov. 1922 $150

Radio News (Sept. 1922)

CROSLEY Radio Apparatus

BETTER—COSTS LESS

The unit outfits shown on this page are priced without tube or batteries

Crosley Crystal Receiver No. 1

This is an extremely efficient crystal detector outfit combining tapped inductance and variable condenser for tuning. Manufactured under the Pickard patents. Furnished complete as shown, with pair of standard double head phones, antenna, insulators, molded rubber dial, formica panel, Adam brown mahogany finish cabinet, complete instructions, etc. (Manufactured under Pickard patents).
Price.. **$25.00**

Crosley Audion Detector Unit

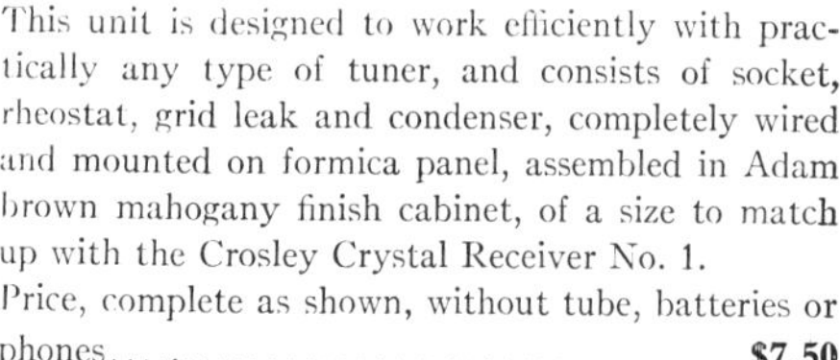

This unit is designed to work efficiently with practically any type of tuner, and consists of socket, rheostat, grid leak and condenser, completely wired and mounted on formica panel, assembled in Adam brown mahogany finish cabinet, of a size to match up with the Crosley Crystal Receiver No. 1.
Price, complete as shown, without tube, batteries or phones.................................. **$7.50**

The Crosley Crystal Receiving Outfit No. 1 includes phones and antenna which are not included with the other outfits

Harko Senior No.5

Consists of tuning element and audion detector unit mounted in neat cabinet, Adam brown mahogany finish, and many refinements over the original Harko Senior models, that have made history for themselves in the radio field.
Through large production the price is only.................. **$20.00**

Crosley R. F. T. A. Unit

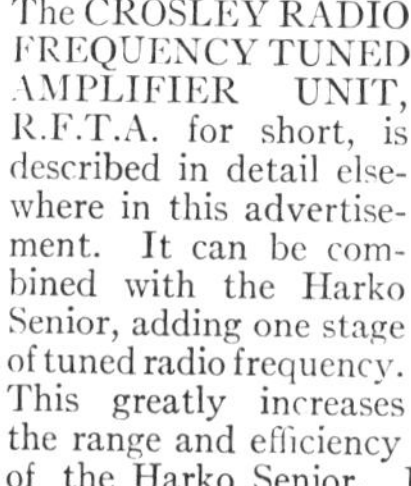

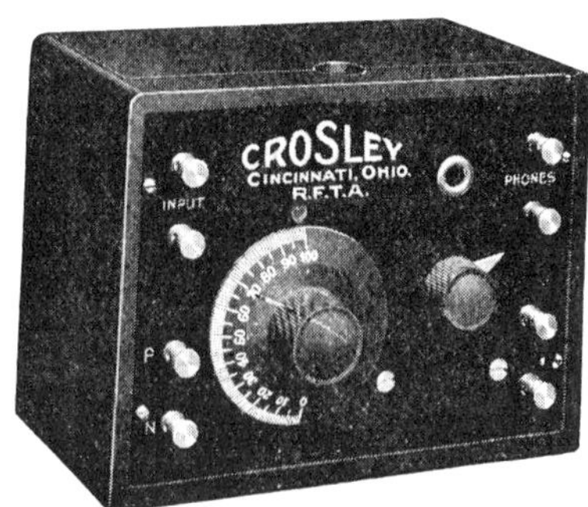

The CROSLEY RADIO FREQUENCY TUNED AMPLIFIER UNIT, R.F.T.A. for short, is described in detail elsewhere in this advertisement. It can be combined with the Harko Senior, adding one stage of tuned radio frequency. This greatly increases the range and efficiency of the Harko Senior. Price is............................ **$15.00**

Crosley Receiver No. 6

This unit in effect combines the Harko Senior and R.F.T.A. unit. Used alone can be recommended highly to get everything that is going up to several hundred miles in the way of broadcasting programmes, code, etc. It has a range from 200 to 600 meters, and will give from six to ten times the volume of sound that the Harko Senior alone will give; it brings in stations loud and clear that would otherwise be inaudible, and at the same time is non-regenerative. It requires one amplifier and one detector tube. Price.... **$30.00**

Crosley Two-Stage Audio Frequency Amplifier

This unit has been sold in large quantities for use with any type of audion detector receiver, and can be used in many combinations with Crosley or other apparatus. It amplifies signals approximately one hundred times, loud enough to be heard all over the room, in connection with any good tuner and audion detector.
Price.................. **$25.00**

Crosley Model No. X

Is a combination tuning element, radio frequency tuned amplifier, audion detector and two-stage audio frequency amplifier, all mounted neatly in one cabinet, simple and easy to tune, has wonderful range and volume of sound. Price................................ **$55.00**

Model No. XV

This cabinet is equipped with the model X panel, the same four-tube outfit, mounted in a special cabinet with amplifying horn incorporated therein, and can be used with head phones or single loud speaking receiver, such as the Baldwin type "C." Price of this four tube outfit is...... **$70.00.**

Crosley Model No. XX

This is the same four-tube panel as is used in model X and XV, and is complete in a floor cabinet, as shown in the illustration. In the cabinet is room for storage battery. It also is equipped with amplifying horn for use with loud speaking receiver.
Price............ **$100.00**

The CROSLEY MANUFACTURING CO.

CINCINNATI, O.

Radio News (Sept. 1922)

CROSLEY

Radio Apparatus

BETTER—COSTS LESS

This page illustrates the flexibility of the small unit idea, and the combinations that can be built up with Crosley apparatus.

This illustration shows how it is possible to start with a Crosley Crystal Receiver No. 1, and add the Audion Detector Unit at a later time, making in effect the Harko Senior.

This shows how the R. F. T. A. can be added. The combination shown in the second illustration is equivalent to the Crosley Receiver No. 6.

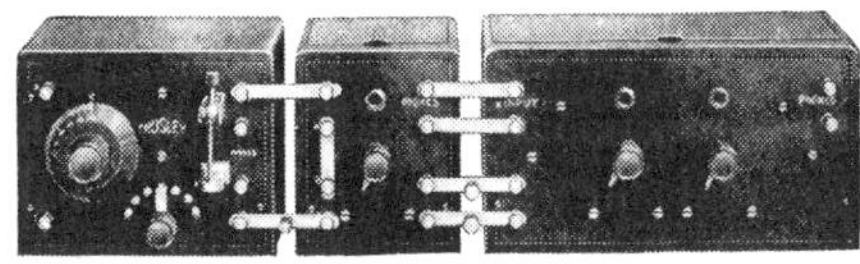

This illustration shows combination of the Crosley Receiver No. 1, Audion Detector and Two-Stage Audio Frequency Amplifier, if same is preferred to the Radio Frequency Tuned Amplifier.

Later, when desired, all four units can be combined as shown in the illustration, of the Crystal Receiving set, Audion Detector, Radio Frequency Tuned Amplifier and Two-Stage Audio Frequency Amplifier, making in effect, the Crosley Model No. X, illustrated on the opposite page.

Now if the purchaser wishes to start with the Harko Senior, he can add the Radio Frequency Tuned Amplifier as illustrated, or if he prefers, he can combine the Harko Senior with the Two-Stage Audio Frequency Amplifier.

The next illustration shows the combination of all three units, which again are equivalent to the model X illustrated on the opposite page.

If the purchaser wishes to start with the Model 6, illustrated on the opposite page, he can add the Two-Stage Audio Frequency Amplifier at a later date, making a Model X. This combination is pictured below.

We believe that the unit idea has been worked out in the Crosley apparatus in a very effective manner. Efforts along this line have been made before, but never so thoroughly or completely. The idea has met with instant enthusiasm wherever shown; its popularity is already assured. The price of all units of Crosley apparatus are way below all competition. Their efficiency is unquestioned, remarkable results having been reported continually on even the simpler units. The low prices are made possible by quantity production in Crosley factories, where practically every piece and part is made, not merely assembled.

The CROSLEY MANUFACTURING CO.

Dept. R. N. 1 **CINCINNATI, O.**

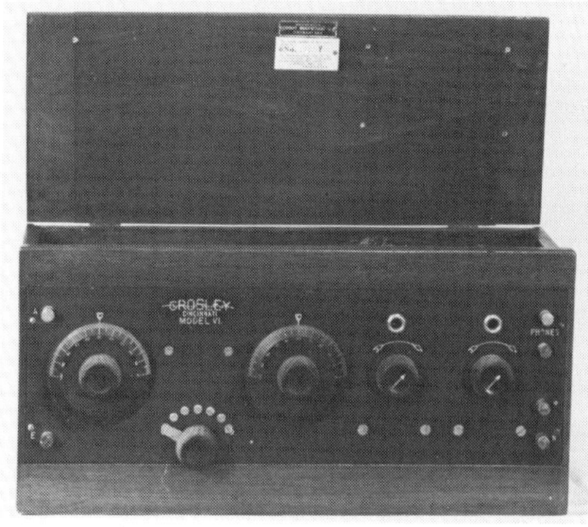

Dave Crocker

VI Special (larger cabinet) Jan. 1923 $30
VI portable also advertised Jan. 1923 $40

Dave Crocker

Late 1923–Early 1924

Dave Crocker

VIII Feb. 1923 $48
VIII Special advertised in Jan. 1923, $50. VIII portable, Jan. 1923, $60.

Dave Crocker

X Early 1923

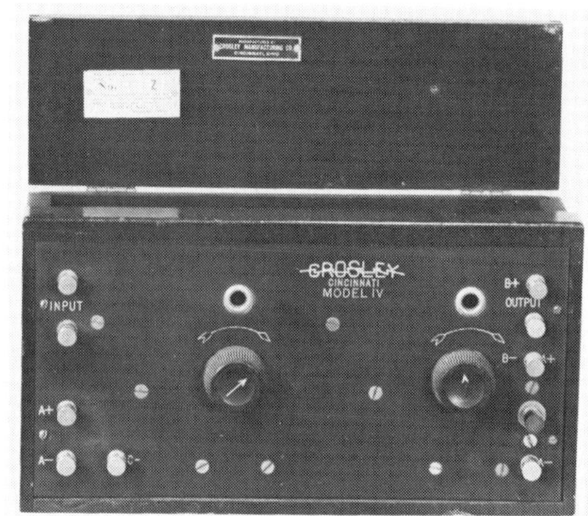

Dave Crocker

Model IV amplifier to match the VI.

Dave Crocker

3B Sept. 1923 $50
3C (similar cabinet to XL) $125. On stand, $150.

XJ Sept. 1923 $65
In Mar. 1924, the XJ and the VI were also made as Superdyne models, with reversed ticklers mounted between the tuning dials, under license from R.S. Miner formerly of Tuska.

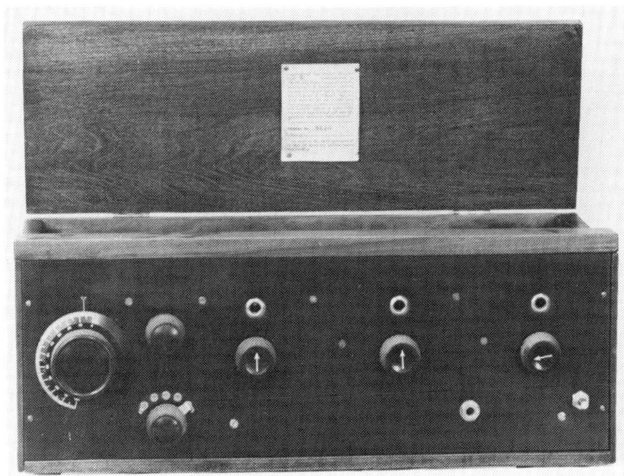

Dave Crocker

Sears, Roebuck Type 6

Dave Crocker

XL Jan. 1924 $140

1,710,966. VARIABLE INDUCTANCE. POWEL CROSLEY, Jr., Cincinnati, Ohio. Filed Nov. 3, 1923, Serial No. 672,491. Renewed Oct. 5, 1928. 4 Claims. (Cl. 171—119.)

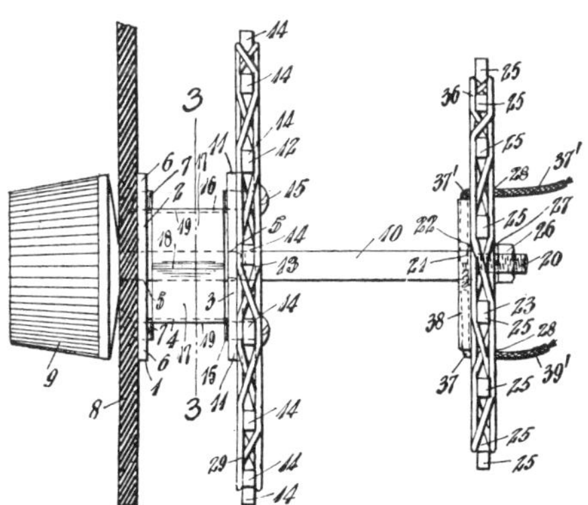

1. In a device of the character described, the combination with relatively movable coils, of a shaft of angular cross-section supporting one of the coils, means fixed to the other coil and having a bearing holding said shaft for non-rotative longitudinal sliding in said bearing, and having edges substantially parallel with said shaft, and a spring formed of a flat sheet bent with members meeting in a ridge bearing along said shaft and with flanges lying along said edges of said bearing means.

Dave Crocker

SR2 Jan. 1924 $23.25 complete. Made for Sears, Roebuck.

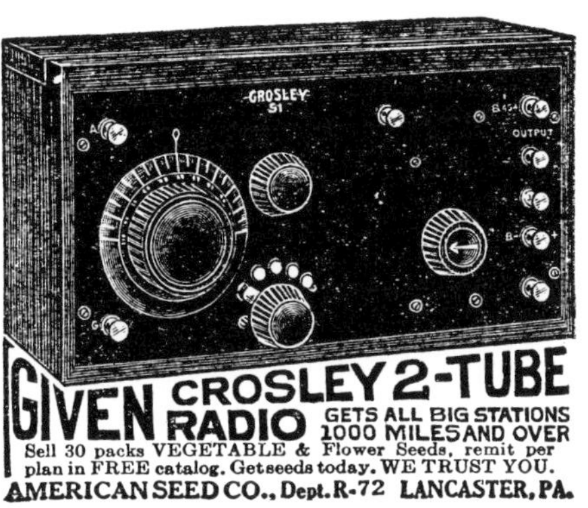

Farm Life (Jan. 1928)

Dave Crocker

50 Portable Aug. 1924 $18

Dave Crocker

Left to right: Early 51, Feb. 1924. 51 DeLuxe, Sept. 1925. Late 51 with wooden panel (after July 1924). Not shown: 51 portable, July 1924, $25. In the first year, Crosley claimed to have sold "over 170,000" or "197,000" (both figures are given in the May 1, 1927 *Crosley Broadcaster*).

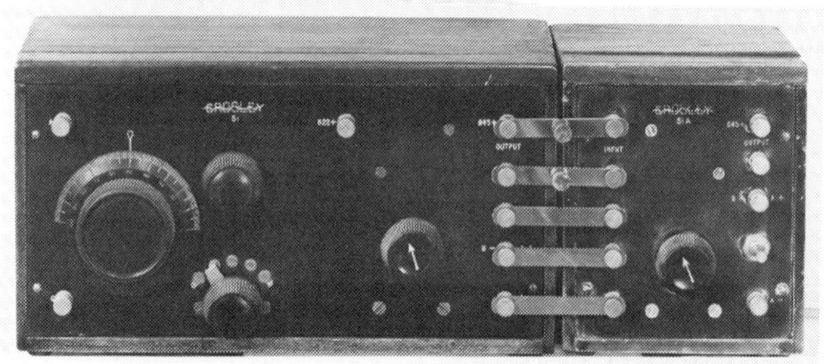

Dave Crocker

late 51 and 51A July 1924 $18.50, $14.

Dave Crocker

DeForest-Crosley 51 and 50

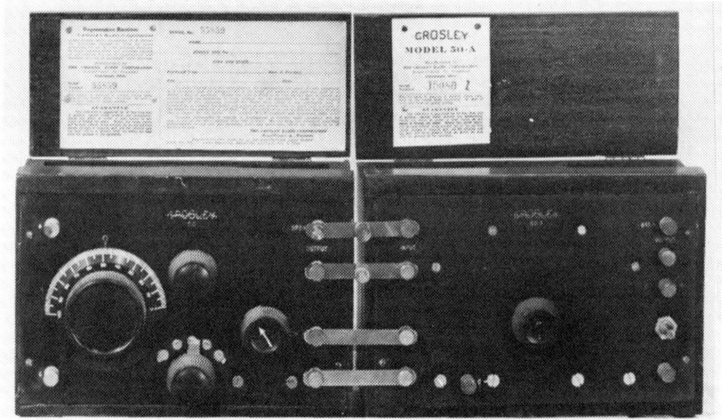

Dave Crocker

50 and 50A July 1924 $14.50, $18.

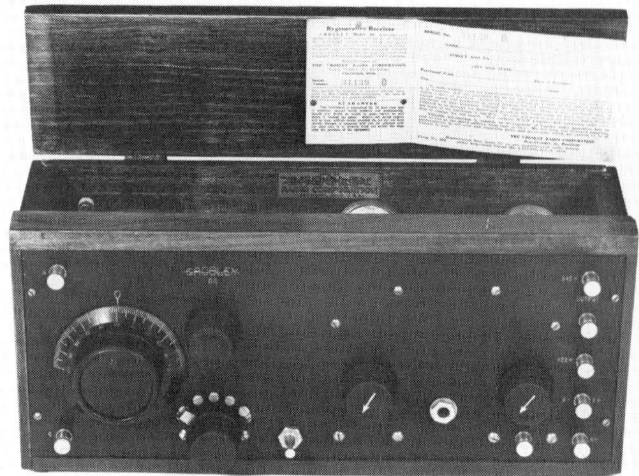

52 July 1924 $30

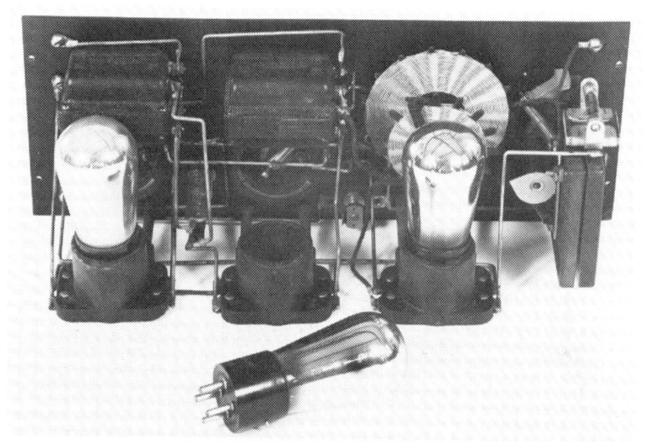

Popular Science Monthly (Mar. 1925), p. 1

51 Special Jan. 1925 $23.50
52 Special Jan. 1925 $35

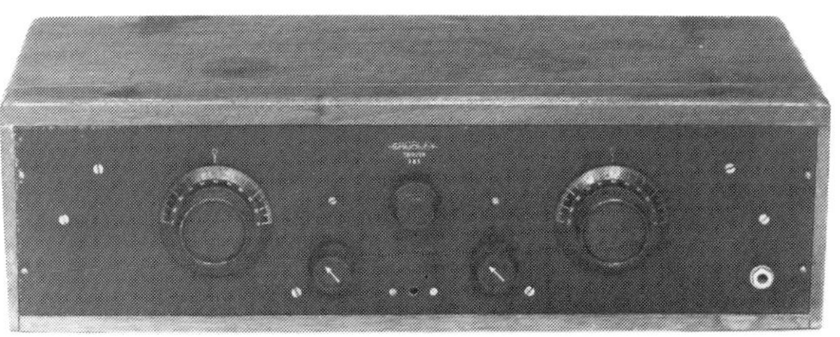

New England Wireless & Steam Museum

Trirdyn 3R3 Apr. 1924 $65

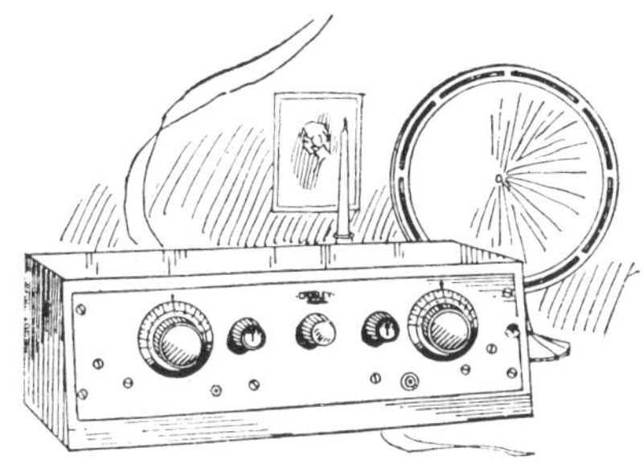

Popular Science (Mar. 1925), p. 1

Trirdyn Jan. 1925 $50

Ralph & Elinor Williams

Trirdyn Regular Apr. 1925 $50
Technical article in *Popular Radio*, Apr. 1925, pp. 340–351.

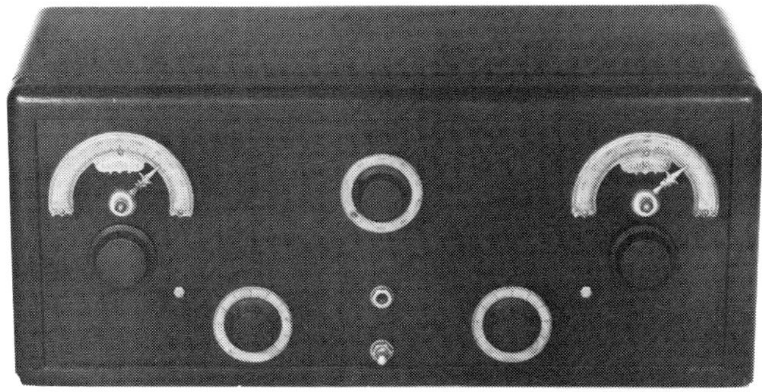

Herb Parsons

Super Trirdyn Regular Sept. 1925 $50

Radio News (Dec. 1924), p. 889

Trirdyn 3R3 Special June 1924 $75

Radio News (Dec. 1924), p. 888

Trirdyn Newport Dec. 1924 $100

New England Wireless & Steam Museum

Trirdyn Special Jan. 1925 $60

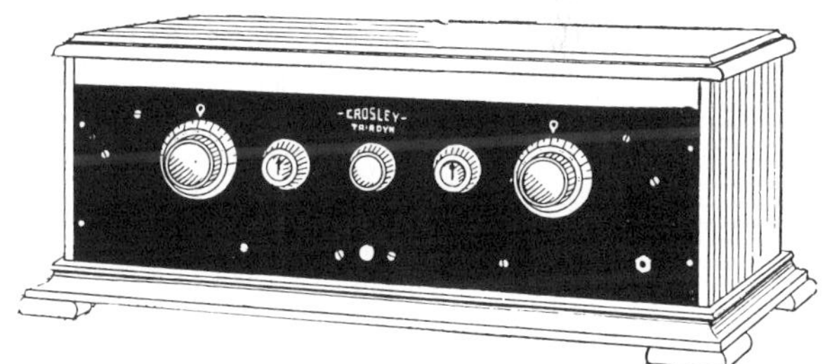

Radio News (Apr. 1925), p. 1835

Trirdyn Special Apr. 1925 $60

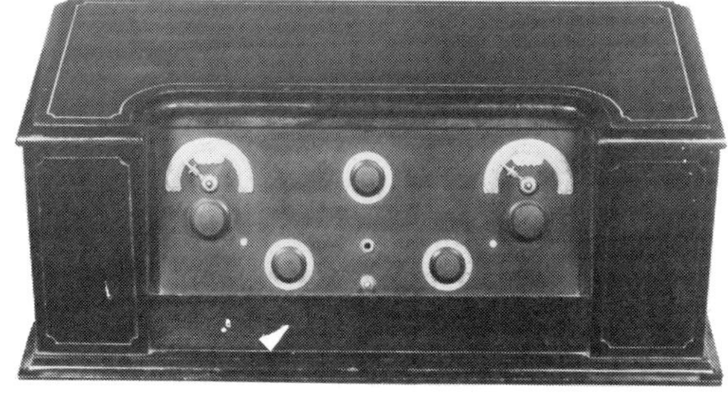

Ralph & Elinor Williams

Super Trirdyn Special Sept. 1925 $60

The Crosley 51 Special DeLuxe is a two tube Armstrong regenerative radio with one stage of audio frequency amplification. Similar to the Crosley 51 Regular, which costs but $14.75. Stations from coast to coast being heard by owners. Operates loud speaker on local and hear-by stations. Price, without accessories

$23.50

The Crosley 52 Special DeLuxe is a three tube receiver consisting of a genuine Armstrong regenerative detector and two stages of audio frequency amplification. Stations from coast to coast heard on loud speaker by many owners. Tunes through powerful local stations. Cabinet has sloping panel and is large enough to house dry cell batteries. Similar to the Crosley 52 Regular, which sells for $25.00, except cabinet is larger. Price of the 52 Special DeLuxe, without accessories

$32.50

51 Special DeLuxe Sept. 1925 $23.50
52 Special DeLuxe Sept. 1925 $32.50

Dave Crocker

Pup Aug. 1925 $9.75
A "repackaged" Model 50

Popular Radio (Oct. 1925)

"Better Results From 3 Tubes Than From 5"

Instead of passing the incoming signal once through each of 5 tubes, Crosley design, in the Super-Trirdyn, passes it through two of the three tubes several times, each time building up its strength and adding to its volume.

Even the technically uninitiated can see the advantages: simplicity instead of complexity; fewer dials to adjust; sharper accuracy in selecting stations; greater clarity; greater volume, greater ease in logging stations.

This simplicity of design, combined with the economies of gigantic production, makes possible a price of $60.00 on the Super-Trirdyn Special, the most efficient and beautiful of all Crosley receiving sets. For Crosley is the world's largest builder of radio sets—owning and operating parts factories, cabinet woodworking and assembly plants.

Listen to a Crosley Super-Trirdyn under the most exacting conditions and you will understand why it represents a genuine achievement in radio performance and value which all America was quick to recognize and reward with increasing sales.

Write for an illustrated catalogue of the complete Crosley line or see them at your Crosley dealer's. Authorized sales and service stations everywhere.

Crosley manufactures receiving sets which are licensed under Armstrong U. S. patent No. 1,113,149 and priced from $9.75 to $60.00 without accessories.

Add 10% to all prices West of Rocky Mountains. Crosley owns and operates WLW first remote control super-power broadcasting station.

CROSLEY·RADIO

BETTER · COSTS LESS

THE CROSLEY RADIO CORPORATION, CINCINNATI, OHIO

2-Tube Crosley 51

Same as wonderful Crosley 50 with additional tube amplifier. Local and nearby stations on loud-speaker always and distance up to 1500 miles under average conditions. Much greater range with head phones.

Special Sloping Front 2-Tube Crosley 51

Same as model 51, with cabinet holding all dry A and B batteries. $23.50.

2-Tube Crosley 51 Portable

The Crosley 51 in a black leatherette case, with nickel trimmings. Space for batteries. $23.50.

Crosley Musicone

A marvelous new development of loud-speaking principles. Diffusion of sound creates perfect reproduction of *all* tones. $17.50.

3-Tube Crosley 52

A larger set for those who want greater reception range on the loud-speaker. Operates on three tubes, using wet or dry batteries. Consistent loud-speaker range 1500 miles or more.

Special Sloping Front 3-Tube Crosley 52

Cabinet contains dry A and B batteries. Same efficient detection and reception as regular 52. $35.

3-Tube Crosley 52 Portable

Same as other 52 models, but in a black leatherette case. Easily carried. All batteries inside. $35.

Prices quoted above do not include accessories. Add 10 per cent west of Rocky Mountains.

This is the latest refinement of the marvelous set that enabled Leonard Weeks of Minot, N. D. to catch the messages of MacMillan's North Pole expedition when sets costing ten times as much failed.

In this set Crosley has developed the famous Armstrong regenerative circuit. This circuit does with one tube what it takes three tubes to do in others.

This set will bring in stations from all over the country. It is simple and easy to operate. With accessories the total cost should be under $25.00. Crosley keeps the cost down with his "radio-for-the millions" ideas in production.

Recent letters from enthused owners of the Crosley one-tube 50 report good reception at these distances:

Mrs. J. E. Martin at East Palestine, O. hears KGO at Oakland, Calif.
O. W. Bryant at Sunset, Texas get WLW at Cincinnati, KDKA at Pittsburgh and Hollywood, Calif.
L. R. Pratt, Hammond, Ind. hears 5NO, New Castle, England.
Eugene Barnhouse at Brookfield, Mo. hears Montreal and Winnipeg, Canada.
Paul J. Hall at Osceola, Neb. hears 2LO at London, England.

Crosley manufactures receiving sets which are licensed under Armstrong U. S. Patent No. 1,113,149, and priced from $14.50 to $65, without accessories.

The Crosley Radio Corporation

Powel Crosley, Jr., President

816 Sassafras Street, Cincinnati

Popular Radio (Aug. 1925)

4 TUBE
PORTABLE

"Take your entertainment with you" by means of a Crosley four-tube 4-29 Portable radio set. A real, long distance receiver in a leatherette covered carrying case. Ample room for batteries and other accessories inside cabinet. An idea radio to take with you on your vacation. A genuine Crosley four-tube portable radio at a ridiculously low price.

$33.00

For Sale By

E-667

NOTE: Mats or Electrotypes of this ad will be furnished Crosley dealers free of charge.

CROSLEY
5-38
for
$38.00
Fulfills a
Promise

The Crosley Crescendon, which permits perfect control of volume and builds up volume on weak signals, makes the five-tube 5-38 greatly superior to other tuned radio frequency radios. This set created quite a sensation last Spring. Recent changes in the panel have made it even more attractive. Stations at very great distances easily tuned in. Let us demonstrate this very low priced five-tube Crosley radio.

For Sale by

E—804

NOTE: Mats or Electrotypes of this ad will be furnished Crosley dealers free of charge.

The CROSLEY
Five Tube 5-50
In a Solid Mahogany
Cabinet
for
$50.00

This is a five tube radio, with its single drum station selector and installed in a solid mahogany, two-toned cabinet. It incorporates two stages of non-oscillating radio frequency amplification, regenerative Crescendon controlled detector and two stages of audio frequency amplification, with means provided for use of a power tube in the last stage. All stations found instantly on this one control, which revolves smoothly under slight pressure. Can be calibrated for wave lengths. Incorporates new shielded compartment metal chassis.

For sale by

E—808

NOTE: Mats or Electrotypes of this ad will be furnished Crosley dealers free of charge.

absolute
balance
in the
CROSLEY
5 tube
R·F·L·-75

$65.00

Popularity of receiving sets using the R. F. L. circuit has been established. This type, offered at the very low price of $65.00, incorporates everything a radio listener could desire. Recent changes in the panels have made it more attractive and still have made it possible to offer the receiver for $10.00 less than its former price. The R. F. L. circuit provides true cascade amplification largely through the use, in each radio-frequency stage, of a complete Wheatstone Bridge whose balance is practically uniform at all wave lengths. Non-oscillating at any frequency, non-radiating, and cannot be made to howl, even if mishandled. Stations can *be* accurately logged. Exquisite two-toned mahogany cabinet, with satin finish.

FOR SALE BY

E—810

NOTE: Mats or Electrotypes of this ad will be furnished Crosley dealers free of charge.

A New and
Beautiful
5 Tube
Console
Radio
With but one
Control and
Built-in Musicone

Beautiful solid mahogany cabinet, with built-in Musicone and compartment for batteries.

$75

CROSLEY
5~75

This is a new five-tube radio, with single drum station selector, all-metal shielded chassis and power tube adaptability. Consists of two stages of non-oscillating radio frequency amplification, regenerative, Crescendon controlled detector and two stages of audio frequency amplification.

For Sale by

E—806

NOTE: Mats or Electrotypes of this ad will be furnished Crosley dealers free of charge.

That Famous
R·F·L·
Circuit
in the
CROSLEY
Six-Tube
R·F·L·90

$90
Double Drum
Station Selector

The tremendously popular R. F. L. circuit is utilized in the new six-tube R. F. L.-90. This is a very beautifully designed console type radio, with built-in Musicone and compartments for batteries. Solid mahogany cabinet. Double drum station selector. Also many other exclusive features that make it the ideal long-distance receiving set. See the R. F. L.-90 on display NOW at our store.

For Sale by

E—812

NOTE: Mats or Electrotypes of this ad will be furnished Crosley dealers free of charge.

4-29 portable July 1926 $33 RFL-75 Sept. 1926 $75

5-75 Sept. 1926 $75 RFL90 Sept. 1926 $90

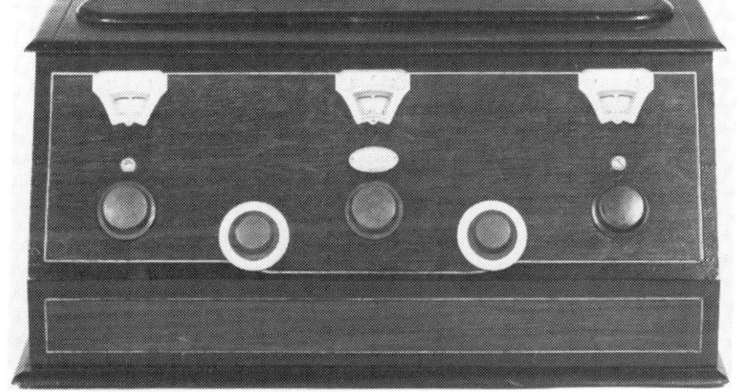

5-38 Sept. 1926 $38 Rich Elskamp

5-50 Sept. 1926 $50 Ralph & Elinor Williams

Technical article, Popular Radio Nov. 1926, pp.656,678–81.

Dave Crocker

6-60 Apr. 1927 $60
6-85 console (like 5-75) $85

$40. *Without Tubes or Batteries*

Lowave adapts any Broadcast Receiver to bring in Short Wave Stations

WLW and other stations are already broadcasting simultaneously, their regular program on low wave lengths below 80 meters. As low wave reception is free from static, it affords great distance increases, improves daylight reception and opns a NEW FIELD for RECEIVING SETS. This receiver is designed so that by changing coils wave lengths from 20 to 80 meters are covered. Using three 201-A tubes or their equivalent it picks up short waves and translates their frequencies to one wihtin the band of your present broadcasting receiver. Utilizes same "A" battery as is used on regular set.

Crosley Broadcaster (Apr. 15, 1927), p. 8

The Newest Achievements of POWEL

Industrialist—Pioneer Radio

The Crowning Accomplishment In a Career of Radio Leadership

If you have waited for a very low priced radio that gave you all the volume, all the purity of tone, all the selectivity and sensitivity of the costliest set you ever heard—

—visit the nearest Crosley dealer and see the four new instruments that constitute the latest and greatest radio achievement of Powel Crosley, Jr.!

Here, at prices so low as to be literally revolutionary, are three 5-tube sets and one 4-tube set—entirely new in principle, design, circuit, and appearance—entirely unique in the results they give on distant and local stations—entirely unprecedented in the value they now introduce.

Emphasizing the amazing performance and value of two of these sets is the Crescendon, a new and exclusive Crosley feature—an extra volume control by which average incoming signals can be built up until the music booms throughout the house, if desired. For the first time now, this basic principle of extra volume control is offered on low priced 4 and 5-tube tuned radio frequency radio sets, its use having heretofore been restricted to one set costing several times as much as the new Crosley on which it is introduced.

An examination of the new Crosley sets—and a comparison of results with any radio on the market—will clearly reveal why their announcement is destined to be regarded as the outstanding radio triumph of Powel Crosley, Jr. Radio, with all its romance, knows no more magic name than Crosley—simply and solely because Crosley engineering, manufacturing, and distribution genius has pioneered the advancement of radio to its present place in American life.

Powel Crosley, Jr., was the first to offer radio parts at prices that enabled millions to build their own sets; first to market a complete crystal set below $25; first to offer sets using tuned radio frequency amplification and first to combine it with regeneration and the reflex principle; first to erect and operate a remotely controlled super-power broadcasting station.

Then came the now famous Crosley Musicone ($14.75)—an instrument reproducing the entire range of the human voice and music—which, in its first year, became the largest selling loud speaker in the world and which today is replacing other types of speakers with a rapidity that is determined solely by the production capacity of one of the largest radio plants in the world.

As a direct result of this unflagging leadership in radio, there are more Crosley sets in use today than any other make. This fact stands first as a tribute to Crosley quality and value. But it also lends emphasis to the unqualified statement that the new Crosleys offer a measure of performance and appearance never before available at anywhere near the price!

Crosley dealers everywhere are demonstrating the new Crosley sets. We invite you to visit the nearest one. We urge you to go anticipating more than the unusual that the public has come to expect when viewing Crosley achievements of this nature. We ask you emphatically and directly to look for the most startling revelation in radio ever announced in the entire history of the industry—and we predict that your expectations will be more than satisfied!

Crosley manufactures radio receiving sets which are licensed under Armstrong U. S. Patent No. 1,113,149, or under patent applications of Radio Frequency Laboratories, Inc.

West of the Rocky Mountains all prices as published are 10% higher.

THE CROSLEY RADIO CORPORATION, CINCINNATI, OHIO

Owning and operating WLW first remote control super-power broadcasting station in America.

CROSLEY JR.

Builder—Master of Mass Production

Four Entirely New Radio Sets—

Definite improvements on ideas heretofore found only in the high priced field!

The Crosley 4-29 and 5-38 New 4 and 5 Tube Radios

With a New Amplifying Device Improving Volume and Selectivity

The Exclusive Crosley Reserve Volume Feature

Crosley 4-tube—4-29

Using only four tubes, this attractive set delivers an extraordinary performance because the Crescendon control is equivalent to one or more additional tubes of tuned radio frequency amplification. Scarcely audible signals from distant low-powered stations can be built up to dancing volume, and local high-powered stations cut down to a whisper without impairing the tonal quality. Attractively cabineted in hardwood, finished in two-toned mahogany, the 4-29 is destined to win a popularity that will be sensational even according to Crosley standards. Without accessories **$29**

Crosley 5-tube—5-38

On the basis of exhaustive laboratory tests during development, the 5-38 is offered as capable of exactly duplicating any standard 5-tube set in volume, selectivity, sensitivity, and tonal qualities. In addition it incorporates the Crescendon for extra volume on those distant stations which do not ordinarily operate a loud speaker to satisfaction.

Imagine any good 5-tube set lifted into the realm of the extraordinary simply by the turn of a dial—and you have the new Crosley 5-38! Hardwood cabinet, handsome two-tone mahogany finish. Without accessories - - **$38**

The CRESCENDON

How often have you been receiving some very attractive distant program in volume insufficient to operate the loud speaker satisfactorily—and wished that there was some knob or dial you could turn to build that volume up and flood the house with music? Heretofore, no set provided such a knob or dial which, almost by magic, would swell that signal into a full, impressive crescendo. Now it is offered as an exclusive Crosley feature—alone a sufficient reason for selecting a Crosley above all other sets. It is the Crescendon. Every radio owner and lover should not fail to have it demonstrated by the nearest Crosley dealer.

RFL-60

Genuine R. F. L. Circuit Five Tube Sets

Incorporating TRUE CASCADE Amplification

Crosley 5-tube—RFL-60

An entirely new and patented circuit that provides true cascade amplification largely through the use, in each radio frequency stage, of a complete Wheatstone Bridge whose balance is practically uniform at all wave lengths. Thus is achieved a cumulative amplification heretofore deemed impossible and closely approaching the theoretical maximum of efficiency per tube. Non-oscillating at any frequency, non-radiating, and cannot be made to howl even if mishandled. A set so marvelous in its performance that its appearance on the market will automatically create a new standard of comparison. Solid mahogany cabinet, oil rubbed finish. Without accessories - - - **$60**

Crosley 5-tube RFL-75

This, the most luxurious Crosley ever offered, is unique in the remarkable degree of selectivity it provides. Absolutely non-radiating and completely fool proof, Crosley RFL sets are recommended for satisfactory performance in the hands of inexperienced operators. Stations can be accurately logged and reappear at the same point on the dials as long as the wave length is unchanged. For accuracy, simplicity, and speed in tuning, for clarity and fidelity in tone, for decorative beauty and value—the Crosley RFL-75 stands unchallenged. Hear it once and no other radio set will ever quite satisfy you. The solid mahogany, duo-toned cabinet holds all batteries. Without accessories **$75**

RFL-75

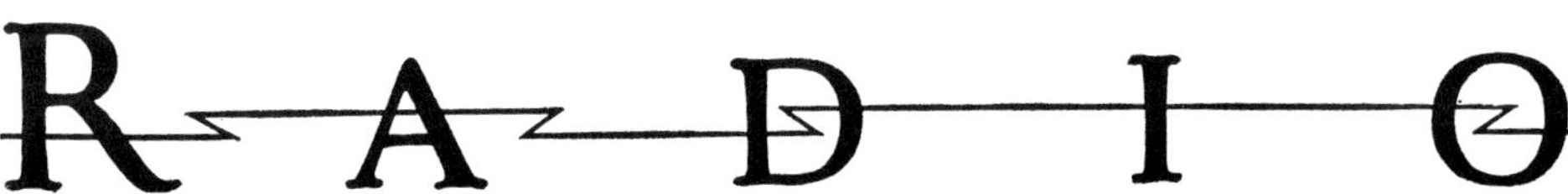

Crosley Broadcaster (Apr. 15, 1927), pp. 8–9

CROSLEY *Six Tube..... BATTERYLESS Single Drum* RADIO

First radio run direct from light socket without batteries for less than $150. Nothing extra needed ....except tubes and aerial..

The Crosley 6 tube A·C·7

$70.

ABC Power Unit $50 extra

AC-7 Mar. 1927 $120
console $145

HERE they are! A table model and console radio without a battery needed. A simplified, compact efficient little power unit supplies necessary A, B and C battery current direct from house lighting outlets or a lamp socket.

No batteries to renew!
No batteries to recharge!
No trickle charger to watch!
No water to refill batteries!
No acid to spill!
No run down batteries just as good program begins!
No mess or fuss in removing batteries!
No chemical action in power supply units!
Nothing going on when set is not in use!

Snap your lamp switch. The set is full powered instantly and constantly at highest peak. Snap it off. Everything is shut off at the electric light connection. No after charging.

This unique power supply unit is designed solely by Crosley, solely for these two Crosley radios. It incorporates new and advanced ideas and principles exclusive to Crosley.

Both the radio sets which it is designed to operate are six-tube tuned radio frequency circuits of amazing efficiency, incorporating three stages of radio amplification, detector tube and two stages of audio frequency amplification.

Radios Designed for A.C. Current

The simplicity and moderate price of these radios lies in the fact that they are *primarily designed* for A.C. current power. The power unit is not a makeshift effort to change A.C. current into power necessary for radio sets designed for battery operation.

The Crosley 6 tube A·C·7·C Console

$95.

ABC Power Unit $50 extra

601 Bandbox Aug. 1927 $55
By October, 3500 per day were being made, by 3000 workers. An AC Bandbox sold for $65, plus $60 for the external power pack (price reduced to $110 on Nov. 1).
602 AC Bandbox Nov. 1927 $110 reduced to $90 in Feb. Technical article in *Radio Broadcast* March 1928 pp. 369–372.

704 Jewelbox Feb. 1928 $95
704B (8-tube) June 1928 $95

Technical article in *Citizens Callbook* vol.11 no.2, Mar. 1930, p.78.

401 Bandbox, Jr. Feb. 1928 $35
401A (5-tube) June 1928 $35

608 Gembox June 1928 $65

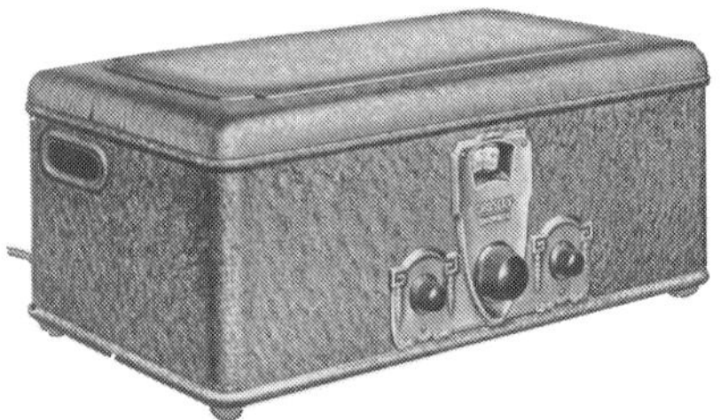

706 Showbox June 1928 $80
705 DC Showbox $85

804 Jewelbox Mar. 1929 $105

5 cuts from *Crosley Broadcaster* (May 15 & June 1, 1928)

NEW GEMBOX AC
ELECTRIC 7 TUBE

The *DYNACONE* is a different type of power speaker that takes its field current from the set which operates it. This employment of the armature principle of actuation has improved reproduction to a marked degree. Each tone is true in its relation to every other tone of the audible scale.

$25

IMPROVED
DYNACONE

THE·WORLD'S
THREE·GREAT
RADIO·VALUES

CROSLEY RADIO

Gembox

The *GEMBOX* has three stages of radio frequency amplification, detector, 2 audio with 171-A power tube in last stage and a rectifying tube—7 tubes in all. Shielded—illuminated dial—power output tube — Mershon condenser in power supply—AC electric operation. All modern, up-to-the minute quick-sale features.

Installed with the Dynacone in the . . .

Gemchest

You have the smartest radio set on the market, and at a price that makes quick sales. The *GEMCHEST* design is adapted from the Chinese Chippendale—three exquisite color combinations—Mandarin red with bronze gold hinges and fittings—Nanking green with rose gold—Manchu black with white gold. Stylish—new—individual—perfectly fitting into modern home interiors.

The *SHOWCHEST* is the same but is equipped with the 8-tube *SHOWBOX* receiver selling at $109.

Both the *GEMCHEST* and *SHOWCHEST* come equipped with the Improved Dynacone.

The Crosley *JEWELBOX* selling at $105 is another wonderful value.

THE CROSLEY RADIO
CORPORATION

Dept. 19 Cincinnati, Ohio

POWEL CROSLEY, Jr., *President*

Owners of WLW—The Nation's Station

Montana, Wyoming, Colorado, New Mexico and West prices slightly higher

Prices quoted do not include tubes

THE
SMART GEMCHEST

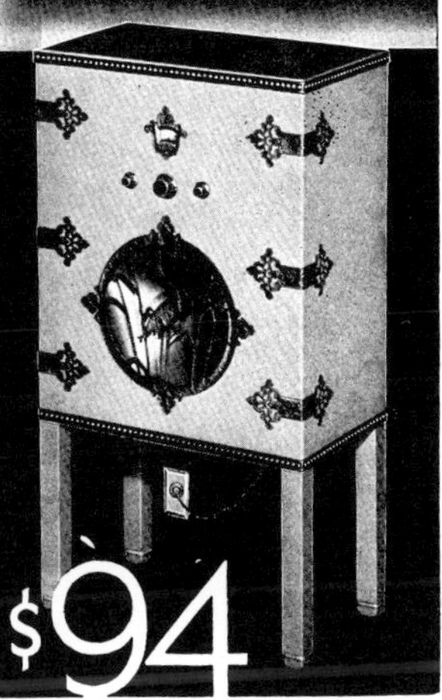

Radio (July 1929)

Steve Conklin

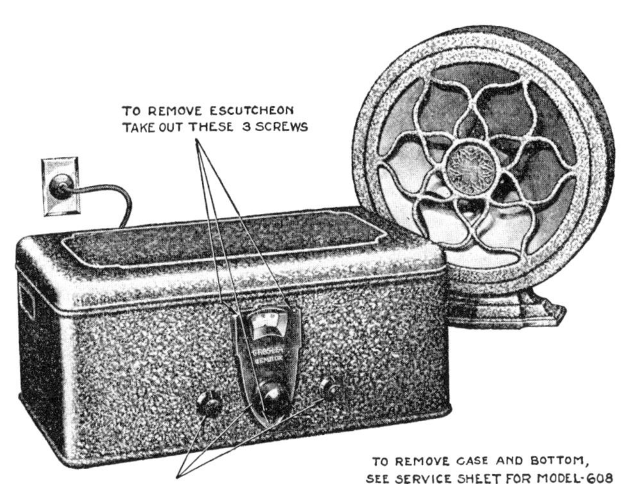

610 Gembox June 1929 $65

609 Gemchest Mar. 1929 $94
Also made with Showbox chassis, as 708 Showchest, $109.

Farm Journal (Oct. 1929)

New Beauty in Screen Grid Battery Radio

25 per cent lower voltage drain than on ordinary battery sets

NOW available for farm homes—all of Screen Grid's marvelous performance *plus* the latest refinements and improvements in radio design.

The clever table set shown at the right, will please you—not only by its smart attractiveness in your home, but quite as much with its rich, pure beauty of tone.

For unwired homes, this model—together with the exquisite console model at the right—incorporates the famous Crosley Screen Grid battery radio.

Not only does this remarkable set give you power, sensitivity and beauty of tone never before attained in battery radio, but it sets new standards of *economy* in operation. Its drain on the batteries is *25 per cent lower than any other standard battery set.*

For homes with electric lights, there is a complete range of Crosley table and cabinet models to choose from. Two of them are shown below. You will see all of them at any Crosley dealer's—*at the lowest prices in radio history!*

If there is no Crosley dealer near you write us for complete details.

THE CROSLEY RADIO CORPORATION
CINCINNATI, OHIO
Home of WLW

You're there with a

CROSLEY

Powel Crosley, Jr., President

$49 WITHOUT ACCESSORIES

Model 21 (above). Metal case with panels in beautiful burled walnut effect. Also available with legs, at $54. A six tube set utilizing THREE Screen Grid Storage Battery tubes, 201-A detector, 201-A first audio, 171-A power output tube. A. C. sets incorporated in the same case, are available in the following models: Model 31, seven tubes, $55. Model 41, eight tubes, $70. Model 41-S, containing the sensational New Crosley Screen Grid Unitrad set, is $85

$49 without tubes speaker or batteries

$88.50 without tubes or batteries

Model 22 (at left). Console cabinet in two tone walnut veneer. Contains same six tube Screen Grid unit as Model 21. Dynacone speaker. Model 32, available in same cabinet, full A. C. operation, at $99.50

CROSLEY A. C. SETS

$150 (without tubes)

Model 82 (left). Eight tubes including rectifier and power output tubes. Model 82-S, containing Crosley Screen Grid Unitrad, $160

(without tubes) **$125**

Model 42 (right). Eight tubes including two power tubes and rectifier. Model 42-S, containing Crosley Screen Grid Unitrad, $140

NOTE: *Western prices slightly higher on all models*

21	July 1929	$49	20 (chassis only)	22	$88.50		
31	July 1929	$55	30 (chassis only)	32	$99.50		
31-S	Oct. 1929	$65	30-S (chassis only)	33-S	$115	34-S (Dec.)	$116
41	July 1929	$70	40 (chassis only)	42	$125		
41-S	July 1929	$85	40-S (chassis only)	42-S	$140	82H	$150

82-S $160. Technical article in *Citizens Callbook*, vol. 11 no. 3, Sept. 1930, p. 79. Performance graphs in vol. 11 no. 2, Mar. 1930, p. 85.

61	July 1929	$85	60 (chassis only)	62	$135

In New York, Alma Gluck, renowned dramatic opera and concert soprano, contributes to the new tone beauty in Crosley Radio. Mr. Henry P. Joslyn, the composer (center), discusses the tone quality with Mme. Gluck. Powel Crosley, Jr., president of The Crosley Radio Corporation, receives the artist's comments, suggestions and advice

ALMA GLUCK

helps to make possible this new tone beauty in CROSLEY RADIO

FROM a distant broadcasting station music sweeps in . . . a Crosley Radio receives the program. . . .

A little group of radio experts "stand by" as Alma Gluck, the great dramatic soprano, "listens in." . . .

The music stops. Then—Mme. Gluck's counsel on the tone quality of the receiving set . . . the suggestions, comments, advice of a great artist, given directly to Crosley engineers.

Thus, in all parts of the country, Crosley Radio is subjected to the expert tone scrutiny of America's foremost composers, directors, musicians, opera and concert artists.

For this purpose, Crosley receiving sets are placed in the homes of these artists.

By a series of unique "tone tests," the greatest authorities on music and voice regularly aid Crosley in developing and improving a new tone beauty never before achieved in radio!

Alma Gluck, Edith Mason, Efrem Zimbalist, José Mojica, George Gershwin and others equally famous regularly contribute to Crosley tone quality . . . a new purity and richness that mechanical tests alone could never attain!

This beautiful tone, developed to the highest technical degree, then "ear tested" by America's famous musicians, is *exclusive* with Crosley.

Hear it yourself at any Crosley dealer's. You'll be amazed and delighted!

You may choose from a wide variety of Crosley table and console models. They embody every modern feature: Screen Grid, Neutrodyne circuits, Power Detection, phonograph pick-up, Dynacoil and Dynacone Speakers, etc. . . . *at the lowest prices!*

Arrange with your nearest authorized dealer to place any Crosley model in your home for a free trial. If you keep it, payments may be arranged on easy terms. Ask the dealer for details.

THE CROSLEY RADIO CORPORATION
CINCINNATI, OHIO
Home of WLW, "the Nation's Station"

One of a wide range of Crosley A. C. Screen Grid cabinet models, priced at $115 (without tubes). Crosley builds sets for direct current, alternating current, and battery operation, at prices ranging from $49 to $160. Western prices slightly higher

Crosley end-table models, similar to above, with 7 or 8 tube A. C. Screen Grid sets, priced from $67 to $85 (without accessories). Western prices slightly higher. No matter where you live Crosley builds a radio set to suit your pocket book

You're there with a

CROSLEY

Radio Broadcast (Feb. 1924)

From Binding Post to Varnish

By HOWARD S. PYLE

UP IN the cold, lonely timberlands of the north—down in the warm, tropical countries—on hundreds of ships on the seven seas, myriad tiny voices of the night bring cheer and comfort and entertainment to eager listeners. And all through proper manipulation of a few simple controls on a small wooden cabinet. It is an accomplishment to inspire awe, and we have a respect for the little cabinet that forms a medium of contact with the world about us. So faithfully and efficiently does the little mystery box serve us, that many do not stop to wonder how and where and by whom the delicate apparatus was assembled and arranged within.

The writer recently made an investigation of the manufacture of modern broadcast receiving equipment, concluding his search with a trip through one of the largest and most modern factories of the independent radio manufacturers.

The entire plant had been laid out after the policy followed by the Ford Motor Company: the raw materials are routed through a definite manufacturing cycle. As each part is completed, it is delayed at an operating station until the arrival of the next part required in the receiver assembly. By this method as a panel progresses through the factory, various completed parts are added until the panel reaches the testing department with a completed set attached to it.

The first photograph shows the individual units being made up. The girl at the extreme right is operating an automatic screw assembling machine—a combination power-driven screw driver and socket wrench. To the left, other operators are seen with automatic rheostat winding machines. A large proportion of the special machinery found to be imperative in accurate work, has been developed within the plant.

Following the completion of the various parts, they are delivered to the assembly department, to be incorporated in the receivers at the proper stations during their progress along the assembly tables.

In the photograph on page 290 we see the first stages in the construction of a four-tube re-

HERE THE PARTS ARE MADE UP

ENGRAVING AND FINISHING PANELS FOR FOUR-TUBE RECEIVERS

ceiver—an extremely popular product of this particular manufacturer. In the operations pictured, the panels appearing in the lower right-hand corner are machined to the desired dimensions, sanded, and then subjected to a finishing operation which completely removes the gloss and gives the panels a pleasing, flat, grained finish. They are then passed through a gang of drills, each operator drilling but one size hole, thus eliminating change of drills and lessening the possibility of error. The next step is through the engraving machines. A battery of five are seen behind the blank panels in the foreground. Here the work is also divided, so that one machine operator engraves only a small amount of "copy." Thus each man becomes expert in engraving some particular character such as a trade-mark or a group of letters or figures.

The panel is then ready for the assembly department. One row of long tables permits the progressive passage of the panel through the various stages of assembly. A separate row permits an uninterrupted line of receivers of five different types to be constantly in production. One worker, for instance, secures the filament rheostat and the two audio-frequency transformers in place, the next places on the binding posts, and so the work progresses down the line until all of the parts are assembled on the panel.

Next in order is the wiring, or "hooking-up." It has been found desirable to limit the number of wires placed by any one operator. Hence we find one worker caring for but two or three wires, the instrument then passing to the next station where two or three more are placed. This procedure enables each operator to work rapidly without sacrificing accuracy. In a plant where the bonus system is in vogue, this division of labor works to the advantage of both employee and manufacturer.

When a number of instruments are completely wired they are passed to the testing department on "wagon-racks." Several of these wagon-racks are shown at the right in the upper picture on the opposite page.

Passing along the right-hand row of tables, each instrument is subjected to an inspection for possible loose connections and given a thorough brushing and cleansing of all soldered joints with benzine and alcohol. This serves to eliminate leakage paths which might be caused by excessive soldering flux.

The instrument is then delivered to the radio-frequency test tables. Two small, continuous-wave oscillators installed in this department furnish energy for testing receiving

HERE THE RECEIVERS ARE INSPECTED AND TRIED OUT

sets on actual radio-frequency signals. One oscillator is adjusted to a wavelength of 200 meters and modulated at a frequency of 1000 cycles. The second oscillator is adjusted to 600 and modulated at 100 cycles. These testing oscillators are arranged on a loop, and the energy is transmitted to the series of loops shown in the photograph. A receiver must function at both wavelengths before passing the testing department.

It has been found practically impossible to locate faults in radio-frequency apparatus unless actual radio-frequency currents are applied to the instruments under operating

CABINETS IN THE EARLY STAGES OF CONSTRUCTION

conditions. In other words, receivers which have satisfactorily passed a mechanical and electrical test, are often inoperative when connected to an antenna system. The testing system just described enables a really satisfactory test to be made, the loop antenna design being such as to represent the electrical characteristics of the average antenna erected to receive broadcasting.

A receiver satisfactorily passing all tests is finally routed to the table at the rear of the testing department where the necessary instructions are placed in the cabinet and the serial number entered on the cards. The set is then packed for shipment.

In order that all of the parts entering into the final assembly may be manufactured in the one plant described, a wood-working department is maintained, which is among the largest and most modern in this country. Here the cabinets and other wood work used in the manufacture of radio equipment are turned out, with a considerable surplus for outside distribution. An example of the modern machinery in use throughout the entire organization is seen in the "sticking-machine"—the paddle-wheel-like machine shown in the lower picture on the previous page. This has replaced the old method of setting up long lines of cabinet work held together by means of "gluing-clamps."

A cabinet, upon leaving the wood-working plant, goes to the spraying department where the finish is sprayed on the cabinet instead of being applied by hand with a brush.

Completing the equipment of this manufacturing plant is an up-to-date printing establishment in which are published all catalogues, instruction sheets, and circular matter, as well as a weekly newspaper.

From the foregoing, it can readily be seen that radio as an industry has assumed a definite place in the business world. Conditions, so chaotic two years ago, have so shaped and stabilized themselves that an establishment such as the one described, representing an investment of many thousands of dollars and employing hundreds of workers, is recognized as a sound and substantial industry. The fly-by-night manufacturers of a year or two ago are now practically extinct. Good workmanship, correct policies, and a desire to give the public dollar for dollar value, have narrowed the great new field of radio manufacturing down to a comparatively few progressive manufacturers.

HERE THE FINISH IS SPRAYED ON

The Glue Wheel Clamps Cabinet Sections Until the Glue Sets, While the Panel Drill, Below, Drills Eleven Holes at Once

Deft-Fingered Girls Assemble the Book-Type Condensers (Above); the Inventor, at Right, Checks the Blueprints for a New Set, Soon to Be Placed in Quantity Production

The Basket-Woven Inductance Coil Follows the Antenna and Tickler Coils into Place as the Growing Set Flits down the Long Assembly Table

Making Five Thousand Radio Sets a Day

Automobile-Factory Methods Are Adopted by Wireless Receiver Producer to Turn Out Outfits in Wholesale Lots

GIVEN a million and seven thousand radio parts, weighing nearly thirty-four tons, how many workers will it take to assemble them into 5,000 radio sets in one working day of nine hours?

This question has been answered by one great radio producer through quantity-production methods, whereby some 550 girls can assemble the thousands of condensers, transformers, rheostats and multistats on their panels and subpanels, place several hundred thousand screws and binding posts, solder something over a half million joints, and pass the completed receivers on to a few men who screw them into their cabinets, add serial numbers and instruction booklets and box the outfits ready for shipment. Starting at one end of a long table with a subpanel and a handful of binding-post parts—three parts for each post—the set moves down the bench with clocklike regularity until it comes out the other end, ninety feet away, complete, tested, and boxed and labeled. A one-tube set makes the journey from a handful of binding posts to a sealed carton with shipping label attached in forty-five minutes, a two-tube outfit in an hour, and a three-tube receiver in an hour and fifteen minutes. Each worker at that table has just enough to do to occupy her for three minutes. Radio outfits are assembled at the Crosley plant in Cincinnati on the same principle by which low-priced automobiles are made at Detroit. A few hours' instruction in the factory school qualifies a new employe to do the one simple task allotted to her. She may never learn the principles of radio or how to build a complete set, but doing the one easy stint over and over 200 times a day, the newest worker quickly acquires a skill that shames a veteran amateur builder.

Everything comes to the assembly tables ready for installation. Each piece of bus wire—and there are twenty-six pieces in the three-tube set—is cut to length and bent in the wire shop, which uses 1,200 pounds, or 60,700 feet, of tinned, hard-drawn copper wire daily. The twenty-six

An Automatic Oscillator, with a Range Up to 600 Meters, Sends Out Signals All Day Long by Which the Final Testers Check Each Detail to Insure the Completed Set Is Perfect

pieces in the big set, ranging in length from one to twenty-four and seven-eighths inches, require seventy-five bends.

The company operates three plants and a printing shop. One factory turns out nothing but cabinets, using five carloads of mahogany and walnut and five carloads of poplar each week. Another plant makes electrical parts — transformers, multistats, condensers, coils, jacks, etc.—while the third factory is devoted entirely to assembling and shipping the completed receivers. Like the popular-priced automobile, these sets eliminate all the unnecessary frills, to stick to the original idea of marketing a medium-priced outfit. Engraving of the panel, for example, is expensive because of the labor used, so it is reduced to the minimum. Even then it takes three minutes to engrave the panel for a three-tube set and longer for smaller outfits, where there is no subpanel and all binding posts must be designated on the front of the set. The binding-post markings on the subpanel of the larger outfits are stamped by a punch press.

By dividing the work into three-minute units, hundreds of separate operations are performed on the three-tube receiver in the seventy-five minutes it takes to assemble it. There are sixty-nine connections to be soldered, and a total of 258 screws, binding posts and parts to be placed, including everything from the smallest washer to the cabinet, shipping carton, instruction sheets and shipping label.

When one of the larger sets starts down the assembly table the first worker deftly places the row of binding posts on the long, narrow subpanel, and at the end of her three minutes passes it on to the second girl, who mounts the tube sockets in another three minutes. Then the third worker gets the strip and assembles the plugs on the rheostats, mounts the cam which operates the book-type rheostat, inspects and whitens the engraving on a front panel and mounts the subpanel, with its posts and sockets.

The fourth step adds the jack and filament switch and mounts the assembled rheostats, ready for the next girl to add the coil bracket, phone condenser and tickler coil. So on down the line, the rapidly growing receiver moves. Here a girl puts on the first transformer and passes the work along to the neighbor on her right, who adds the second one; another girl adds the plate condensers; and then a whole row of workers—eight of them—begin adding wires and soldering the connections, each having just enough to do to occupy her for three minutes. Add an antenna coil, the tickler shaft, induction coil and black spaghetti insulation on some of the bus wires, the knobs for the rheostat and tickler shaft, and the growing set is ready for its first inspection.

The inspector checks the details of the assembly to make sure that everything is in place and then adds the barrel of the rheostat and checks the position of the indicating arrow on the knob, while her neighbor installs the knobs and dials of the plate condensers. Another mechanical test, and then the twenty-fifth girl slips the row of binding posts under a corresponding row of brass fingers, which automatically establish all the connections, and begins the final electrical test. In an adjoining room an automatic oscillator sends out signals all day long, ranging from 200 to 600 meters, to be picked up on the inspector's loop aerial, while she manipulates the knobs and switches to make sure that every circuit is correct and every part functioning.

An Automatic Machine Winds the Basket Coils (Above), and a Pantograph Engraver (Below at Left) Marks Panels from a Master Plate

The Shipping End of One of the Long Assembly Tables, Each Side of Which Turns Out a Set Every Three Minutes, and (Below) a Soldering Job on One Subpanel Connection

Past the test, and another nine minutes suffices to mount the set in its cabinet, add the serial numbers, instruction sheets and booklets, box the receiver in a corrugated-cardboard carton, seal the joints with paper tape and paste on the label. Each of the tables averages around 200 sets a day. Most of the parts are made by the company, the principal exceptions being the molded sockets, dials and knobs. The parts factory and assembly unit use 15,000,000 screws and 25,000,000 nuts each week in building 5,000 sets a day, as well as 360,000 binding posts. The quantities of the small parts are so great that automatic computing scales are used to weigh them out of the bins, requisitions for hundreds of thousands being translated into pounds. Seven and one-half tons of formica panels—a total of 2,500,000 square inches—must be cut, drilled, engraved and polished each week for the 30,000 panels and 15,000 subpanels. The parts plant is expected to turn out weekly 60,000 book-type condensers; 75,000 multistats; 75,000 completed sockets; 75,000 basket-weave coils, and some 36,000 audio-frequency transformers. The latter use up 360,000 silicon-steel laminations, each about three inches long, and keep five punch presses busy turning them out.

A recently developed motor-driven automatic machine is used to wind the lattice turns of the basket weave on the inductance and tickler-coil "spiders," stopping automatically after specified numbers of revolutions for the operator to draw out the various tap leads. The new machine replaces hand-turned coil winders with which the operator was required to keep a mental count of the turns made.

DIRECTS PRODUCTION

Lewis M. Crosley, vice-president of The Crosley Radio Corporation, supervises the production of all radio apparatus made by the concern headed by his brother, Powel Crosley, Jr. Thousands of pieces of apparatus are made every week in the Cincinnati plants and are sent to all parts of the world.

Crosley Radio Weekly (Mar. 1, 1926) p. 3

The Wire Benders Turn Out Thousands of Pieces Each Day, All Cut and Bent Ready to Be Soldered in Place; 600 Pounds of Wires a Day Is Not an Unusual Output

A short "factory tour" is in *Radio News*, May 1926, pp. 1536–1537.

A CORNER IN THE TESTING LABORATORY

Radio Broadcast (Nov. 1924) p. 64

The "English roundhouse" an unique booth where 3,000 receiving sets are balanced and tested each day at the plant of the Crosley Radio Corporation, Cincinnati, O.

Radio Retailing (June 1928), p. 121

Test equipment at the Crosley plant is described by K.W. Jarvis in Proc. I.R.E., vol. 17 no. 4, Apr. 1929, pp. 664–710.

Crosley Buys Large Factory

IN ORDER to meet the demand for Crosley-made radio receiving sets, Powel Crosley, Jr., president of the Crosley Manufacturing Company, has has purchased the factory building now occupied by the Thomas J. Corcoran Lamp Company, on Colerain avenue at Sassafras street, in Camp Washington, Cincinnati.

This real estate transaction, involving more than $150,000, meets the question of whether the radio industry is an established business or a passing fad, for preparations are being made to manufacture nearly 5,000 radio receiving sets every day in this new plant, which will be ready for occupancy by early spring.

The large building will house, in addition to the general offices, manufacturing and assembling departments of the Crosley radio products, the radio broadcasting station WLW, which will have all of the latest improvements of this particular field in the radio world, making the station one of the finest in the world.

There is a floor space of over 100,000 square feet in this new four-story building, as compared with 30,000 in the present Crosley factory, at Alfred and Colerain, and this large space will be fully utilized with the manufacture of radio receiving apparatus. It is the intention of Mr. Crosley to manufacture radio parts in the present building and to use the new one for the making of the complete outfits. There is a B. & O. and Southwestern R.R. Company siding which goes to the plant, facilitating the shipping of the raw and finished products.

Large New Factory

The new factory is four stories high, with plenty of daylight on all floors, supplied by hundreds of windows, which will make the place ideal for ventilation and working conditions. New machinery will be installed which will aid the rapid but careful production of radio sets. It is this study of production methods and the great number of receiving sets manufactured that has gained for Mr. Crosley the title of "The Henry Ford of Radio."

The history of Powel Crosley, Jr., is most romantic. About eight years ago he conceived the idea that a mail order business would be profitable and he organized the American Automobile Accessories Company, a corporation of which the Crosley Manufacturing Company is now a branch. Within a few years his automobile accessories business amounted to more than $1,000,000 annually. However, during the winter months the automobile mail order business was somewhat dull, and in order to stimulate his trade, Mr. Crosley entered the phonograph field, at first simply buying and selling his instruments. Later, he purchased a woodworking plant and made his own phonographs. Then as his mail order business grew he found it necessary to purchase a large printing plant to handle the tremendous amount of printed matter he mailed to the trade. Thus, Mr. Crosley was operating an automobile accessory, phonograph and printing business to great financial advantage. A short time later he started the manufacture of a toy, or rather a utility, for children from six months to two years old. This is known as the Go-bi-bi, and hundreds of thousands are in use now throughout the United States.

But three years ago, when the demand for radio receiving sets became noticeable, Mr. Crosley decided to enter this new field, carrying on at the same time the other huge tasks he had undertaken.

Crystal Set Was First

The radio business was operated, as everyone knows, under the name of the Crosley Manufacturing Company. A crystal set was first made; then came vacuum tube outfits. Extensive advertising soon resulted in the organization taking a leading place in the radio field, but until May, 1922, the business was conducted in the plant of the American Automobile Accessories Company.

One corner of the factory had been set aside for the manufacture of radio equipment. In May, 1922, however, the business had grown so extensively that larger quarters were required, and the plant at the corner of Colerain and Alfred streets was acquired.

The radio broadcasting station the company had operated at the automobile plant was replaced by a larger and more powerful station, WLW, when the company moved into the new quarters, and within a short time even this new broadcasting station was replaced by a still more powerful Western Electric plant. On January 6, 1923, Mr. Crosley bought the Precision Equipment Company, Walnut Hills at Peebles Corner.

Since the Crosley Manufacturing Company has been housed in its present place and the demand for radio has increased so rapidly it became evident that the present quarters were not large enough to take care of the business which has grown from a hundred sets a week to nearly a thousand a day. In the new building the organization will be able to increase the output to nearly five thousand a day.

Radio Topics (Mar. 1924)

Combined Craftsmanship of 42 Men and Women Required In Assembling of One Crosley Five Tube Model 5—50; Perfect Work Certain After Set Passes Four Testers

Individual Tasks Are Performed By Well Trained Girls

Rigid Inspection Prevents Defective Set Leaving Factory, Despite Fact That Thousands are Shipped Daily.

Combined craftsmanship of thirty-six girls and six men is required in the assembling of one of the five tube 5-50 radio receiving sets now being produced in such large numbers by The Crosley Radio Corporation. Each of these 42 girls and men has a special task to perform —and must do his or her work so perfectly as to pass the rigid inspection of four carefully trained testers. In a previous issue of the Radio Broadcaster we stated the work was done by 33 girls, and in order to correct the error we decided to interview John R. Butcher, supervisor of the assembly department, to obtain first-hand information as to just how the work is accomplished.

Mr. Butcher, an expert radiotrician, has been in the radio business for five years and has supervised the assembling of Crosley radios during the past three years. In addition to performing this task, he has had charge of inspection and testing, which experience has enabled him to master the task of building Crosley radios according to the standards of the organization.

In order to make his explanation of the work more understandable, Mr. Butcher suggested that we start at the very beginning of an assembly table, and watch carefully the intricate tasks being performed by each worker. He stated there are 53 separate pieces of wire in each set, not counting, of course, the wires used in the various parts. Each piece of wire is numbered, and will be referred to by number as the work is explained.

"Here at the beginning," said Mr. Butcher, "is girl No. 1. She rivets the sockets to the all-metal shielded chassis used in these sets, and then passes it on to Girl No. 2. The work of the other girls will be described in order in which it is done.

"Girl No. 2—Checks serial number. Mounts nine binding posts, jumper and "C" wires.

"Girl No. 3—Mounts three binding posts (Vol. A to G inc.). Bends lugs. Mounts binding post strip with phone condenser. Mounts two balancing condenser studs.

"Girl No. 4—Mounts two auxiliary condenser cams. Mounts two auxiliary condenser adjusting screws.

"Girl No. 5—Puts spaghetti on wire No. 30 and 33. Puts spaghetti on wire No. 27 and 38. Solders wire No. 30 and 33 to last auxiliary condenser. Solders wire No. 27 and 38 to second auxiliary condenser. Puts spaghetti on wire No. 44. Solders wire No. 44 to third socket second clip. Solders No. 10 to fourth socket third clip. Solders wire No. 17 to second socket third clip.

"Girl No. 6—Mounts two coil supporting studs. Mounts studs on two auxiliary condensers. Mounts two auxiliary condensers.

"Girl No. 7—Mounts one by-pass condenser. Mounts detector socket. Mounts grid condenser.

"Girl No. 8—Puts spaghetti on wire No. 6-12 and 45. Solders wire No. 6-12-22 and 45 to balancing condenser. Solders wire No. 39-43 and 47 to second by-pass condenser. Puts spaghetti on wire No. 11. Solders wire No. 11 to fifth socket third clip.

"Girl No. 9—Mounts varind bearing and bracket. Mounts balancing condenser. Mounts rheostat.

"Girl No. 10—Mounts two transformers.

"Girl No. 11—Prepares two transformers for mounting. Puts spaghetti on wire No. 42. Solders wire No. 42 to first socket second clip. Solders wire No. 17 to first transformer S-1. Solders wire No. 37 to first by-pass condenser. Solders wire No. 3 and 37 to second socket fourth clip.

"Girl No. 12—Solders wire No. 10 to second transformer S-1. Solders wire No. 8 to fifth socket first clip. Solders wire No. 40 and second transformer lead S-2 to C-0 to 6 lug. Solders wire No. 31-40 and 50 to first by-pass condenser. Puts spaghetti on wire No. 31.

"Girl No. 13—Solders wire No. 21 to C-0 to 40 lug. Solders wire No. 21 to first transformer S-2. Solders wire No. 19 and 43 to B+90 lug. Solders wire No. 19 to first transformer P-2. Solders wire No. 14 to B+45 lug. Solders wire No. 14 to second transformer P-2. Solders wire No. 23 to fifth socket fourth clip.

"Girl No. 14—Solders wire No. 23 and 24 to fourth socket fourth clip. Solders wire No. 20-24 and 25 to

Crosley Broadcaster (Dec. 15, 1926), p.9

For additional information see:
"Radio Sets Put On Production Basis," J.B. Nealey. *Iron Age, vol.124*, no.27, Dec. 26, 1929, pp.1717–1721, 6 figs. "Description of method of assembling and testing radio units at plant of Crosley Radio Corp., Cincinnati, which has resulted in manufacturing economies; entire process of manufacture is as near straight-line mechanical progression as possible; duplicated testing mechanism developed by company is located at end of line where set is tested for loose connections; converters added to receiving unit."

third socket fourth clip. Solders wire No. 25 and 26 to second socket first clip. Solders wire No. 26 to first socket fourth clip. Solders wire No. 15 to C+A—Lug.

"Girl No. 15 —Solders wire No. 16 to 4th socket 2nd clip. Solders wire No. 16 to 1st transformer P—1. Solders wire No. 18 to Musicone lug. Solders wire No. 18 to 2nd socket 2nd clip. Solders wire No. 5 and 49 to 4th socket 1st clip. Solders wire No. 5 and 48 to 2nd by-pass condenser.

"Girl No. 16—Solders wire No. 2-3-4 to G lug. Solders wire No. 36 to Sel. A lug. Solders wire No. 35 to Vol. A lug. Solders wire No. 7 and 9 to phone condenser. Solders wire No. 9 to 2nd transmormer P—1.

"Girl No. 17—Solders wire No. 51 to 1st socket 1st clip. Solders wire No. 52 to 3rd socket 1st clip. Solders wire No. 1-2, 51 and 52 to 1st transformer lug.

"Girl No. 18—Solders wire No. 53 to Phone condenser. Solders wire No. 1-8-49 and 53 to 2nd transformer lug. Solders wire No. 46 to 5th socket 2nd clip. Solders wire No. 29 to 1st coupler coil lead No. 3.

"Girl No. 19—Ties cable. Mounts three condenser supporting blocks. Cuts and Scrapes 1st coupler coil lead No. 4. Cuts and scrapes 2nd coupler coil lead No. 4. Puts spaghetti on 1st coupler coil lead No. 4. Puts spaghetti on 2nd coil lead No. 4.

"Girl No. 20—Same as No. 19.

"Girl No. 21—Assembles tickler coil, worm nut and guides to supporting arm. Mounts tickler assembly and worm shaft to varind bracket. Cuts and scrapes antenna coil lead No. 5. Puts spaghetti on antenna coil lead No. 5.

"Girl No. 22—Puts spaghetti on wire No. 28. Solders wire No. 28 to antenna coil lead No. 4. Puts spaghetti on wire No. 34. Solders wire No. 34 and antenna coil lead No. 5 to 1st socket 3rd clip. Solders wire No. 4 to antenna coil lead No. 3 Solders wire No. 36 to antenna coil lead No. 2. Solders wire No. 35 to antenna coil lead No. 1. Solders wire No. 46 to tickler coil lead No. 2.

"Girl No. 23—Solders wire No. 15 to Rheostat. Solders wire No. wire No. 20 to Rheostat. Solders wire No. 47 to 1st coupler coil lead No. 3. Solders wire No. 42 to 1st coupler coil lead No. 1. Solders wire No. 44 to 2nd coupler coil lead No. 1. Solders wire No. 22 to 2nd coupler coil lead No. 3. Solders wire No. 39 to 2nd coupler coil lead No. 2.

"Girl No. 24—Assembles stud on 2nd coupler coil. Mounts antenna coil. Mounts 1st balancing coil. Mounts 1st coupler coil. Mounts 2nd balancing coil. Mounts 2nd coupler coil.

"Girl No. 25—Puts spaghetti on wire No. 41. Solders wire No. 41 and 1st coupler coil lead No. 4 to 3rd socket 3rd clip. Solders 2nd coupler coil lead No. 4 to balancing condenser. Puts spaghetti on wire No. 32. Solders wire No. 32 to 1st balancing coil lead No. 2. Solders wire No. 50 to 1st balancing lead No. 1. Solders wire No. 45 to 2nd balancing coil lead No. 2. Solders wire No. 48 to 2nd balancing coil lead No. 1. Solders wire No. 7 to tickler coil lead No. 1.

"Girl No. 26—Mounts set screw on dial. Mounts stop on dial. Mounts balance weight on dial. Assembles soldering lug on dial. Mounts 3rd condenser with dial.

"Girl No. 27—Mounts resistance unit on 2nd condenser. Mounts short coupling. Mounts 2nd condenser.

"Girl No. 28—Mounts resistance unit on 1st condenser. Mounts long coupling. Mounts 1st condenser.

"Girl No. 29—Solders wire No. 12 and 13 to grid condenser. Solders wire No. 13 to 3rd condenser stator. Solders wire No. 41 to 2nd condenser stator. Solders wire No. 32 to 2nd resistance unit. Mounts four felt strips on chassis. Puts in thumb screws.

"Girl No. 30—Solders wire No. 11 to grid condenser. Solders wire No. 34 to 1st condenser stator. Solders wire No. 31 to 1st resistance unit. Solders wire No. 28 and 30 to 1st resistance unit. Solders wire No. 33 to 1st condenser stator. Solders wire No. 29 and 38 to 2nd resistance unit. Solders wire No. 27 to 2nd condenser stator. Solders wire No. 6 to 3rd condenser rotor.

"Girl No. 31—Mounts two screws on dial. Sets condensers. Sets dial.

"The set is now ready to be tested and inspected. The next three girls, referred to as No. 32, 33 and 34 subject the set to the most careful examination and after it passes from their hands it is as perfect as human hands can make it. Girl No. 35 gets it next. She is in an enclosed test booth and it is her task to see that the set operates properly under actual working conditions. She tunes in as many broadcasting stations as she can, and when none is in operation she uses an especially arranged buzz system with which she can put the receiver through the most rigid test.

"No. 36 is a man, whose task it is to place the set in its cabinet. He then passes it on to No. 37 who screws on the name plate and attaches the bolts that hold the chassis in the cabinet. No. 38 puts on the auxiliary condenser windows and the dial windows and No. 39 puts on the dial covers and knobs. No. 40 is a girl, who inspects the work done by these men and if it is satisfactory she passes it on to No. 41, who sees that the dial moves freely in the window and tacks on the serial card. It is then ready for the final operation—that being the work of No. 42, who puts the set in its carton, puts on the labels and places the completed set on a chute bound for the shipping department."

Assembled By Forty-two Trained Workers

The set illustrated above is the Crosley five tube 5-50, which is assembled and tested by 42 carefully trained men and women. Every set is tested and inspected by four people before it is passed on to the packers.

STOP—LISTEN

Judge—"What's the matter with your headlights?"

Driver—"I used wood alcohol in the radiator and the darn thing has gone blind."

To get the most from a single tube it must be connected in a regenerative circuit.

High Efficiency Production Methods Place

A Peak Behind The Scenes Reveal Systems, Finest aud Most Equipment and Skilled

Where Crosley Parts Are Cadmium Plated

A corner of the Plating Department showing how the metal parts are carried along through the plating baths on metal hangers. As the end of each vat is reached, the hangers automatically lift out, dipping into the succeeding vat and moving along continuously. Each part must travel twice the length of the room and pass through several baths before the process is complete. Cadmium plating is the most modern and efficient method known for protecting metal parts from corrosion.

Assembling Tuning Condensers

Here the die-cast condenser rotors are being mounted in their "bathtub" supports. As each man finishes assembling a condenser "gang" he puts it on the moving belt, which carries it quickly to the testers, farther down the line. The condenser assembly is then subjected to a rigid inspection and carefully adjusted before being passed on to the Set Assembly Department.

This issue of The Crosley Broadcaster is dedicated to those who would like to take a trip to Cincinnati and see the Crosley factory in operation, but who cannot. We are endeavoring in these pages to give you in the very best way we know how a slight conception of what is going on behind the scenes.

Let us say right in the beginning that we can only hope to do this in a very small way, because the power of painting word pictures is limited and the space to reproduce pictures of everything that is happening out in the factory is also limited.

We hope, however, that the few pictures which we are able to reproduce in this issue will give some conception of what is taking place in this huge modernly equipped and modernly operated factory to provide our dealer and distributor organization with the merchandise needed to take care of the tremendously increasing demand for Crosley merdise.

We trust that some day each one of you can take the time to visit us and see with your own eyes what it is so impossible for us to describe

Every Minute A Finished Chassis

At the far end of this long table the individual units of the embryo set starts on their journey. Each of the forty workers along the line has a particular operation to perform. Gradually the set takes form, until, as it reaches this end of the table, it is completely assembled on the chassis, ready to be tested. The inspectors and test operators in the foreground carefully give it the "once over", sending it back up the line unless it is, without question, up to standard.

An O. K.'d Gembox Passes On

The Gembox chasis being hooked onto the conveyor has just passed through a long line of test booths where it has been given almost every conceivable kind of reception and operation test. A special department, housed in shielded metal booths and centrally located, handles the reception tests. These tests are in addition to those given to finished chassis at the end of the assembly tables.

Crosley Merchandise First In Radio Market

Tremendous Activity---Latest Conveyor Modern Machinery, Critical Test Labor All Play Their Part

to you. It is a thrilling sight to see an organization of twenty-five hundred persons all busily engaged in the manufacture of radio apparatus. It is a more worderful sight when it is realized that the introduction of a most modern conveyor system and most modern machinery makes it possible for everyone of this vast army to perform the work of 4 or 5 persons under ordinary methods.

We trust that these pictures in this issue of the Broadcaster will only be a starter and that from time to time we can take you behind the scenes and show you more and more of what is taking place. We feel sure that when you more fully realize the tremendous and practically unlimited resources of the organization which is backing you up, that you will with even greater confidence go after the business which is rightfully yours in your community.

Crosley is setting the pace in the merchandise being manufactured for the 1928-29 radio season. It is only the natural thing, therefore, that every Crosley dealer and distributor should set the pace in his territory.

Into The Cartons They Go

At each tick of the clock another Gembox is carefully wrapped and packed into its carton, ready for shipment. The belts carry an endless line of sets on their way to the Shipping Department on the floor next below. One has but to watch for a few moments to count hundreds of them. This, of course, is but one of several such packing lines.

A Glance At The Conveyor System

This modern conveyor system speeds up production in the Crosley factories. This continuous conveyor runs from department to department, all over the Crosley plant. The unfinished parts start on the top floor of the plant, and are gradually assembled as they go down from floor to floor, until they finally reach the shipping department in the basement, complete, tested, and packed in their cartons, ready to send out. The chassis seen on the conveyor are Gemboxes, on the way to the Testing Department.

An Imposing Array of Showboxes

These Showboxes are being mounted in cases, prior to the final inspection and test. It is estimated that these sets and the parts that go into them each pass through more than a thousand inspections and tests before being finally approved. These multiple safeguards are a guarantee to every Crosley dealer and distributor that his customers will be more than satisfied.

Two More Carloads

Two carloads of Crosley sets ready to start on their journey. They represent happiness for hundreds of homes, and worth-while profits for Crosley distributors and dealers. Every hour the endless chain continues, more parts starting in at the beginning and more carloads of sets leaving this platform. Efficiency and reliability are the keynotes of the process—two fundamentals which explain the quality and value represented in Crosley merchandise.

Crosley Broadcaster (Apr. 15, 1929)

ARMATURE ASSEMBLY. The young women in the picture are assembling the most important part of the Dynacone, the armature. The armature oscillates between the poles of two powerful electro-magnets and the length of the oscillation determines the depth of tone.

BOBBIN WINDING OPERATION. On each arm of the armature is a small bobbin. Just as the armature is the most important part of the Dynacone, the bobbin is the most important unit of the armature. The bobbins carry the wiggles of signal current which are converted into sound.

CONE ASSEMBLY. The cone is stamped out of special paper impregnated with waterproofing material. It is shaped and sewed in one operation. This moisture proof feature combined with the fact that all metal parts of the Dynacone are cadmium plated insures ample protection against the moisture found in damp climates. These features make the Dynacone the ideal power speaker for use in damp or humid climates—in the tropics, near ocean, lake, or stream, and on the ocean itself.

TESTING OPERATION. The young lady with the keen ear has connected the assembled Dynacone to the A-C supply. If she gets the proper hum out of the cone, all is well. If not, something is rotten somewhere.

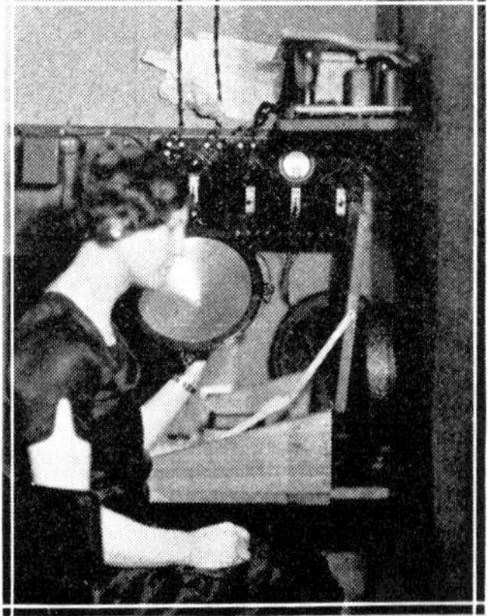

Addition under construction, Aug. 1929 *Crosley Broadcaster* (Sept. 1, 1929)

Crosley Broadcaster (Apr. 15, 1929)

Architect's drawing of 8-story addition, completed Sept. 1929, 6-story addition of Sept. 1926, and original factory bought about Feb. 1924. At far left: cabinet assembly and shipping plant, Sept. 1929.

Crosley Broadcaster (Sept. 1, 1929)

Cabinet assembly and shipping plant.

Late 1929 photo, in new addition

CUTTING & WASHINGTON

Cutting & Washington Radio Corporation

Fulton Cutting, born in New York City December 27, 1886, came from an old and socially-prominent family. He received his AB from Harvard in 1910, followed by an MA, MEE, and PhD in short order. In April 1916 he had already contributed a paper to the *IRE Proceedings*, a highly-technical and thorough treatment of power-transformer design for spark transmitters. He served as IRE president in 1922.

Bowden Washington was born on July 7, 1892 at Bar Harbor, Maine, and attended the Browning School, New York and spent two years studying electrical engineering at Columbia. He built his first transmitter and receiver in 1903, was listed in the 1908 *Modern Electrics* Blue Book as "BW" and twice had photographs of his station published in *Modern Electrics* in 1909. He helped set up two Mexican stations in 1913, was assistant engineer at Clapp-Eastham in 1913–14, then went to Harvard in 1915 where he assisted G.W. Pierce in equipping Cruft Laboratory, and met Fulton Cutting.

During 1915 Washington and Cutting worked at Cruft to develop a radio transmitter using the quenched spark gap invented in 1911 by E. Leon Chaffee, then an instructor in physics at Harvard. About this time Cutting and Washington formed a partnership, and, losing no opportunity for free publicity, Washington wrote a descriptive article for the *IRE Proceedings* in February (published in August).

In April 1917 the firm of Cutting & Washington, Inc. of Cambridge, Mass. was incorporated in New York with a capital of $200,000; the three incorporators all being Cuttings. C&W developed a production model (4A) of its "impact excitation" transmitter and sold the government 1000 of them at $750 apiece. The company also made 60 motor boat sets using CN114 receivers in 1917, and a number of lightweight aircraft transmitters, all using the Chaffee gap. Some of this gear was described by Washington in another IRE article in December 1918.

In the Navy during the war, Washington claimed in a 1924 autobiographical sketch to have had charge of the direction-finders on 350 destroyers and to have been on the staff of the Commander-in-Chief. His path crossed that of Expert Radio Aide A.F. Van Dyck in February 1919 on the Pennsylvania in Cuba, who made these entries in his diary:

"Ensign Washington has done a good job on the compass, eliminated the collector rings as I have wanted to do many times. Not a standard installation at all, but it works. He is pretty wise but works hard to convince you he is wiser than he is. Talked with him for hours the other night. I noticed that most *of the Bostonese accent wore off during the evening—not all. . . . Ens. W. is a great BSer & tries to give the impression that he knows all abt* any *subject. When in swimming he keeps asking if you can make a 'so-and-so' dive (which most of us haven't heard of) etc. to give the impression he's great. So far I have found that he has had a* great deal *of experience with guns, cameras, radio, swimming, navigation, boat handling, laboratory installation, factory management, women, in fact all we have talked about. Some of the officers are wise to him, as I overheard today."* (It's worth noting that Van Dyck was a Yale man.)

DR. FULTON CUTTING
Chairman of the Board

Bowden Washington
Vice-President and Chief Engineer Cutting & Washington Radio Corp.

In June of 1919 C&W moved to 6 & 8 W.48th St., New York City, re-incorporating as the Cutting & Washington Radio Corp., capital $36,750. They advertised commercial radio gear regularly in the *IRE Proceedings*, and maintained their own coastal radio station WSA in Easthampton for ship traffic, having combined with Independent Radio Telegraph Co. Apparently C&W was a wholly-owned subsidiary of Independent for the next five years.

There are several mysteries connected with C&W's Armstrong regenerative license, which was received on July 7, 1920. In addition to the usual provisions to sell to radio amateurs, experimenters, and scientific schools, this license had a third clause: "to purchasers in the U.S. for use in their own non-commercial land stations; i.e., stations used for the private purposes of their owners, and which do not receive or transmit for others commercial messages for money or other valuable consideration." This was too early to refer to broadcast reception, so it must have meant point-to-point communication. But C&W's whole business had been, and continued to be, commercial, specifically shipboard communication. So what were they doing with "non-commercial land stations"? And why did they not take out a license for their commercial business (they were admittedly equipping some of their stations with Audion detectors by that time, though not necessarily regenerative ones)? Why then did they take out an Armstrong license at all? Typically, such licenses were saddled on companies already making and advertising regenerative sets, not by their request, but by demand of Armstrong's patent attorneys, who saw a chance of getting some revenue to pay their considerable fees without jeopardizing a future sale (the amateur market being thought negligible). Cutting & Washington, however, were not advertising amateur equipment, only commercial. Perhaps they contemplated entering the amateur field, as most of their competitors were doing, the commercial market experiencing a severe slump after the war.

Whether by design or accident (probably the latter) they did not enter the amateur market for nearly two years. Their first broadcast set, the type 11, appeared in May 1922, just in time for the "radio bust," but was surely intended to have been ready much sooner. Its production was subcontracted to a New York City manufacturer (C&W never had a factory of their own) who may have been De Forest. If so, it would explain the delay, as the De Forest people were unable to get even their own models into production, let alone subcontract for someone else. Cutting & Washington were so disgusted that they pulled up stakes and moved to Minneapolis to hook up with the Minneapolis Heat Regulator Co. (later Honeywell) which was willing to subcontract production. MHR may have made the model 12 which appeared in June. C&W also ran broadcast station WLAG, beginning on September 4, 1922.

Unfortunately their proposed arrangement didn't work out, but they did find a second manufacturer willing and able to make radios, but with no self-interest in the market: Automatic Electric Co. of Chicago, makers of telephone equipment. All subsequent C&W radios were made by Automatic.

Cutting & Washington's troubles were only beginning. As soon as the "negligible" amateur (broadcast) business began to wag the dog, Westinghouse, now owner of the Armstrong patent, filed a series of lawsuits to pick off its licensees. C&W got theirs in the Second Circuit Court of Appeals on December 10, 1923, when two of the three judges ruled that while their subcontracting arrangement was legal, sales through distributors and jobbers were not; in other words, they would have to sell directly to the final customer. This, of course, was not the usual industry practice, and while they could get around it as De Forest did, by consigning their sets to dealers (title remaining with the manufacturer until the final sale was made) it was an awkward arrangement. Even the new model Teledyne couldn't pull them through (particularly at the $190 list price) and C&W went into receivership by August 1924.

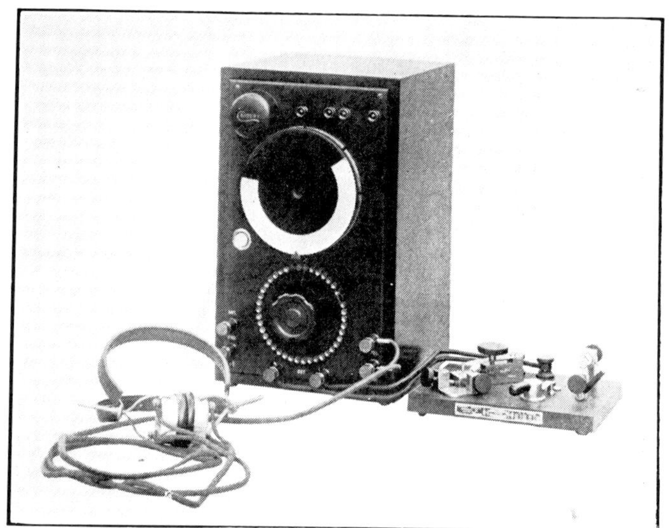

IRE Proc. (1918), p. 311

8A Mar. 1918

WLAG closed down, to be reopened with the backing of the Washburn-Crosby Co., makers of Gold Medal flour, who asked only that it identify as the "Gold Medal Station" and change its call letters to WCCO.

Independent Wireless Telegraph Co. sold C&W to the "Radio Engineering Co." of Minneapolis, which probably was a legal entity only, an intermediary, for Mssrs. Cutting and Washington reappeared in September in New York City with their Colonial Radio Corp., capital $4 million. They opened a factory in Long Island City, New York employing 200 people and by November were showing their models 16 and 17 at the Chicago Radio Show. In February 1925 after having completed 500 model 16s, Cutting built another 600 into six-tube sets using Dubilier fixed-RF transformers; however the production models didn't work at all like the five prototypes and the factory shut down while the trouble was corrected. At this time, Colonial merged with the Multiple Electric Products Co. of Newark, New Jersey makers of Atlas speakers, who unloaded the troublesome radios on the public, in return for future distributorship of Colonial models. But by December the two companies had parted, on poor terms.

Theoretically, Colonial continued to introduce new models each year, but very few must have been made until 1929. In that year a minor flurry of articles and ads hit the radio magazines. Colonial absorbed Valley Appliances of Rochester, New York in February 1930, and King of Buffalo in October, apparently closing all but the Buffalo plant in mid-1931. When it bought the King plant from Sears, Roebuck & Co. it probably contracted with Sears to become the major manufacturer of Silvertone radios, and continued to make radios for Sears even after discontinuing sales under the Colonial name around 1934 or 1935. Colonial was sold to Sylvania in 1944, and merged into that company in 1950 as its Radio and Television Division.

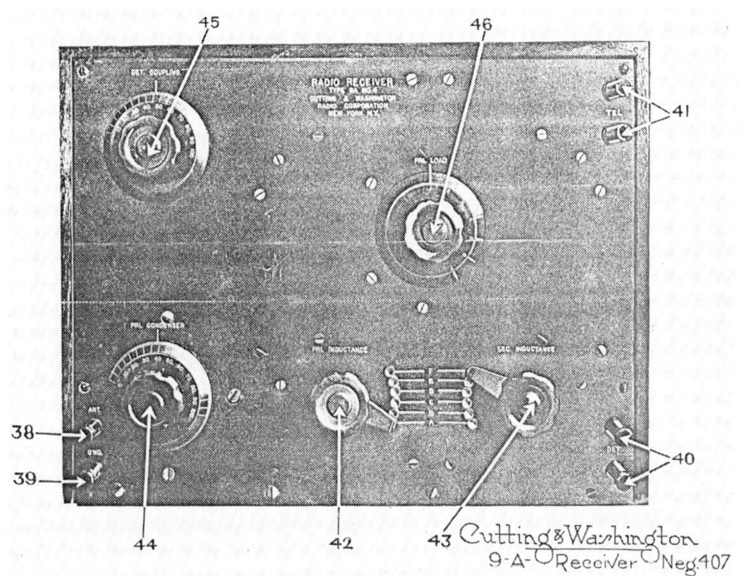

9A June 1919

Mel Rosenthal

11 May 1922 $125

Type 14, for use with a loop, was announced in Aug. 1922 but probably not made. It was described as using four tubes (RF, Det., 2 AF) and having three controls. A four-foot rotating loop and a fixed coil eight feet square were supplied.

12 June 1922 $65

John Drake

John Caperton

12A Oct. 1923 $97.50

Popular Radio (Feb. 1924), p. 1

Today's Cinderella needs no fairy godmother. She calls up Prince Charming and together they waltz at her Radio Dance, while out of the air, over her C & W Receiver, comes the rhythmic swing of Schubert melodies, Strauss waltzes, fascinating modern two-steps, played by the nations best orchestras. . . . .

Get the *Best* Program with a C & W Receiver

Tested with other broadcast receivers — on the same antenna — listening to the same stations, Model 11-B, the Cutting & Washington Receiver illustrated above, proved to give greater volume from the desired station, with less interference from all other stations.

Highly selective, a 3-tube, double-circuit regenerative receiver, with remarkably long range and clear reception. Uses dry cell A-battery; 3-UV 199 tubes; special sharp tuning C & W Circuit, shock absorbing tube mounts, automatic rheostat switches, shielded panel. Leader of the C & W Line, the set that will get your station if the station is to be had Price complete, ready to operate, $160.

Compare — then choose. Ask for a demonstration by the nearest C & W Dealer

Dealers—Distributors: Write at once for details of the Cutting & Washington Selling Plan — a real opportunity in Radio.

Cutting & Washington Radio Corporation
Minneapolis :: :: :: Minnesota

Cutting and Washington

A Genuine C & W Receiver Only $50 Complete

(Including Tube and Batteries)

Really two sets in one—this ingenious new receiver designed by Dr. Fulton Cutting and Mr. Bowden Washington.

In your home, on a permanent antennae, it's a sharp tuning double circuit set. In the home of a friend—in your car, it's a single circuit receiver, ready to operate on any sort of temporary antennae.

A special C & W circuit, the result of the designers' long experience, makes this receiver as selective as the most expensive sets.

Let it introduce you to Radio at its best. Now on display at C & W Dealers.

No Other Set at This Price

offers you these special features. C & W Model 15 (Armstrong Patent), single and double circuit; 1-UV-199 tube; shock-absorbing mount, preventing vibration howls; screened panel; automatic rheostat switch; all wires back connected; dials safety mounted under silver finished rings; standard C & W high grade construction throughout; complete, with all batteries and tube inside cabinet, $50.

Cutting and Washington

Compare—then Choose. Ask for a Demonstration by nearest C & W Dealer

THE SAYRE-LEVEL RADIO COMPANY

Philadelphia Distributors

Dealers write for discounts and territorial rights.

PHILADELPHIA
Henricks Hardware Co., 17th and Venango Sts.
Geo. W. Snell Company, 6522 Woodland Ave.
E. K. Tryon Company, 912 Chestnut St.
Eidlow's Electric Shop, 5518 Chester Ave.
AVONDALE, PA. John Evans, Jr.
BELLEFONTE, PA. R. C. Witmer
BERWICK, PA. J. N. Harry
CONSHOHOCKEN, PA. Dennison's Electric Shop, 68 Fayette St.
LANSDALE, PA. Wm. R. Weikel
PALMYRA, PA. Palmyra Music Store
POTTSTOWN, PA. Nuebling's Sporting Goods Co.
WILKES-BARRE, PA. Ludwig & Co.
LEWES, DEL. A. W. Joseph & Co.
CAMDEN, N. J. Munger & Long
PLEASANTVILLE, N. J. Mark Sanders
SWEDESBORO, N. J. W. H. Hannold & Sons
TOMS RIVER, N. J. Traco Music Store
TYRONE, PA. Electric Battery Service Station
WILDWOOD, N. J. R. W. Ryan

Phila. Inquirer (Nov. 4, 1923)

Radio in 1914

1914 model, Cutting & Washington Receiver, one of the first commercial sets in America, and—

A present Cutting & Washington model, the standard for simplicity and sound design.

and today

THROUGHOUT their long experience of Radio, Dr. Fulton Cutting and Mr. Bowden Washington have kept to one idea—make Radio simple—make it dependable—make it useful.

How well they have succeeded is proved by their famous record as designers of U. S. Naval and Marine Equipment—Radio apparatus that *must* not fail—that must be serviceable and efficient.

Today you can have a genuine Cutting & Washington Set for your home. Dr. Cutting and Mr. Washington have at last given to Home Radio the benefit of their long experience and expert engineering.

Ask your dealer to point out the superiorities of Cutting & Washington equipment—or write direct for illustrated catalog.

Cutting & Washington Radio Corp.
Minneapolis, Minnesota

Dealers

C & W Sets are so simple and efficient that you can easily *prove* they are the sets for your customers to own. They are:

1. Designed under Armstrong Patents.
2. By America's Pioneer Radio Manufacturers.
3. Built by world's largest makers of automatic electric equipment.

Write for "The Future of Radio Retailing", a book that outlines the liberal franchise offered C & W Dealers, and the C & W plan for a profitable business in Radio.

Cutting and Washington

America's Oldest Manufacturers of Commercial Radio

Popular Radio (Aug. 1923)

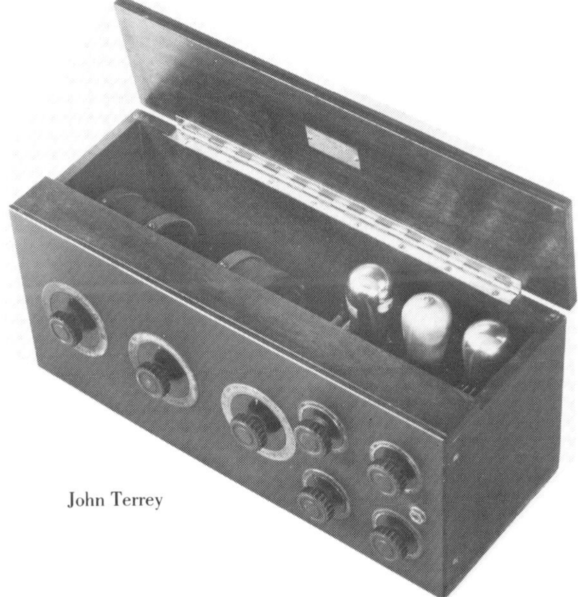

John Terrey

11A June 1923 $135 (uses 01As, shown on opposite page)

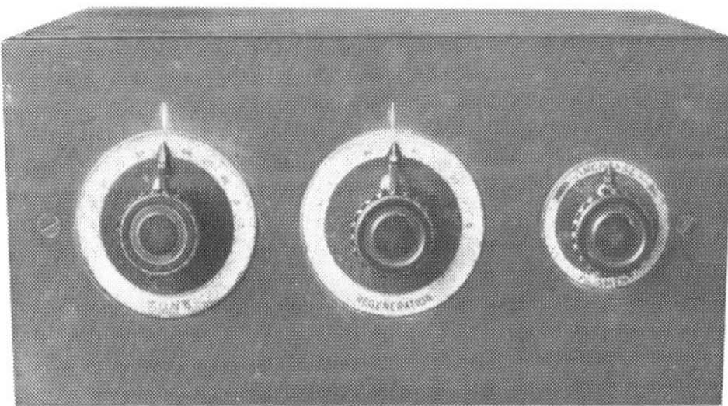

15 Nov. 1923 $50

Richard Hostler

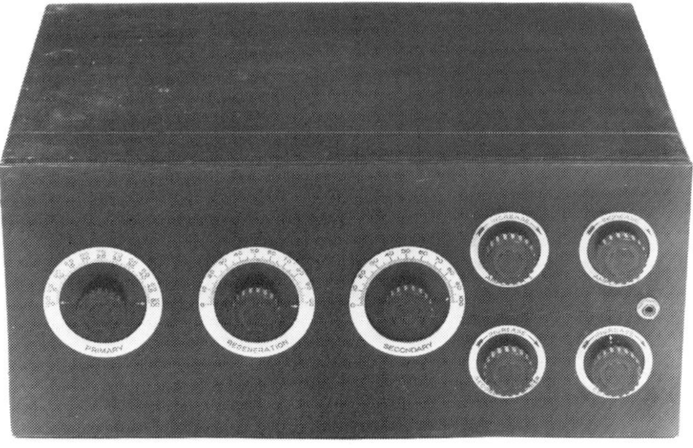

John Drake

11B Oct. 1923 $325 (console model, uses 99s)
11B Jan. 1924 $160 complete

Teledyne Mar. 1924 $190 complete

In a descriptive article in *Radio Broadcast* for May 1924, pp.47–51, Washington shows how he added a fourth tube to his old circuit, ahead of the regenerative detector, preventing radiation and interference to other sets.

Other engineers that may have had a hand in the design: Henry C. Forbes, assistant chief engineer in 1923–1924, who joined Zenith and designed the Super Deluxe models, and H.S. Williams, who wrote a construction article in the *New York Sun*, Mar. 15, 1924, pp. 1, 4; Mar. 22, p. 2.

Thousands in Chicago Heard Colonial Radio Last Week

The Show has closed, but One Hundred Chicago Dealers are ready to supply Colonial Radio to all seeking the finest in local and long distance reception.

Thousands heard Colonial during Show Week, at the Congress Hotel and at Dealers' Demonstrations. Note that Colonial Dealers are continuing their Show Week Demonstrations indefinitely. Thousands more saw the Colonial Salon at the Coliseum, saw the handsome cabinets containing Dr. Fulton Cutting's and Mr. Bowden Washington's combined success—the Colonial adaptation of the Weagant Circuit. But these people *did not hear Colonial Radio.*

If you heard Colonial Radio last week, you know more now than any advertisement can tell you—more about *wonderful reality* in reception of music and speech, more about distance-getting and the logging of many stations scattered over the map.

But if you only saw the Salon and did not hear one of the Demonstrations, be sure to do so by visiting your nearest Colonial Dealer during broadcasting hours. As you listen, bear in mind that this thrilling radio is not confined to people who can pay a high price. The Colonial principle has been engineered to meet the needs of all.

Colonial Sets can be purchased as low as $58.50. Other Colonial Sets range upward to $175. Find just the one that fits your ideas. Colonial Sets are operated on wet batteries, or on dry batteries only, or simply attached to the light socket.

When you know the enjoyment that comes from this greater radio, you will not think of being without it. The whole world of music, romance and education is on the air and Colonial will bring this world into your home hour after hour. It will bring in each syllable and each note with amazing fidelity.

If you have not heard Colonial Radio, go to one of these demonstrations—take the family.

DEPENDABLE COLONIAL PERFORMANCE

COLONIAL RADIO CORPORATION · LONG ISLAND CITY, N.Y.

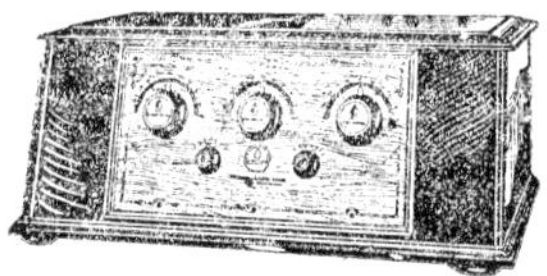

Exclusive Distributor

Music Trade Radio Corporation

328 No. Michigan Ave. Chicago State 2628

MODEL 21-5 WITHOUT ACCESSORIES $87.50

CHICAGO
Baldwin Piano Co., 323 S. Wabash Ave.
Bryn Mawr Music Shop, 2101 E. 71st St.
Hartman Furniture Co., 6158 S. Halsted St.
Joseph Jiran, 1333 W. 18th St.
Lazar & Son, 1378 Milwaukee Ave.
Schroeder Piano Co., 6309 S. Ashland Ave.
Kimberly Radio Corporation, 528 N. Michigan Ave.
Station Tire & Spring Co., 9500 Ewing Ave.
Von Lengerke & Antoine, 33 S. Wabash Ave.

ILLINOIS
Atlanta—J. N. [illegible]
Arlington Heights—[illegible] Music Co.
Augusta—Earl L. Smith
Aurora—Wiswell Music Co.
Amboy—C. [illegible]
Blue Island—F. A. [illegible], 301 Western Ave.
Belvidere—Manley Motor Co.
Canton—Canton Music Co.
Chester—[illegible] Motor Co.
Chadwick—Henry Frank

ILLINOIS [Continued]
Carlyle—[illegible] Bros. Garage
Des Plaines—Des Plaines Mercantile Co.
Dixon—Kennedy Music Co.
De Kalb—Lewis & Palmer Music Co.
Decatur—Linn & Scruggs Dry Goods Co.
Evanston—Homer Piano Co., 1521 Sherman Ave.
Evanston—Central Radio Shop, 1315 Chicago Ave.
Elgin—Crocker Radio Shop
Freeport—Haller & Bauder, 15 N. Van Buren St.
Fairfield—Fleming Furniture Co.
Harvard—Manley Motor Co.
Lawrenceville—Valley Battery & Radio Co.
Joliet—West Music Co.
Marengo—Geo. Rappien
Morris—W. J. Goode
Milford—Home Furniture Co.
Mt. Morris—C. H. Newcomer Electric Co.
Nauvoo—Ochsner Motor Co.
Moline—Carlson Bros., Inc., 1404 5th Ave.
Nashville—Kirchhofer Motor Co.
Princeton—Roy Rapp
Peru—U. S. Clothing Co.

ILLINOIS [Continued]
Paris—Fritch Drug Co.
Peoria—Lacey's, Inc., 424 Main St.
Lake Forest—Lake Forest Music Shop
Quincy—R. E. Roach Music House
Robinson—Geeseman & Lorrance
Woodstock—W. R. Burns
Sandwich—Carroll Plumbing & Heating Co.
Savanna—R. G. Fuller
Newton—Gilmore's Music & Jewelry Co.
Waterloo—Jacob Bersche
Wayne City—C. F. DeZouche & Son
West Chicago—F. A. Goetz

INDIANA
La Porte—Sonneborne Sons Co.
Hammond—Straube Piano & Music Co., 631 Hohman St.
Gary—The Colonial, 710 Broadway
Winchester—A. T. Engle & Son
Crown Point—P. T. Hand
Valparaiso—J. Lowenstine & Son

WISCONSIN
Baraboo—Willey's Music House
Ripon—L. A. Wilson Co.
Sheboygan—A. Zohlen
Beaver Dam—Dan M. Freitag, 413 West St.

Chicago Tribune (Nov. 22, 1925)

21-5 Aug. 1925 $87.50

Not shown: 23-5 phonograph panel, Aug. 1925, $85

24-5 portable, Aug. 1925, $85

N. Y. Herald-Tribune (Dec. 20, 1925)

Through a typographical error the above picture was not identified in the group of illustrations of radio receivers published in the Herald Tribune Radio Magazine for Sunday, December 13. It is the Colonial Cabinet Model 20-6 receiver. It employs six tubes in a three-stage tuned radio frequency circuit and sells for $175. The table unit on which it stands sells for $40 and the loud speaker for $30.

Described in *Radio News*, Mar. 1926, pp. 1273–1274.

News and Notes of the Radio Trade

N. Y. Herald-Tribune (Oct. 17, 1926), p. 11

This is probably the model 25, said to be ready for production in May 1926, and advertised in Oct. for $225. A model 25 was also offered in Aug. 1927 at $250, and the 26 DC model in Sept. at $250.

Announcing the

COLONIAL

Colonial 16

The Colonial 16 is a five-tube, dry battery operated receiver. It employs two stages of tuned *compensated* radio frequency amplification, detector, and two stages of audio amplification. All circuits are shielded and compensated to give uniform response on all waves. No regeneration; no radiation; no howling.

Cabinet by Brewster, the finest builder of custom built automobile bodies in the world. Beautiful in design; executed in the best of taste. Panel in bronze, done by a new etching process. Maximum of selectivity and sensitivity. Can be logged with greatest accuracy. Absolutely no body capacity. All batteries enclosed in cabinet.

Colonial 17

The Colonial 17 is a four-tube, dry battery operated receiver. It employs one stage of tuned *compensated* radio frequency amplification, detector, and two stages of audio amplification. The cabinet and panel is a novel combination of beautifully decorated metal and natural grained, lightly polished wood. It is highly sensitive and its selectivity is well above the average. Can be logged with ease and accuracy. All batteries are enclosed in cabinet.

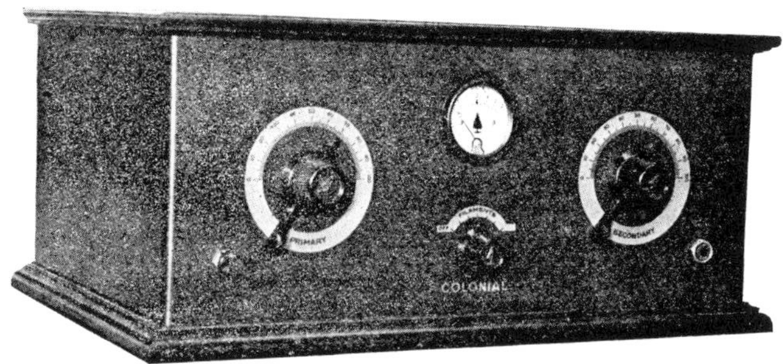

A Revelation in Radio

Experts in radio have described Colonial 16 and 17 as "a revelation in radio." They are the latest contribution to radio science of Dr. Fulton Cutting and Mr. Bowden Washington, and represent the culmination of years of experimenting and testing. We waited until we were absolutely sure that the sets were mechanically and electrically perfect before we placed them before the trade. Now we know they are absolutely right and we guarantee every Colonial receiver that leaves our factory to give perfect satisfaction.

Merchandising Policy

Exclusive franchise and absolute territorial protection are two of the outstanding features of what Colonial gives its jobbers or distributors. We believe the jobber is entitled to this protection from the manufacturer.

Colonial has developed a dealer co-operative sales service that is unique and that is sure to be of signal importance to the retail merchant. Colonial offers direct to the dealer, through the jobber, this sales plan personally inaugurated by Colonial "dealer salesmen." This has at once been recognized as the best, most direct and most efficient manner of securing dealer sales that has yet been devised for radio.

We have a booklet completely describing the Colonial Merchandising Plan. Write for your copy today.

COLONIAL RADIO CORPORATION

East Avenue and Tenth Street
Long Island City, N. Y.

The model 16 was announced in Nov. 1924 ($200) but first advertised in May 1925, as the 16-5 ($150) and 16-6 ($175). The model 17 was announced in Nov. 1924, $85.

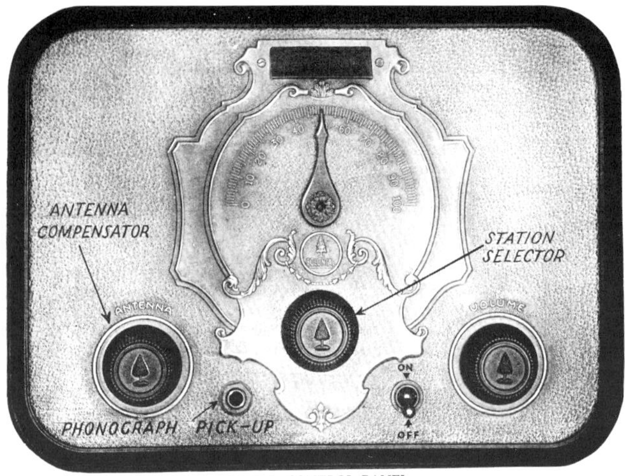

FIG. 6—CONTROL PANEL

31AC Nov. 1928 $268 & $278
31DC Nov. 1928 $288 & $298
A short technical article is in *Citizens Callbook*, vol.11 no.1, Jan. 1930, p.73.

Radio Retailer & Jobber (Mar. 1930)

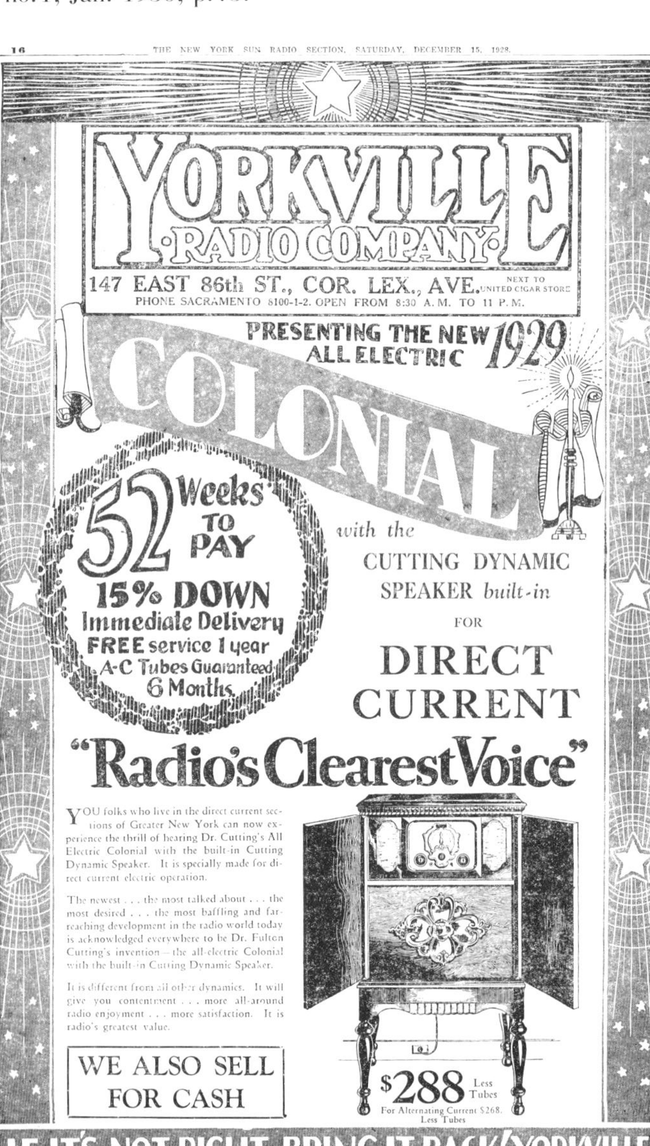
16 THE NEW YORK SUN RADIO SECTION, SATURDAY, DECEMBER 15, 1928.

YORKVILLE
RADIO COMPANY
147 EAST 86th ST., COR. LEX., AVE., NEXT TO UNITED CIGAR STORE
PHONE SACRAMENTO 8100-1-2. OPEN FROM 8:30 A. M. TO 11 P. M.

PRESENTING THE NEW ALL ELECTRIC 1929
COLONIAL
with the CUTTING DYNAMIC SPEAKER built-in FOR DIRECT CURRENT

52 Weeks TO PAY
15% DOWN
Immediate Delivery
FREE service 1 year
A-C Tubes Guaranteed 6 Months

"Radio's Clearest Voice"

YOU folks who live in the direct current sections of Greater New York can now experience the thrill of hearing Dr. Cutting's All Electric Colonial with the built-in Cutting Dynamic Speaker. It is specially made for direct current electric operation.

The newest . . . the most talked about . . . the most desired . . . the most baffling and far-reaching development in the radio world today is acknowledged everywhere to be Dr. Fulton Cutting's invention—the all-electric Colonial with the built-in Cutting Dynamic Speaker.

It is different from all other dynamics. It will give you contentment . . . more all-around radio enjoyment . . . more satisfaction. It is radio's greatest value.

WE ALSO SELL FOR CASH

$288 Less Tubes
For Alternating Current $268. Less Tubes

IF IT'S NOT RIGHT-BRING IT BACK! YORKVILLE

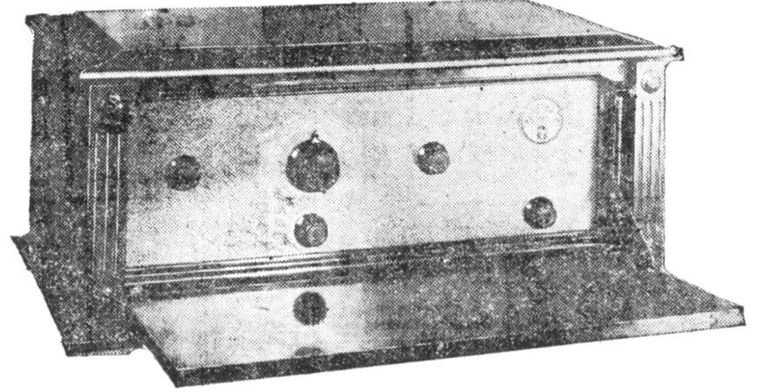

Table-style Colonial power receiver.

N. Y. Sun (Sept. 17, 1927)

Colonial

All-Electric Radio...Models for Direct and Alternating Current

"...it is the most naturally toned radio in the world"

Searching, experimenting, testing...Colonial engineers gradually evolved the ideas and applied the principles which brought about the perfection of this great receiver.

Their aim was to produce an all-electric Radio which embodied every quality feature...and, by advance production methods, place it within easy buying reach.

Now...in the most critical laboratories of all...*the homes of thousands of enthusiastic owners*...The Colonial is justifying the purposeful years of labor which brought it into being.

Beyond this perfection of light-socket radio, beyond its beautiful appearance and beyond its marvelous performance, there is still another source of enduring satisfaction in the possession of it. *Colonial Radio, in each of those homes represents an intimate, friendly association between manufacturer and owner.*

When you buy your Colonial All-Electric, you will know that your radio enjoyment is the concern of a local manufacturer whose reputation and growth are founded upon the dependability of its products ...who recognizes that continuing success depends upon the gratification of your wishes for the utmost in radio value.

Why don't you come in and see...and hear...how wonderfully Colonial meets your every wish? We shall be glad to arrange for a home demonstration and leave the decision to you.

COLONIAL FEATURES: All-metal chassis—completely electrically shielded...Uses the New A. C. Tubes...Single-dial tuning control, with 3-dial accuracy...Bridge volume control..."A" metal (patented alloy) transformers...Fieldless LITZ coils. *...to mention just a few!*

One Dial Receivers Licensed under U. S. Patent 1014002

$147.50

Table and Speaker with Tubes Extra
(Direct current models slightly higher)
Easy Terms if Desired

IF YOU CAN'T CALL, TELEPHONE MURRAY HILL 2448 OR SEND THIS COUPON

I am interested in Colonial Radio. Please have your representative call ....(date) at ....o'clock
Name ....
Address ....

COLONIAL RADIO SALES CO., INC.
3 East 43rd Street 20 Vesey Street
OPEN EVENINGS UNTIL 9:00

COLONIAL RADIO CORPORATION, Long Island City, *Manufacturers of Radio's Greatest Values*

N. Y. Herald-Tribune (Feb. 26, 1928)

28 (?) Sept. 1927 $147.50

32 June 1929 $235 & $270 Technical article by Cutting in *Radio Broadcast*, Oct. 1929, p. 358. *Radio Retailing* (Oct. 1929), p. 145

Day-Fan Electric Co.

The Dayton Fan and Motor Co. was established in Dayton, Ohio in 1889. It made fly and ventilating fans, electrically or water-powered, for use in hotels, bars, restaurants, and stores. In 1921–1922, M.D. Larkin of the M.D. Larkin Co., electrical and mill-supply distributor, was very successful in handling radio products and wanted to become a manufacturer. As he was also president of Dayton Fan and Motor, he hired Capt. Orin E. Marvel of the McCook Field Radio Laboratory, in charge of aircraft radio research there, to design a line of components. First advertised in December 1922, these were very much like their Atwater Kent counterparts, and could be mounted on breadboards or panels, per factory plans, though Day-Fan built no complete receivers.

In September 1924 the reflexed OEM-7 and OEM-11 made their appearance, modestly priced and fairly successful, and continued with some changes for the following season too.

At this point a new president entered the scene: Charles F. Kettering, founder of Dayton Engineering Laboratories (Delco), inventor of automotive self-starting and lighting systems, and by now vice-president of General Motors and general manager of its research organization. He invested new capital in Day-Fan, moved it to a 400,000 sq. ft. plant on Miami Chapel Road and Wisconsin Boulevard in June 1926, and changed its name to the Day-Fan Electric Co. At this time 600 were employed, producing 700 sets per day. Marvel remained chief engineer for at least another year, but later left for Sonotone; he died in 1941 at the age of 55.

A group of former employees estimated in 1979 that Day-Fan had produced 500,000 sets by 1929, though that figure seems high.

In October 1929 General Motors bought Day-Fan and created the General Motors Radio Corp., owned 51% by GM and 49% by RCA, GE, and Westinghouse. A quote from the September issue of *Radio Retailer and Jobber* sums up the situation:

> *"(Day-Fan) has been laboring along for several years and losing something like $200,000 per year, of the competence, gradually-depleting, of a vice president of General Motors, who has been the sole and long-suffering angel and the one substantial stockholder in the Day-Fan Co. to date; but he has been trying for months manfully to persuade the General Motors crowd to lighten his financial burden by shouldering a pro-rata portion of it. By all indications, the General Motors crew has relieved its financially-involved vice president entirely, by swallowing the Day-Fan concern, hook, line and sinker."*

To give the GM management their due, Day-Fan's past performance was of little concern to them; what they needed was an already-existing company with an RCA license which they could take over, without arousing suspicion of collusion with RCA (the joint stock ownership was a tightly-kept secret for some time), as RCA was already under fire for its monopolistic practices. The Department of Justice did file an anti-trust suit in May 1930, and as part of the response, RCA and GM liquidated the General Motors Radio Corp. about November 1931.

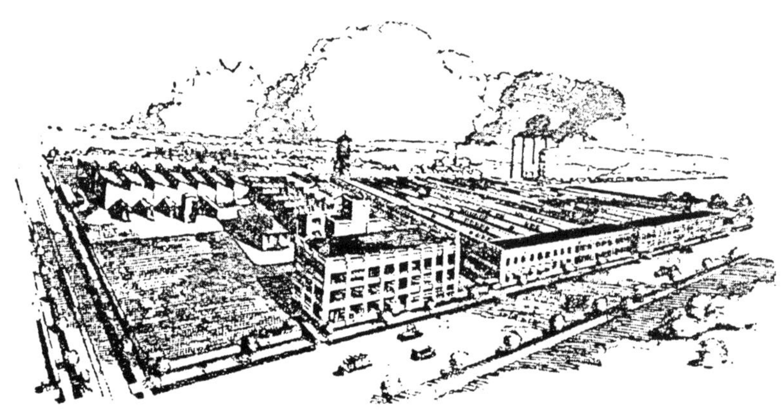

Day-Fan Factory and Laboratories

Radio Retailing (July 1925), p. 74

O. E. M. 12—4-Tube, $75.00

This is the famous O. E. M. 7 Duoplex stripped of all cost not absolutely essential. Mahogany finished cabinet; four tube; triple dial control. A truly remarkable value.

O. E. M. 7—4-Tube Super-Selective, $98.00

Duoplex circuit, two stages of radio, detector, and two stages of audio amplification, with one tube doing double duty; prelogged at the factory.

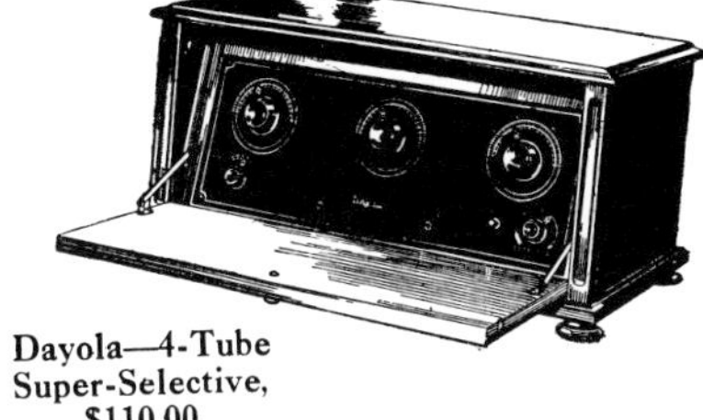

Dayola—4-Tube Super-Selective, $110.00

This is the same as O. E. M. 7, but housed in a beautiful two-toned waxed and polished drop front mahogany cabinet with inclined panel and room for "B" batteries. Duoplex circuit prelogged.

Day-Fan 5—5-Tube, $115.00

This is the newest Day-Fan achievement. Five tubes, a new principle of tone reproduction, extreme selectivity, and a real and practical single dial control.

Day-Fan Speaker, $30.00

Mahogany finish cabinet and grill—adjustable diaphragm on unit—and a tone quality that does justice to the Day-Fan Receiver.

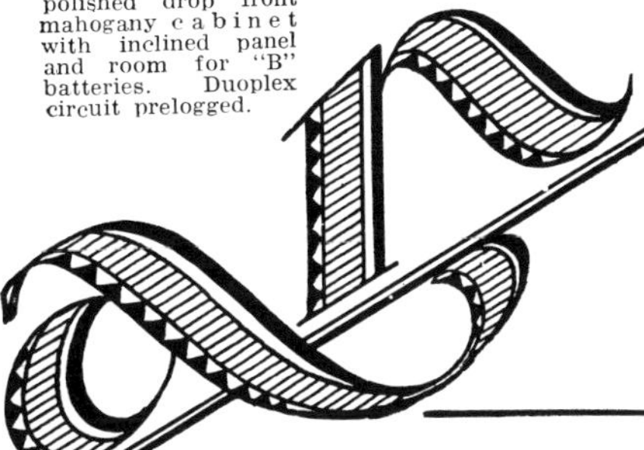

1925 Day-Fan RADIO 1926

Daycraft—5-Tube, $145.00

This set combines the Day-Fan 5 and the new Day-Fan Loud Speaker, built into the beautifully figured mahogany cabinet, providing room for "B" batteries. It has all the qualities of the new 5-Tube Day-Fan circuit, with the exceptional tone quality of the Day-Fan Loud Speaker. DAYCRAFT—WITH LEGS — 5 TUBE, $165.00. This is the Daycraft set with the addition of demountable legs. These legs are also furnished separately to fit any Daycraft set at a list price of $20.00.

Daygrand—5-Tube $195.00

This is a simpler Console model than the Daytonia, with room for "B" batteries, but without any equipment and with the new Day-Fan Speaker. It contains the same panel and set as the Day-Fan 5.

Dayroyal—5-Tube $300.00

This is the Day-Fan 5 panel and set, housed in an exceptionally beautiful mahogany vertical type desk cabinet. Built-in Loud Speaker. Room is provided for all equipment and cabinet is completely wired for batteries and charger.

Dayphone—5-Tube, $105.00

This is a phonograph model with the same circuit as the Day-Fan 5. It is designed to fit into any cabinet, such as the combination phonograph and radio cabinets furnished by some talking machine manufacturers. Panel size, 12x15½ inches.

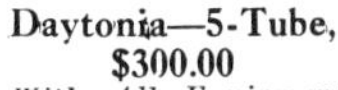

Daytonia—5-Tube, $300.00

With All Equipment Except Tubes.

This is a Console model with built-in Loud Speaker, batteries and a special silent recharging apparatus. It contains the same panel and circuit as the Day-Fan 5.

THE DAYTON FAN AND MOTOR COMPANY

DAYTON, OHIO

For More Than 36 Years Manufacturers of High Grade Electrical Apparatus

1,807,995. ELECTRICAL APPARATUS. ORIN E. MARVEL, Dayton, Ohio, assignor, by mesne assignments, to General Motors Radio Corporation, Dayton, Ohio, a Corporation of Ohio. Filed Mar. 30, 1925. Serial No. 19,370. 17 Claims. (Cl. 250—40.)

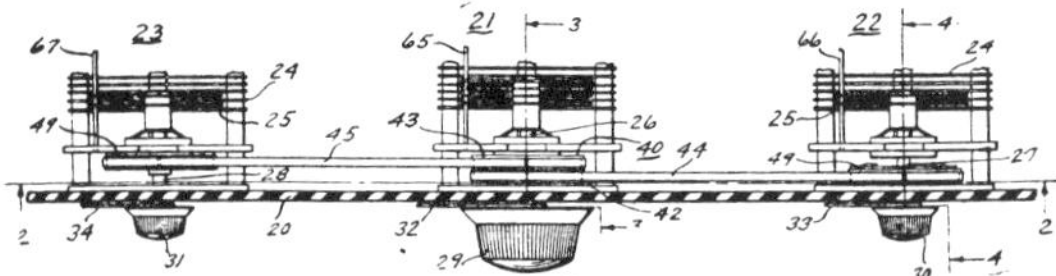

5. An apparatus of the character described comprising a pair of control devices each having a movable element, primary means for moving said elements in unison, a secondary means for moving one of said elements independently of the other, and stop means for a rotatable element adapted upon movement of the primary means to a predetermined position to coordinate said movable elements to a definite normal relationship with respect to one another.

Mechanism used in the Day-Fan 5.
This model is described in *Radio News*, May 1926, p.1544.

5106 OEM-7 (4 tubes) Sept. 1924 $98
5106 OEM-11(3 tubes) Sept. 1924 $90

4 consoles or models with speakers available by Nov.: Dayola $125, Daycraft $160, Dayradia $225, Daytonia $285.

Warren Falls

5049 Day-Fan 5

July 1926 models:
5049 DayFan 5 $89
5048 DayFan 6 $100
5050 DayFan 7 $115

Saturday Evening Post (Dec. 18, 1926)

Assembling the wiring bridle in the plant of the Day-Fan Electric Company, Dayton, O. Every wire except grid leaks is placed within this cable.

Radio Dealer (Sept. 1927), p. 98

Day-Fan

Electric and Battery Radio Receivers are used by Great Broadcasting Stations to Test the Tone Quality of their Programs

Literary Digest (Nov. 12, 1927), p. 3

THAT tells the story of the famous Day-Fan *tone*—its fidelity to the broadcasted music and speech—better than a thousand words of ours.

The top picture on this page is the Day-Fan Junior, a six tube receiver, which at $65.00 is one of the outstanding tone achievements in radio. The two other pictures show Day-Fan's electric receivers, the De Luxe Model that uses a motor and generator for radio current, the Day Cee 6 that uses AC tubes. With either of these you just plug into the light socket and listen—and what you get is like opening the door of the broadcasting station while the program is on. Single Dial Control—of course.

In all the maze of radio claims and counterclaims, when you wonder what radio to buy, here are two things to do. First, notice which receiver the broadcasting stations themselves use for their listening. They ought to know all there is to know about receivers. Second, go to a Day-Fan dealer and hear a Day-Fan.

The coupon will bring you fuller information about all the Day-Fan Models. Send it now. You will never have to wonder again what radio will serve you best.

DAY-FAN ELECTRIC COMPANY
DEPT. LD 5 DAYTON, OHIO

DAY-FAN ELECTRIC CO., Dept. LD5, Dayton, Ohio.
You may send me free circular on Day-Fan Radio Receivers.

Name____________________

Address____________________

City____________ State________

Licensed under R. C. A. Patents.
One Dial Control Licensed under
U. S. Patent 1,014,002.

Steve Conklin

5057. Uses Kellogg tubes and separate power supply.

June 1927 models:

5051	DayFan	6	3 models $115 to $350
5053	DayFan	6B	
5052	DayFan	6B	110 volts DC
5060	DayFan	6B	32VDC
5055	DayFan	6 Jr.	$65
5061	DayFan	6 Jr.	console chassis.
5066	DayFan	6 Jr.	AC. Dec. 1927, $95
5057	DayFan	67	(Kellogg tubes) $170 (?)
5065	DayFan	67	(RCA tubes)

Radio Retailing (July 1928), p. 121

console: 5077 or 5080 (200v or 110v speaker) model 26
AC table model: 5069, model 25 (June 1928) $150. 5070, similar, made for Winchester-Simmons
battery table model: 5072. model 35. $79.50.

PRESENTING

DAY-FAN

UNHURRIED DEVELOPMENT

ADVANCED DESIGN

Day-Fan 9-tube receiver in Walnut Table Model. Beautifully fashioned cabinet of high grade walnut... buried walnut panel...hinged top...constructed with care for every detail.

Day-Fan 9-tube receiver in large Console Cabinet with built-in Dynamic Speaker, specially designed for this set. Sliding doors. High grade walnut; matched-burl walnut panels. A piece of furniture that will win approval of the discriminating.

$225.00 Less Tubes

THE CHAMPION of Sound Radio Merchandising

STABILIZED QUALITY for a YEAR ahead

Radio Retailing (June 1929), p. 158

model 73 (not shown)	July 1929	$65	Console:	74	$119.50
model 67 (not shown)	Sept. 1929	$45			
A5005, model 93	Oct. 1929	$159.50	Console:	94	$210
5091 or A5003, model 66	June 1929	$115	Console versions:	68	$169.50
				69	$225
				72	$175

DE FOREST

De Forest Radio Telephone & Telegraph Co.

While Lee de Forest's connection with radio began around 1900, his Radio Telephone Co. first sold equipment to amateurs and experimenters in 1909, "RJ" for "Radio Junior." Its predecessor De Forest Radio Telephone Co. had equipped the U.S. Navy fleet with 26 arc radiotelephones in late 1907 for its round-the-world cruise, but after its unsatisfactory experience with them, the Navy was not about to buy any more. With no further Navy orders, and a very small market for RJ equipment, the company's main income was from stock sales (the promoters' intention all along). Once these tapered off, the company was bankrupt by March 1911, officially on August 23. De Forest himself left for California where he worked for Federal Telegraph until he was hauled into court for stock swindling, along with four of his colleagues. Just before being acquitted, in December 1913, he revived his old company and transferred its assets to a new organization, the Radio Telephone & Telegraph Co., started with the $50,000 received from AT&T in payment for repeater rights to his Audion.

The Radio Telephone & Telegraph Co.'s first product was an Audion control box similar to the one that Wallace & Co. had been making for two years in de Forest's absence (covered very thoroughly in Gerald Tyne's "The RJ-4 Mystery" in the *AWA Old Timers Bulletin* 19-1, June 1978). This was rechristened the RJ-4, and shortly, in May 1914, the more elaborate RJ-5 was added to the line.

On October 3, 1914 the Radio Telephone & Telegraph Co. became the De Forest Radio Telephone & Telegraph Co., operating under that name until 1924.

The Marconi Co. filed suit for infringement of the Fleming-Valve patent, and got a favorable decision in September 1916. At the same time, it admitted infringing the de Forest triode patent. Marconi obtained an injunction against de Forest's manufacture of detectors, which was lifted on condition that de Forest post bond and provide a continual accounting of Audions made and sold. De Forest appealed the decision; by January 1917 he had apparently violated the court-imposed conditions, and Marconi's injunction took effect. Since the De Forest Co. could now no longer make Audions, it ceased advertising in *QST*, and changed its *Electrical Experimenter* ads to amplifiers, transmitters, and auxiliary apparatus only. It hardly mattered, since the hams would shortly be shut down anyway, for the war's duration.

To keep occupied, perhaps, de Forest stepped up his broadcasting activities, which had begun during the summer of 1916 and reached a high point with his sending out election returns on November 7 to an estimated audience of several thousand. In January he began broadcasting phonograph records regularly, and asking for reception reports (*QST*, Jan. 1917 p.26; April, pp.72–74). This activity kept pace with his development of higher and higher-power "oscillion" tubes which had begun in July 1914 (the first tube transmitter was pictured in *Electrical World* of July 18, 1914). Eventually de Forest too was

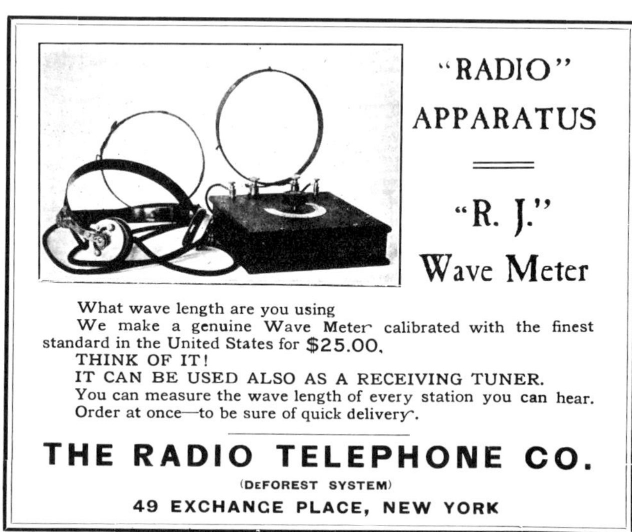

Modern Electrics (Jan. 1909), p. 370

forced off the air by wartime restrictions. By this time, however, he had his hands full with lucrative Navy contracts.

The contracts went back to at least May 21, 1914, when the Navy ordered 27 Audion amplifiers, and 900 double-plate Audions with Hudson filaments at $5 each. In the same year it bought 34 Ultraudions, and more later (contract data from the De Forest papers at Foothills Electronics Museum, and from Howeth, *History of Communications-Electronics in the U.S. Navy*, 1963, p.215). AT&T paid $90,000 for a non-exclusive license to use the Audion in radio, on August 7, 1914. But expenses were heavy, and by September 1916, if not before, the De Forest Co. was back to its usual hand-to-mouth existence. According to private correspondence of Robert F. Gowen, then applying for a job, De Forest had outstanding Navy contracts of $65,000, but with a payroll of over $1000 a week, had to get some gear out the door fast, and get paid for it, to keep going on the rest. Gowen took the job, at $15 a week, on the promise of perhaps twice that amount "in a month or two" when finances permitted, and eventually became chief engineer.

By early 1917, in spite of the injunction against making Audion detectors, things were much better. At least three Navy contracts came through: December 21, 1916, 19 CF349 airplane sets, $22,173; March 9, 1917, 16 CF118 dirigible receiver/transmitters, $30,400; March 17, 1917, 30 CF99 motor boat receiver/transmitters, $48,000 (the last two used the CN115 and CN116 receivers, respectively, made by National Electrical Supply Co.). From October 1916 to October 1917 De Forest also produced 2156 CF185 Audions, and probably quite a few spherical Audions (*Saga of the Vacuum Tube*, Tyne, 1977, p.131). Presumably the Navy accepted liability for this apparent violation of the court injunction; it did so officially on March 29, 1918.

Furthermore, on March 16, 1917 de Forest sold all his remaining Audion rights, other than his personal license to make and sell, to AT&T for $250,000. This windfall, being far more than the company needed to operate, was largely returned to the stockholders as a $1.45-per-share dividend. De Forest himself, as the owner of 120,000 shares, received $174,000. Aside from his regular salary, this was probably the first money he had received, personally, for one of the most valuable patents in the history of radio. He received a further $102,000 in 1919, an 85¢-per-share dividend made possible by profits from wartime contracts. All this money de Forest eventually spent in developing his sound-on-film apparatus in the 1920s.

De Forest may have originated the Audion, and built them for years, but once the large companies like GE and AT&T entered the field with well-trained people and well-equipped facilities, making standardized models, he was left in the dust. The Army Signal Corps offered him a contract in 1917 for the VT21, but it had to be uniform in characteristics, and interchangeable with the VT1 already coming into use. Quoting from another letter of Robert Gowen, to a co-worker who had joined the Army:

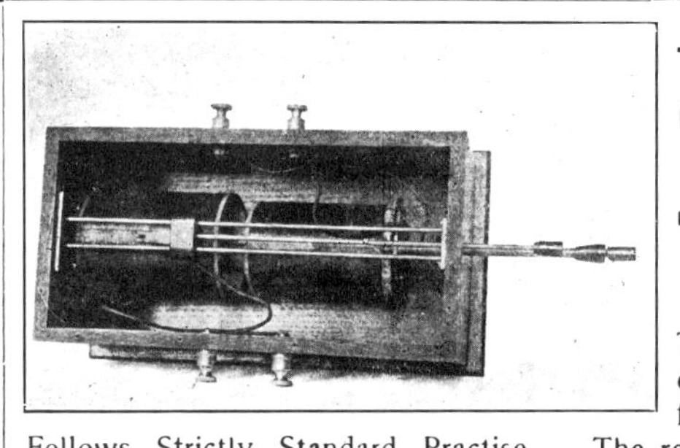

—THE—
"R-J"
Tuner

The best Tuner ever designed for the price.

Follows Strictly Standard Practise. The result of expert engineering experience

"VARIOMETERS" ARE SPHERICALLY AND SPECIALLY WOUND COILS FOR VERY ACCURATE TUNING. NO CONTACTS.

BE MODERN AND USE THEM

The Radio Telephone Co. 49 EXCHANGE PLACE NEW YORK

Modern Electrics (Mar. 1909), p. 450

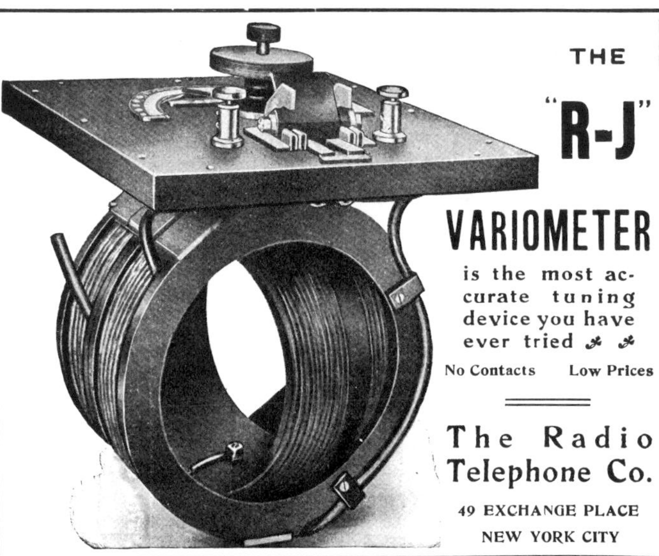

Modern Electrics (April 1909), p. 38

De Forest Apparatus

DESIGNED BY EXPERTS

WIRELESS TELEGRAPH AND TELEPHONE

RECEIVING OR TRANSMITTING

HIGH CLASS APPARATUS OF ALL SORTS AT REASONABLE PRICES

Variometers, Loose Couplings, Variable Condensers of all sizes, Helices and Spark Gaps, large and small, Heavy Transmitting Keys, Audion and Radion Detectors, Wavemeters, Telephone Receivers of extreme sensitiveness, Complete Commercial Tuners, etc., etc.

R. J. VARIOMETER—A variable Tuning Coil without sliding contacts, or a loose Coupling using rotating coils.......... **$16.00** Net

R. J. WAVEMETER— A **calibrated** Receiving Set that will measure the wave length **you** are using or of **any station** you can hear. Its variable condenser can be used separately.......... **$20.00** "

R. J. HELIX & SPARK GAP—Equal to that used in most ship stations, **$15.00** "

R. J LOOSE COUPLING—Rotating coil type to eliminate most extreme interference.......... **$17.50** "

HEAD TELEPHONE RECEIVERS—De Forest type. Most sensitive known, light, fit the ears.......... **$16.00** "

Technical advice and assistance gladly given to all purchasers. Call and see our apparatus at our new offices in the Metropolitan Tower. What we sell is GOOD.

If you wish a REAL Wireless Station go to those who KNOW HOW! Address

SALES DEPT.

RADIO TELEPHONE CO. 1 MADISON AVE. NEW YORK CITY

Modern Electrics (Nov. 1909), p. 391

Nov. 18, 1917

"Doc has finally given up his cut and try methods on the Army tube after wasting two weeks and told me yesterday he would present me with a check for a hundred dollars if I would get one that would pass by next Saturday. I am not much interested in the check as I would have to give a third to Brad and a third to Coyer as it is only fair that they should get theirs but I am afraid that the time is too short as I have not been allowed to get a bit of data as yet to work on except that when Doc was not looking last week I finally managed to get through special tubes enough to get a curve of B battery against Plate spacing. It sure is discouraging to be expected to produce a miracle without any data and also not to be allowed to get any data. Here's hoping that it will be different this week for that grid-relation-to-tube-operation-and-control curve is going to be a long hard job even if one is not interfered with."

And from another letter of Sept. 8, 1918 (to a different person):

"I am now Production Engineer and General Manager of the Vacuum Tube Division of the De Forest Radio Telephone and Telegraph Co. here in New York and we are manufacturing a new type of radio receiving tube for the army. This tube I designed and patented and I therefore have a very personal, as well as financial interest, in its production. The manufacturing difficulties, especially in the developmental work, have been particularly great and although we have shipped more than half of the Government's preliminary order for 15,000 of these tubes, there are important questions as to details of manufacture arising daily, which require my continuous attention.

*"The problem has been so difficult that, after four months of development work, the manufacturing department gave it up and insisted upon my leaving the Research and Inspection Laboratories to supervise the manufacture and devise ways and means for the production."**

*Gowen's correspondence, in the collection of Will Jensby WØEOM

These RJ variometers are quite different internally from the advertised model, using wood rather than hard rubber.

According to Tyne, 15,057 VT21s were shipped between March 8 and September 18, 1918. Apparently the "preliminary" order that Gowen referred to, was the *final* order. One has only to compare any two VT21s in existence today, and try to find two alike, to see why. The Navy's experience wasn't much different; on a contract dated May 8, 1918 for 2000 tubes, 1802 of them were rejected for lack of uniformity (Howeth, p.218).

With the Armistice in November 1918, the profitable wartime contracts came to an abrupt end. And Lee de Forest was dissatisfied with his work. His Audion had grown almost beyond recognition; books about it had been published, which he could barely comprehend, and while his company had trouble making any two tubes alike, GE and Western Electric were satisfying the government's demand with tens of thousands of excellent specimens. His competitors sold far more radio receivers than de Forest did. Even in his special field of Oscillion transmitters, his total production was swamped by better models from GE, WE, and even Marconi.

Lee de Forest needed a new field to conquer, one which was suited to his cut-and-try methods, free from competition by well-funded-and-staffed commercial interests. Recording sound on motion-picture film was the area that intrigued him, conceived at first as a replacement for the phonograph, but soon narrowed to talking pictures. This work occupied him during most of the 1920s, and has little bearing on this story, except that it effectively removed Lee de Forest from the De Forest Radio Telephone & Telegraph Co. Although his name continued to be used in advertising, he had essentially no contact with the company after 1921, nor influence on its product designs.

De Forest might have been content to remain with his company if he could have occupied himself with radio broadcasting. He enjoyed this, and of course was a recognized pioneer in radio-telephony, having broadcast continually since 1907. He set up his broadcast station 2XG again immediately after the war, but soon ran afoul of the New York Radio Inspector when he relocated his transmitter without a change in license. Ordered off the air, he moved to the west coast and set up another station, but the continuity was broken. His spot in the New York ether was filled by engineer Gowen, who broadcast from his Ossining home every night and soon eclipsed his boss, being heard as far as North Dakota by February 1920.

Gowen, an enthusiastic amateur, stated that he tried to get the De Forest Co. interested in the ham market after the war, with his newly-conceived "unit" set and self-supporting plug-in coils, but was rebuffed. Two months later, however, in January 1919, things were different. Government contracts were cancelled—particularly those for the BC14A artillery-spotting crystal receiver—and with the commercial market at a low ebb too, the amateurs were about all that were left. Gowen was authorized to develop his designs, travelling to Providence, Rhode Island to work out manufacturing arrangements of the "Honeycomb" coils with the Coto-Coil Co. and its head, Thomas Giblin (Coto later made coils under its own name, and Giblin's also, and Giblin made a line of radios).

THE NEW R. J. 5 DE FOREST AUDION DETECTOR

Licensed for private, amateur or experimental use only. The only amateur Audion Detector manufactured under the patents of Dr. Lee de Forest.

Manufactured by the

Radio Telephone & Telegraph Company

309 Broadway, NEW YORK

No other Detector on the market will compare with it at any price.

Renewal Audion bulbs may be secured, in exchange for old or broken ones, for \$3.50 and \$5.00 each. All bulbs are tested before shipment, but the "X" grade, or \$5.00 bulbs, are tested for the maximum possible sensitiveness. With the Audion you can easily increase your range from 50 to 103 per cent.

Price \$25.00

The de Forest Audion Detectors and Bulbs may be obtained from the following authorized distributors.

The J. J. Duck Company,	Toledo, Ohio.	Superior Wireless Instrument Co.,	Buffalo, N. Y.
Geo. S. Saunders & Co.,	Boston, Mass.	Adams-Morgan Co.,	Upper Montclair, N. J.
J. H. Bunnell & Co., Inc.,	New York City.	F. B. Chambers & Co.,	Philadelphia, Pa.
John Y. Parke & Co.,	Philadelphia, Pa.	The Stanley Company,	Salem, Mass.
Clapp-Eastham Co.,	Cambridge, Mass.	Woodill & Hulse Electric Co.	Los Angeles, Cal.
J. Elliott Shaw Co.,	Philadelphia, Pa.	H. A. Moore,	Nutley, N. J.
Manhattan Electrical Supply Co.,	New York City.	Rogers Electric Co.,	Toronto, Ont., Canada.
McCreary-Moore Co.,	Kansas City, Mo.	Harry W. Reinhart,	Scranton, Pa.
Ralph Krows Electric Co.,	Seattle, Wash.	Newark Elec. Supply Co.,	Newark, N. J.
C. Brandes, Inc.,	New York City.	Aylsworth Agencies Co.,	San Francisco, Cal.

When writing, please mention "Modern Electrics and Mechanics."

Modern Electrics and Mechanics (May 1914), p. 671

DeFOREST AUDION DETECTORS

Incomparably Superior to Any Other Known Type

Type RJ4 Audion Detector

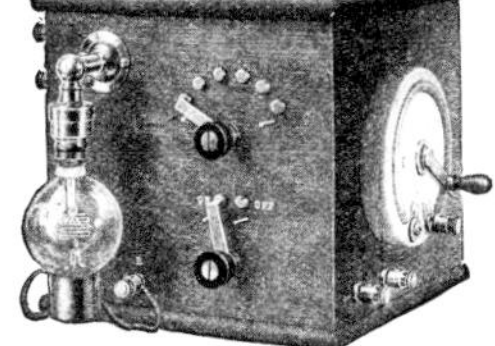

Type RJ5 Audion Detector

If you desire long distance reception of messages, you must have an Audion Detector. Tests of the Bureau of Standards show it to be the most sensitive and reliable detector ever invented. It stays in adjustment even where a transmitting set is used, and can be depended upon, absolutely, at all times.

The above types are regularly furnished with regular, or "S" grade, tantalum filament Audion Bulbs. Fitted with other grade bulbs at the difference of price of bulbs as listed below. A 4-volt storage battery or three dry cells are needed to light the filaments of the detectors, but are not furnished at these prices.

Bulbs are sold only for purposes of renewal, and then only upon return of the old bulb.

Price List, F. O. B. New York

Type RJ4 Audion Detector, with regular "S" grade bulb.........................\$18.00 net
Type RJ5 Audion Detector, with regular "S" grade bulb......................... 25.00 net

***Renewal* Bulbs**

Type "S" Regular Grade Audion Bulb (Tantalum Filament).....................\$3.50 net
Type "X" Extra Sensitive Audion Bulb (Tantalum Filament)..................... 5.00 net
Type "S" Regular Grade Audion Bulb (Hudson Filament)......................... 5.00 net
Type "X" Extra Sensitive Audion Bulb (Hudson Filament)....................... 7.50 net

Hudson Filament Bulbs have a very long life, between 800 and 1,000 operating hours, which is about three times the life of tantalum filament bulbs.

All reliable wireless dealers handle Audion Detectors and renewal Bulbs. If you do not know your local dealer, we will give you his name.

For further information, see your dealer, or write us.

DeForest Radio Telephone & Telegraph Co.

101 PARK AVE. NEW YORK

For our Mutual Advantage mention Popular Electricity and Modern Mechanics when writing to Advertisers.

Popular Electricity (Dec. 1914), p. 58

54 POPULAR ELECTRICITY *and* MODERN MECHANICS *for* NOV.

The New DeFOREST AUDION AMPLIFIER

Licensed for private, amateur or experimental use only. The only amateur amplifier manufactured under the patents of Dr. Lee deForest.

Type PJ 1, One-Step Audion Amplifier, Price \$65.00

This instrument is particularly designed for telephone and telegraph amplification, stethoscopic work for physicians' use, and all purposes where an amplification of sound up to 5 to 10 is sufficient. It is particularly suited to the needs of the average wireless amateur, enabling him to easily read signals which are quite inaudible to others not so equipped.

ASK YOUR DEALER OR ADDRESS

DeFOREST RADIO TELEPHONE & TELEGRAPH CO.

101 PARK AVENUE NEW YORK

For our Mutual Advantage mention Popular Electricity and Modern Mechanics when writing to Advertisers.

Popular Electricity & Modern Mechanics (Nov. 1914), p. 54

The RJ4 was first advertised (with an illustration of a Wallace detector) in Nov. 1913. The later version appeared in Apr. 1914.

The RJ7 sold for $200 in early 1916 (catalog B16). The RJ6 was the same, minus amplifier, while the RJ10 was only the tuner portion, at $65. By late 1916–early 1917 the RJ6 Special and RJ7 Special, with switches for the Ultraudion circuit (regenerative) added to the panel, sold for $120 (later $150) and $225.

RJ9 Feb. 1916 $14

QST (Sept. 1916)

UJ1 Aug. 1916 $27.50

Combination Audion Detector and Single Step Audion Amplifier.

Licensed for Private and Amateur Use Only.

"Incomparably Superior to Any Other Known Form of Detector."

Catalog B15, early 1915, $110. A two-step EJ2 looked similar and sold for $150, in early 1916.

EJ1 early 1916 $60

By May 1919 the new line of radio equipment was ready, and catalogs were distributed. Before this, the De Forest Co. advertised a large quantity of components and materials left over from the cancelled BC14A contract; in fact some of these parts were still listed in the catalogs 2½ years later, giving some idea of the number on hand.

On November 19, 1920, the factory was partially destroyed by fire, particularly the glass department where Oscillions were made. No new products were advertised until April 1921, when Catalog E announced the Interpanel line. Robert Gowen was made chief engineer and plant manager some time in 1921, replacing Lawrence C.F. Horle who had held the post in 1920 (Horle joined Federal in 1924) but which of them designed the Interpanel is not clear. Gowen went to China in October, his experiences being recounted in *Radio Broadcast* magazine in March 1923.

"Doc" departed for Berlin in October, having visited that summer and concluding that he could develop his Phonofilm system there with reduced living expenses and freedom from commercial pressures and interruptions. He left the company in the hands of his lieutenants Charles Gilbert and Randall Keator, Gilbert becoming president. Throughout 1921, all advertising featured Interpanels. The company lost $25,827 in 1921, on gross sales of $108,439.

The new proprietors' necks were saved by an unforeseen development in late 1921: the broadcast boom. In January 1922 the Everyman crystal set was advertised, and in April the Radiohome. But to judge from the numbers surviving today, neither of these models sold very well. The reason may lie in the move to a new factory in Jersey City, New Jersey around April, which would have disrupted production at a critical moment (the boom collapsed in May). Yet 1922 was a banner year for De Forest, which posted profits of $288,532 on gross sales of $538,771, and paid a stock dividend of $1 per share. The chances are that many of these sales were Interpanels already in stock at the beginning of 1922, when virtually anything would sell to the hordes of crazed fans. And, once tooled up, the factory could have made more Interpanels on short notice. So, if Gilbert and Keator sat around in their new plant for the rest of the year, twiddling their thumbs, at least they had plenty of money in the bank.

The De Forest executives were painfully aware that their models, without regeneration, ran a poor second to those of the lucky manufacturers with Armstrong licenses. They tried to get around this deficiency by making their models easily convertible by the customer into regens by switching a pair of wires, placing the third honeycomb coil in the detector plate circuit. This earned De Forest a show-cause order by the New York District Court from Westinghouse, eventually resulting in a contempt judgment and a fine.

Meanwhile, De Forest bought an Armstrong license, or rather, it bought a defunct company that owned one, in early 1922, with the idea of transferring it. This was the Radio Craft Co., Inc. of Brooklyn, who had advertised in mid-1920 a somewhat pretentious line of receivers for hams, and had thought its future rosy enough to purchase an Armstrong license for $10,000 (incredible, since most licensees paid only a 5% royalty on sales, but the entire contract is reprinted in the 1923 *FTC Report*, pp.209–11). After offering a regen model in December, Radio Craft's ads disappeared after June 1921.

DE FOREST RADIO TELEGRAPH AND TELEPHONE APPARATUS

AUDION RECEIVING CABINET
TYPE RJ 6—PRICE $150.00

Incorporating in one cabinet the famous de Forest Audion-Ultraudion detector and special Audion Tuner. For spark and arc signals. Most efficient receiving set ever offered at less than $500.00.

THE value of radio equipment for motor cruisers and yachts is undeniable, both for the convenience of communicating and receiving intelligence under ordinary circumstances, and for its use in emergencies. But in either case the value is entirely dependent upon the quality and reliability of the instruments and apparatus of which the equipment is composed.

de Forest Radio Apparatus is everywhere conceded to be the finest available for private use. It has been developed and refined through the persistent efforts of Dr. Lee de Forest and his staff until it has reached a degree of efficiency which cannot be expected in ordinary amateur apparatus. Used extensively by prominent yacht owners and by U. S. and foreign navies.

WIRELESS TELEPHONES OPENS UP A NEW AND INTERESTING FIELD FOR THE BOAT OWNER, requiring no special operator. Their perfection and practicability have been demonstrated beyond the shadow of a doubt.

Write for free bulletins or let us recommend radio apparatus suitable for your requirements.

TYPE OJ3—$400.00 COMPLETE

Oscillion Telegraph, capable of transmitting the voice 15 miles, or telegraphic messages 40 miles. Larger transmitters for greater ranges.

FAMOUS DE FOREST AUDION DETECTOR TYPE RJ8

Three to five times more sensitive than any other known form of detector. (U. S. Bureau of Standards Report.) PRICE $25.00.

COMBINATION AUDION-ULTRAUDION DETECTOR
TYPE UJ2—PRICE $30.00.

DE FOREST RADIO TELEPHONE & TELEGRAPH CO.

1391 SEDGWICK AVENUE, NEW YORK CITY

Sole Makers of the Genuine Audion, Oscillion and the Highest Grade Receiving Equipment in the World

NEW AUDION AMPLIFIER FOR INCREASING STRENGTH OF RECEIVED SIGNALS 25 TIMES.
TYPE EJ2—PRICE $32.00.

You benefit by mentioning "The Electrical Experimenter" when writing to advertisers.

Electrical Experimenter (Mar. 1917), p. 788

The RJ6 appeared in late 1915 and sold for $95 with oak panel, $110 with hard rubber. The RJ8 was first advertised in Feb. 1916.

At Last!

Electromagnetic waves of any length from an incandescent lamp.

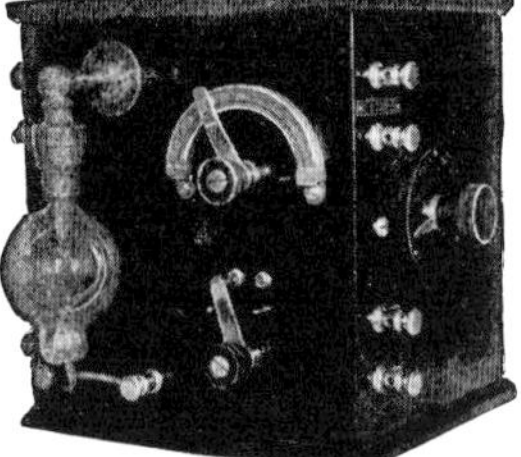

TYPE "S"—$60.00

DeForest "Oscillion"
(Oscillating-Audion)

Generator of absolutely undamped oscillations of any frequency. Permits Radio Telephone speech surpassing in clearness that over any wire. For Laboratory and Research Work has a field utterly unfilled. Patents issued and pending.

TYPE OJ3—$400.00 COMPLETE

Oscillion Telegraph, capable of transmitting the voice 15 miles, or telegraphic messages 40 miles. Larger transmitters for greater ranges.

TYPE EJ2—PRICE $32.00.

NEW AUDION AMPLIFIER FOR INCREASING STRENGTH OF RECEIVED SIGNALS 25 TIMES.

It is not a Detector in any form.

TYPE RJ11—2500—12000 METERS, $35.00

THE DEFOREST LOADING INDUCTANCE

TYPE VC4—PRICE $20.00

VARIABLE CONDENSER

This Condenser is similar to our commercial type but is enclosed in an oak cabinet. It has 35 semi-circular aluminum plates. The maximum capacity is approximately .0025 M. F.

MANUFACTURED BY

DEFOREST RADIO TELEPHONE AND TELEGRAPH COMPANY

NEW YORK CITY

Office and Factory
1391 SEDGWICK AVE.

Cable Address:
RADIOTEL, N. Y.

Electrical Experimenter (Apr. 1917), p. 920

Ford Museum, Dearborn, Michigan

CF753 1917

French Type A

New England Wireless & Steam Museum

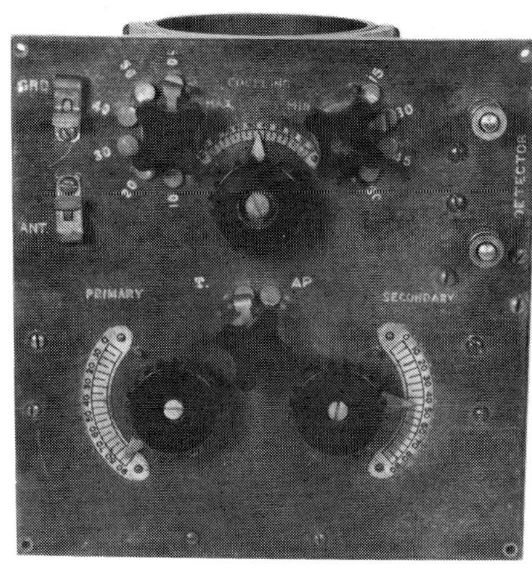

BC14 (part of SCR54) panel only.
"BC" = *B*asic *C*omponent.
"SCR" = *S*et, *C*omplete, *R*adio.

De Forest's Catalog H, first edition of February 1922, claimed that a regenerative tuner, matching the SP series, would shortly be added to its line. The second edition, two months later, read "non-regenerative tuner." Doubtless Westinghouse politely informed De Forest that its newly-acquired license was *not* transferable from Radio Craft, and that it would *not* be adding any such tuner to its line. So Radio Craft was maintained, in a partitioned half of the factory, and reappeared in ads in December 1922, for the regen models that De Forest couldn't make under its own name.

When De Forest bought Radio Craft, it also got the firm's founder, Frank M. Squire (2UY in 1916 , a draftsman at A.H. Grebe before starting his own company). Squire immediately set about designing a receiver radically different from anything that De Forest, or anyone else, had ever made. It would be totally self-contained, aside from headphones. Once the Fleming-Valve patent expired in November 1922, De Forest would be free to make receiving tubes, and its engineers were apparently confident that they could make a small tube with an oxide cathode that would take so little power, that a single dry cell could run three of them. This was the DV1 (a tungsten-filament DV6 was also planned, for 6 volts).

BC14A made by Liberty, identical to the De Forest model. The General Radio version was quite different, with internal buzzer and engraved dial scales.

The BC14 artillery-spotting receiver was patterned on the French type A. It received messages from an airplane circling overhead, equipped with a small spark transmitter, watching the fall of artillery shells, so the guns' aim could be corrected.

Remarkably efficient and handsome audion control panels within the price limit of **EVERY AMATEUR, BUILT AND DESIGNED FOR THE AUDION, AS ONLY THE MAKERS OF THE AUDION CAN BUILD IT.** IN THE PAST, INSTRUMENTS OF THIS CALIBRE HAVE NEVER BEEN OFFERED AT TWICE THESE PRICES.

GENERAL SPECIFICATIONS, ALL MODELS: Beautifully engraved and grained genuine bakelite panels; 4 prong heavy nickeled tube receptacles; smooth running, positive acting, adjustable rheostats; impregnated efficient mica stopping condensers; adjustable or fixed grid leaks with polished nickel covers; bakelite covered binding posts with new slotted positive connecting feature; positive segmentally connected nickel plated switches; 40 volt B battery of two 20 volt type units cast en block with operating life of 6 to 12 months—a remarkable advance in B batteries; handsome quartered oak cabinets, round corners, standard de Forest "Early English" finish; perfect workmanship and expensive design throughout.

TYPE P500—$22.00

Type P-400 Audion Panel—Consists of tube receptacle, grid leak and condenser, rheostat and binding post for connection to tuner, A and B batteries on bakelite panel 7½" x 7½" without cabinet and B batteries........**$9.50**

Type P-401 Audion Panel—Same as above with cabinet, **$11.75**

Type P-402 Audion Panel—Larger panel with 40 volt B battery in cabinet..............................**$18.85**

Type P-500 Audion-Uitraudion Panel—Type P-402, Panel with two switches for audion-ultraudion connection and either 20 or 40 volt B battery in cabinet...........**$22.00**

TYPE P401—$11.75

NOTE: All de Forest audion apparatus is sold without the tubes. Audion detector oscillator or amplifier tubes, $7.00 each additional. Catalog sent on receipt of 10c in stamps to cover postage.

DE FOREST RADIO TEL. & TEL. CO., 1415 Sedgwick Avenue, New York City

You benefit by mentioning the "Electrical Experimenter" when writing to advertisers.

Electrical Experimenter (Aug. 1919), p. 372

A Concert by Wireless

Lee De Forest Gives Amateur Operators a Treat Over the 'Phone.

Thousands of amatuer wireless operators within a radius of 100 miles of New York heard a wireless telephone concert given recently at the De Forest experimental laboratories at Highbridge. The entertainment lasted for more than half an hour, and operatic selections and popular music were poured into the telephone to be sent out in wireless waves to every listening ear in and about the city. Phonographic records were used and a special record was put on to oblige an operator "somewhere in Flushing."

Notice of the concert had been sent out several days ago, and so that amateurs were waiting with receivers clapped to their ears for the signal that would tell them that the performance was about to begin. All that the operators had to do to enjoy the music was to tune up to the wave length of the sending station.

Walter Schare was in charge of the concert, and after the first few selections had been played on the phonograph expressions of thanks from the unseen audience began to sputter into the receiving instrument. From Yonkers came a hearty vote of thanks, and one enthusiastic Staten Islander insisted on sending messages of appreciation several times.

The concert is one of a series planned at the laboratories. Indeed, it is the plan of Lee De Forest to establish a sort of wireless newspaper to which every amateur with an instrument can subscribe. In this way news can be telephoned and the interesting happenings of the day can be sent to listening ears "hot off the wire."

We are informed that the test will continue every evening from Monday to Friday inclusive, at 8:00 o'clock on a wave length of approximately 800 meters. The De Forest Company would deem it a great favor if those hearing the concert would report by mail.

QST (Jan. 1917), p. 26. See also Apr. 1917 , pp. 72–74

Catalog (May 1919), p. 19

Front view of 15 Panel Typical "De Forest Unit Receiving Set," consisting of a Tuner with wave length range of 150 to 25,000 meters, a crystal and audion detector, and a one step amplifier. The most efficient complete receiving apparatus ever put out under $150.00.

Catalog (May 1919), p. 20

Rear view of 15 Panel "De Forest Unit Receiver" shown on the other side of this page.

Squire arranged a license under the patent applications of William Priess, who had worked on amplifier circuits and reflexing during the war at the Washington Navy Yard. Priess later claimed to be the "father of reflex" and must have presented an impressive bargaining position to De Forest, though as it turned out, never had a single U.S. patent issued to him on reflexing. He eventually came to work for De Forest, but in 1922 and a part of 1923, Squire was chief engineer, and claimed responsibility for the new model D7 (*Radio Broadcast*, February 1923, pp.297–302). Designed by August 1922, it was on exhibit at the Chicago Radio Show in October. Very shortly, with little apparent change in appearance or circuitry, the D7 became the D7A. Possibly the set was modified to take DV6As, when the DV1 didn't work out as planned (it was eventually made as a thoriated-tungsten 3-volt tube, patterned after the UV199 which came out at just this time).

Along with the D7, the D4, D5 and D6 made their entrances. These were regenerative models, made for legal purposes by Radio Craft, but advertised side-by-side with the De Forest models. This ruse lasted a few months, but since Radio Craft models are rare now, probably not many were sold. De Forest also continued to advertise the MR6 Interpanel, non-regenerative. Radio Craft could, and did, legally advertise the *regenerative* MR6. It was the discovery of panels engraved for the regenerative circuit, on the De Forest side of the factory partition, that supposedly led to the contempt of court judgment noted earlier.

Dr. De Forest trying out one of his more recent inventions, a portable wireless telephone operated entirely by current from a lamp socket.

DIFFERENTIALLY BALANCED DOUBLE PRIMARY MULTI-WAVE TUNER

Catalog C, p. 46

T100 May 1920 $135

Catalog C, p. 45

T-200 Dec. 1919 $77.50

TYPE P-100 AUDION CONTROL PANEL.

Radio Broadcast (Dec. 1924), p. 216

Catalog (May 1919), p. 46 & 65

TYPE P-200 TWO-STEP AMPLIFIER

Catalog (May 1919), p. 49

P200 May 1919 $69.50

Catalog C, p. 38

P-300 Dec. 1919 $73.50

TYPE P-300 COMBINATION AUDION-ULTRAUDION AND 1-STEP AMPLIFIER

Most examples have two binding posts at top center for tickler connection.

Frank Squire continued to improve his reflex. By March 1923 he had a larger model, D10, capable of driving a loudspeaker, on exhibit at a New York amateur radio convention, where it was wheeled around on a tea wagon "with the moosic pouring out in great shape." Priess claimed to have designed the D10, and probably joined De Forest at about this time as chief engineer, while Squire went to Brandes.

In early April 1923 de Forest, Gilbert, and Keator sold 86% of their stock for $670,000 to a group of Detroit industrialists connected with the Paige-Detroit Motor Car Co., headed by Edward H. Jewett of the Jewett Radio & Phonograph Co. Gilbert and Keator remained on the payroll; Priess got a lucrative royalty contract (reportedly $3.50 per set). Simultaneously, RCA argued its case before the New Jersey Chancery Court and obtained a temporary injunction on April 13, preventing De Forest from selling to jobbers and retailers. RCA claimed that De Forest's right to sell to amateurs, which had been personally reserved when Lee de Forest sold the rest to AT&T in 1917, did not extend to broadcast listeners, and demanded that De Forest secure a written agreement from every purchaser that its tubes and apparatus would not be used for commercial communications. While this demand was refused, De Forest still had to devise some way of getting around the injunction, finally adopting the "consignment" method used by GE years before in similar circumstances. As announced by Jewett in August, when his stock purchase was completed, De Forest would retain title to the merchandise until the final sale; the dealers' payments being held in a bank meanwhile. It worked: for the first nine months of 1924, De Forest netted $116,482 on gross sales of $563,240.

DE FOREST "INTERPANEL" SETS

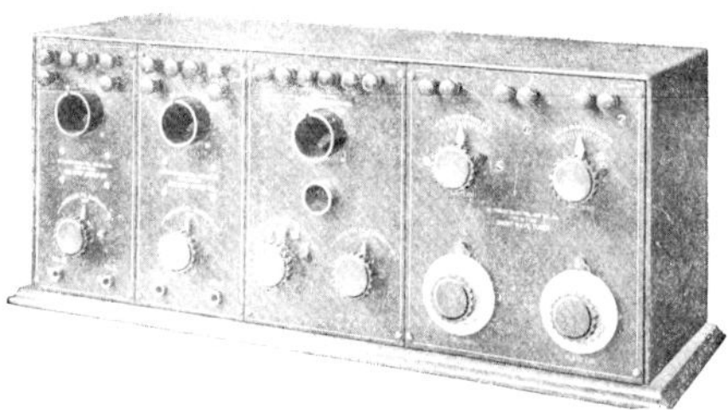

Combination Set MR-3

Price $106.50 Code Word—Tulpen (Shown above)
Shipping Weight—34 lbs.
Consists of the following:

MP-100 Audion Control	$13.50
MP-200 First-step Amplifier	17.75
MP-200 Second-step Amplifier	17.75
MT-100 Tuner	43.00
AP-100 Cord and Plug	1.50
Cabinet	13.00
	$106.50

Combination Set No. MC-1

Price $40.25 Code Word—Tulpenart (Not shown)
Shipping Weight—15 lbs.
Consists of the following:

MP-100 Audion Control	$13.50
MP-200 One-step Amplifier	17.75
Cabinet for both	9.00
	$40.25

Combination Set No. MC-2

Price $60.50 Code Word—Tulpsteel (Not shown)
Shipping Weight—22 lbs.
Consists of the following:

MP-100 Audion Control	$13.50
MP-200 First-step Amplifier	17.75
MP-200 Second-step Amplifier	17.75
AP-100 Cord and Plug	1.50
Cabinet	10.00
	$60.50

Combination Set No. MP-400

Price $45.00 Code Word—Tuluxi (Not shown)
Shipping Weight—15 lbs.
Consists of the following:

2 MP-200 Amplifier Panels giving two steps of amplification $17.75 per panel	$35.50
AP-100 Cord and Plug	1.50
Cabinet for both panels	8.00
	$45.00

MR-6 Interpanel Feb. 1922 $112
The MR-5, with one amplifier, was $91; the MR-4, with no amplifiers, $72.50.

Photo by Lou Lindauer

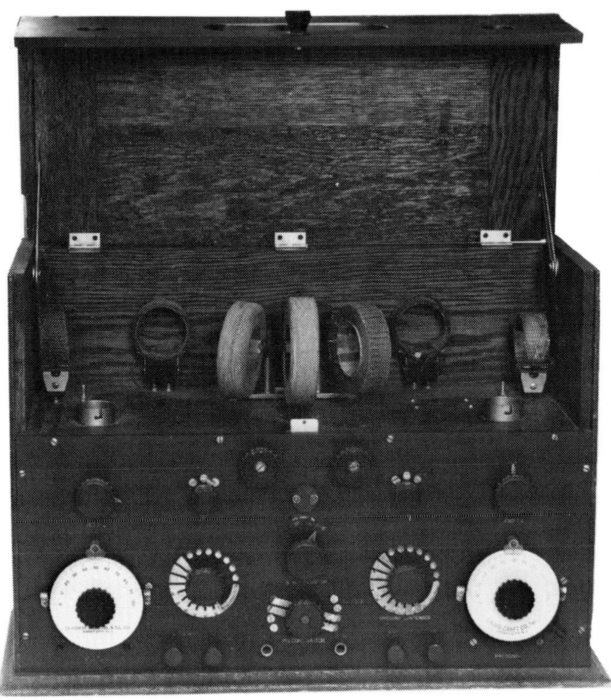

New England Wireless & Steam Museum

RS-500 May 1920 (advertised in 1921) $275

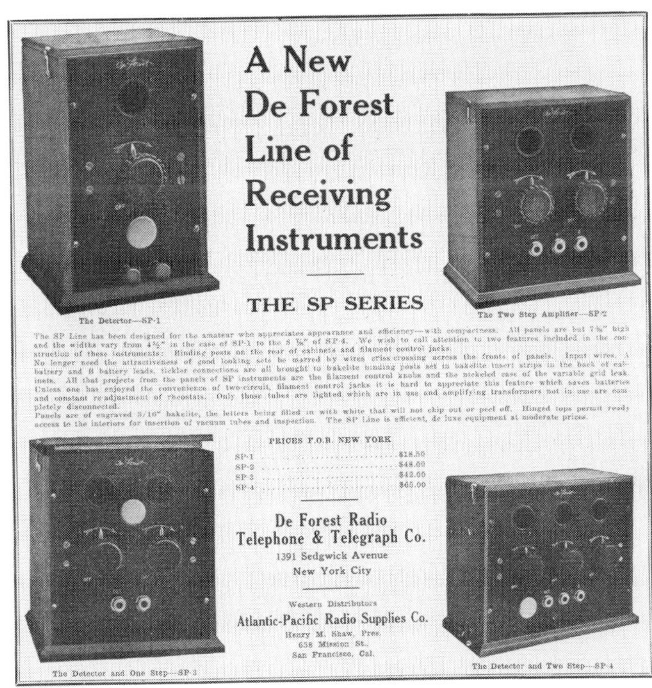

Radio (Feb. 1922)

TYPE RS-100 JEWELERS TIME RECEIVER

Catalog C, p. 43

RS-100 Dec. 1919 $25
The slightly larger RS-200 was listed in Catalog C in May 1920 for $45.

Gilbert and Keator left by September 1924 to form their own radio distributing firm in New York City. Priess withdrew also, incorporating in his own name in October; before he left, he revised the old D10 by making one of the RF stages tunable for greater selectivity. This was the D12, in production by August 1924 and advertised in September; probably several thousand were made.

RCA's injunction was lifted on October 21, allowing De Forest to sell directly once again, though it continued (to the puzzlement of one trade paper) to consign its products for six months more.

Late in 1924 De Forest hired Roy Weagant away from RCA, where he had been since Marconi days. Weagant updated the D12 by re-styling its cabinet and panel, and replacing the crystal detector with a fifth tube. In one fell swoop, this obsoleted every D12 in dealers' stocks, but by the terms of the De Forest consignment contracts, it, not the dealers, owned these unsold models, and was obliged to take them back. To avoid a financial loss, Weagant modified these D12s with the D17 circuit: a tube detector, and one fewer reflexed stage (therefore two audio transformers instead of three). These modified D12s were disposed of by April 1925.

Once Weagant had a free hand, he scrapped the reflex principle entirely, perhaps to avoid further royalties to the now-competing Priess. By September 1925 he had designed two basic chassis, both straightforward TRFs: the simple F5 and the more elaborate W5 and W6. Some of the F5s went into leftover D17 cabinets, of which De Forest apparently had carloads.

Aside from the CS-5 short-wave receiver introduced in April 1930, De Forest made no more receivers. Nominally, it produced vacuum tubes for the next few years, but in fact existed mainly for the stock-promotion schemes

Radio News (Apr. 1921), p. 727

Interpanel Sets

The Most Advanced Idea in Radio Telephone Transmitting and Receiving Apparatus

THE DeForest RADIOPHONE INTERPANEL Set establishes a new standard of design and efficiency for DeForest Apparatus, and provides the most convenient and all round satisfactory method of purchasing Radio Apparatus yet invented.

The INTERPANEL Set consists of a series of panels, each constituting a complete piece of apparatus in itself, and designed to be combined with other panels, thus forming a Set as complete as may be desired, the operating possibilities depending only upon the total number of panels used. The Set for both Telephone and Telegraph Transmission and reception consists of four panels, as follows:

Complete Set of Four Units, in Cabinet, each panel sold separately for mounting in home constructed cabinet; or completely assembled in cabinet as shown above. Also for mounting in Horizontal or Vertical Table-style. Complete Set as above, without batteries or tubes, type MS-1, $189.25

Type MT-100—A complete short wave Tuner of highest possible efficiency;

Type MP-100 A new Audion Control panel designed especially for tubes of the gaseous type, now considered as standard;

Type MP-200—A one-step Amplifier panel complete in every respect; and

Type OT-3—A complete Radiophone Transmitter, capable of transmitting speech at least 30 miles, and up to 500 miles.

(Additional steps of Amplification may be added as desired)

Panels are all 9 inches high; varying widths. Designed for placing side by side, with binding posts in line and convenient to wire. Adaptable to any operating requirement. Panels may be bought individually and mounted in operator's own cabinet; or bought completely mounted in cabinet. Or panels alone may be mounted on table in either horizontal or vertical style.

Horizontal Table-style mounting. Legs attached to corners of each panel. Any number of panels can be mounted in this style. Ample space under panels for batteries. A very convenient and inexpensive method of mounting.

Vertical Panel-style mounting, without cabinet. Two legs hold each panel upright. Any number of panels may be joined and mounted this way.

30 Mile Range for the Telephone Transmitter on Average Amateur Aerial.

Tests show a 30 mile telephone transmitting range for the Set, which can be exceeded under favorable conditions. Telegraph range from 60 to 100 miles with unlimited reception possibilities. One 6-volt storage battery required for all filaments and microphone; Motor-generator, "B" Battery or rectifier supply may be used.

This INTERPANEL Set provides the ultimate in RADIOPHONE apparatus; ease and convenience in installation and operation; minimum space, handsome appearance, great efficiency and extreme economy.

Send Now for Catalog "E" and Prices

Get the full details of this new INTERPANEL idea and get your order placed early.

DEFOREST RADIO TELEPHONE & TELEGRAPH COMPANY

Inventors and Manufacturers of High Grade Radio Apparatus

1415 Sedgwick Avenue **New York City**

DT-600 Everyman Feb. 1922 $25 John Terrey

of its backers, and its chief business was lawsuits. Jewett bailed out and began advertising his own radios by August 1925. Lee de Forest petitioned the company into receivership on April 19, 1926, investment banker Arthur Lord being elected president on May 13 and appointed equity receiver July 19. On August 4 de Forest sued the company and its former officers, charging they had squandered $1¼ million between August 1923 and May 1926, by paying excessive salaries and conducting the business inefficiently. "Utterly shiftless absentee management" and an "asinine contract with . . . Priess" were de Forest's words in his 1950 autobiography.

Powel Crosley contracted to manage De Forest on January 5, 1927, to supply $300,000 of operating funds in return for 39,000 shares of stock and a 3% commission on net sales. His interest in the company, however, was apparently limited to obtaining bargaining leverage in his patent-license negotiations with RCA, and he had no intention of fulfilling this contract, even after Lord sued him on September 21. Instead, a group of bankers headed by Wiley Reynolds who had been a director long before, reorganized the company in mid-1928. Quoting de Forest again in 1950, "During the stock boom of 1928–29, including a wicked, utterly indefensible, purchase by De Forest Company of 'Jenkins Television,' Reynolds had

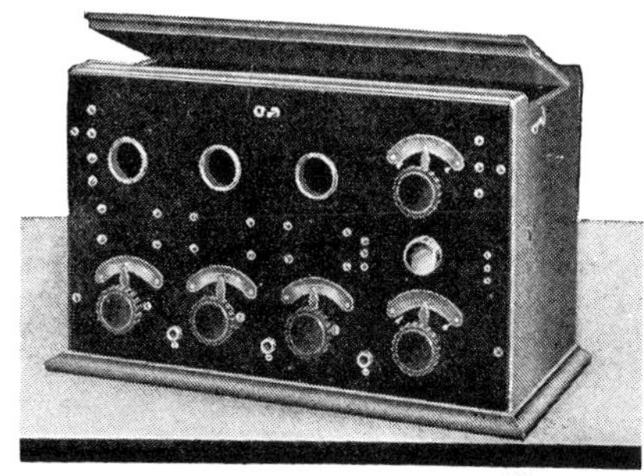

DT-900 Radiobest Apr. 1922 $85

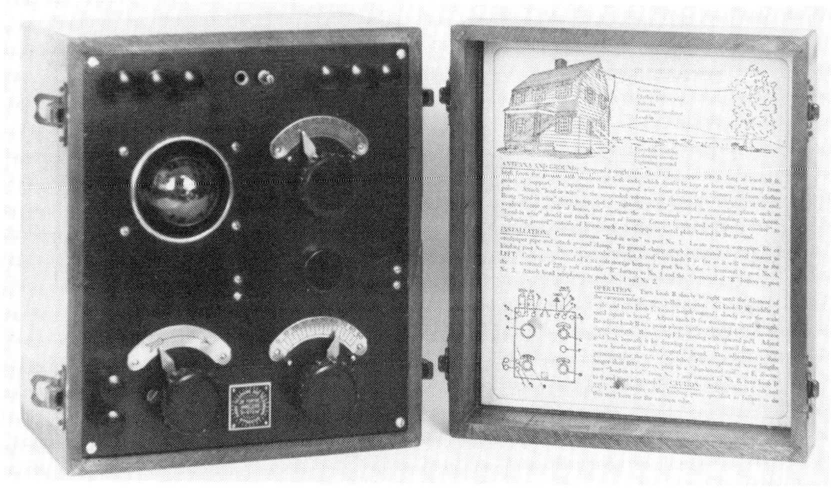

DT-700 Radiohome Feb. 1922 $36 Rich Elskamp
The matching DT-800 two step amplifier was $35 in April.

Earle Drake

Radio (Dec. 1922), p. 55

De Forest
D-7
Reflex
Radiophone*

Listen to Half the Continent This Christmas!

HERE'S the latest De Forest triumph, the D-7 Reflex Radiophone* Receiver. It's the newest and most sensitive set of them all, with a thousand mile range on a two-foot indoor aerial! That's what you've been waiting for. No outside aerial is needed. The whole set is as you see it here.

Easy to control with its single knob—small, compact, super-efficient—and an ornament to any library table!

Economical to operate, too, because you get five stages of amplification on three tubes, and correspondingly longer life for your storage batteries.

If you want to bring into your home the news, the music, the lectures of half the American Continent—with no trouble—clearly, without interfering noises—this is the set for you. Remember it's a Radio Christmas—and here's one set that is all you ever hoped a receiver could be. Ask your De Forest dealer about D-7 and the other De Forest sets—today.

De Forest Radio Tel. & Tel. Co.
Jersey City, N. J.

*Reg. U.S. Pat. Off.

What is the Finest Receiving Set Made?

To be worthy of this distinction, the set must have three characteristics—

World-wide receiving range—
Reception in all wave lengths—
Reception without distortion.

RadioCraft D-6 Regenerative Radiophone* (by permission of De Forest Radio Tel. & Tel. Co.) will receive even European stations, and of course all those on this continent.

It receives on all amateur, broadcasting, and transatlantic wave lengths.

In clearness and avoidance of distortion, it must be heard to be adequately described.

D-6 uses outside aerial only, and head sets or loud speaker.

It claims to be the finest receiving set now manufactured and any authorized De Forest dealer (who also carries the RadioCraft line) will be glad to prove this to you by actual demonstration.

RadioCraft Regenerative Radiophones* (by permission of De Forest Radio Tel. & Tel. Co.)

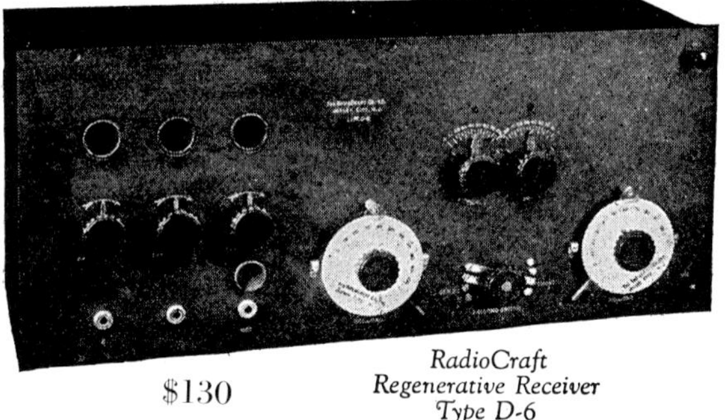

$130

RadioCraft Regenerative Receiver Type D-6

range from the simplest to the most elaborate.

It is unnecessary to add that this entire line of De Luxe radio equipment is most exquisitely finished in every detail.

The RadioCraft Co., Inc.
139 Franklin St., Jersey City, N. J.

*Reg. U.S. Pat. Off.

Radio Broadcast (Feb. 1923), p. 299

FRANK M. SQUIRE

The Regenerative Principle is Only Half the Battle

YES, it's regenerative. But there's one other point you want to know. What parts make it up? De Forest parts go into RadioCraft sets—and that tells the whole story. Here is regenerative equipment inexpensive in price and so efficient that it will bring in broadcast within a radius of 150 miles and upward. It's the fastest selling line because it gives your customers the greatest satisfaction at low price.

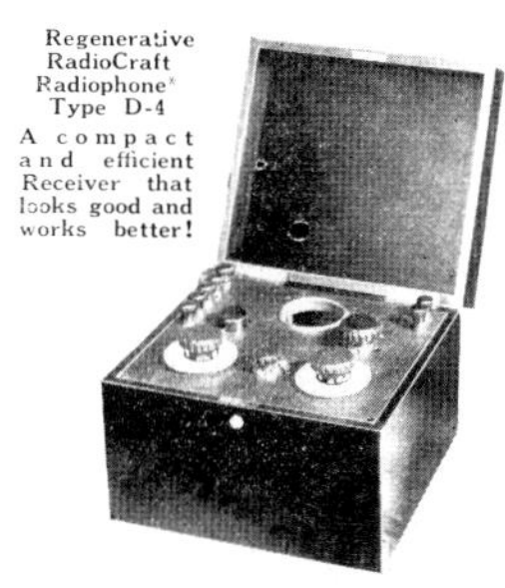

Regenerative RadioCraft Radiophone Type D-4 A compact and efficient Receiver that looks good and works better!

D-4 is designed for the user who wants a compact portable outfit for camping or touring. It is also ideal for the family which desires to purchase a tuner and detector unit, and after becoming accustomed to its use adding the 2-Step Amplifier, **D-5, shown below,** for the purpose of actuating a loud speaker or greatly increasing the reception range. The two units are so designed, being exactly of the same size, that they can be placed one beside the other and connections made by means of short wires.

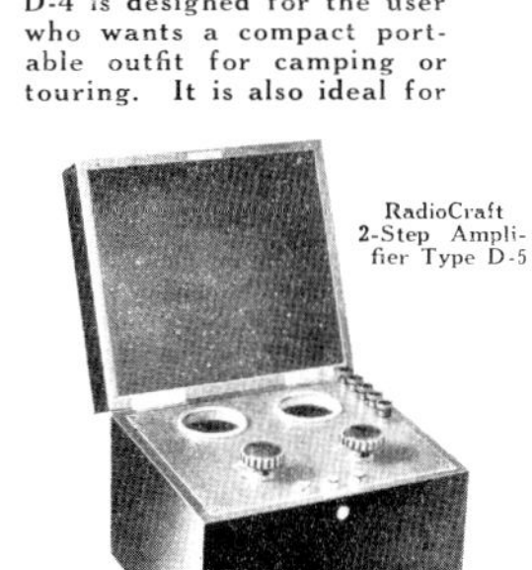

RadioCraft 2-Step Amplifier Type D-5

By permission of DeForest Radio Tel. & Tel. Co.

See Our Exhibit at Permanent Radio Fair, Hotel Imperial, New York City

The RadioCraft Co., Inc., 139 Franklin St., Jersey City, N. J.

Say you saw it in RADIO MERCHANDISING.

Radio Merchandising (Feb. 1923), p. 43

D-4 Jan. 1923 $36
D-5 Jan. 1923 $35

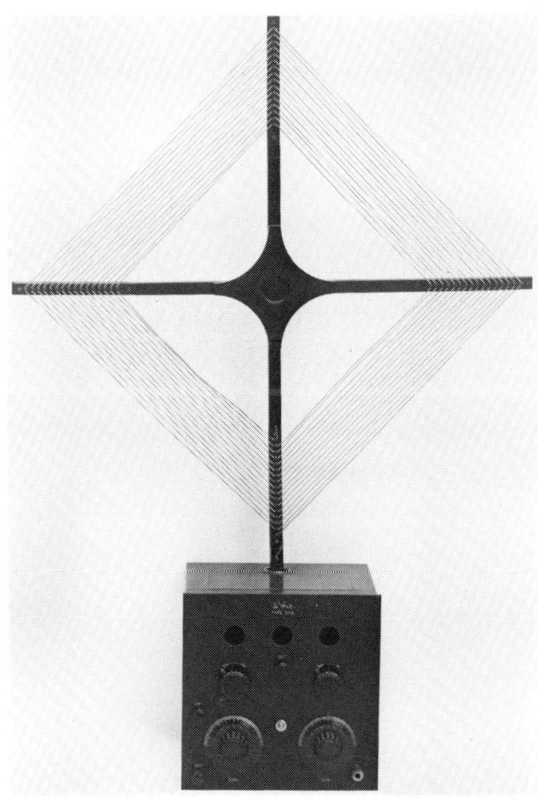

D-7A

John Wolkonowicz

D-7 Dec. 1922 $125

It became the D-7A in Feb. 1923, with shielding, a vernier condenser, and new transformer contacts. Later, the tap switches and coupler were removed.

cleaned up (if one may use that fine word for such foul business) some eight million dollars. And the stockholders, of course, were cleaned out." (*Father of Radio*, p.406). Jenkins, acquired on a stock swap, conducted research and operated TV broadcast stations, but had no income, nor hope of any.

In the end, not even $1 million from RCA could keep De Forest afloat. (This was the settlement in the "Clause nine" suit. Clause nine in the RCA license agreement forced manufacturers to equip their sets initially with RCA tubes; never enforced because of a storm of protest from the industry, it was struck down in 1931). An equity receiver was appointed on June 21, 1932, RCA purchasing the company in July 1933 for about $400,000. Lee de Forest personally formed several other companies on the west coast, the first of them in October 1930, making mostly diathermy machines.

D-10 July 1923 $150

 Also available in mahogany cabinet. Technical article in *New York Evening World*, Jan.26, 1924, pp.3–4

The newest radio triumph of the greatest name in radio—*De Forest*

Saturday Evening Post (Feb. 28, 1925), p. 176

THIS new instrument is extremely simple to operate (simplicity is one of the chief elements of greatness).

It uses a De Forest Tube as a detector—detection is immediate and accurate. Four tubes besides the detector—three stages of radio frequency amplification and two stages of audio. The very practical De Forest loop antenna is used of course. De Forest D-17 is complete and self-contained in one unit—readily movable from room to room—no wires to put on the walls—no connections inside or out.

Easy to tune

Extreme delicacy of adjustment gives the D-17 the two qualities of selectivity and distance-getting to a degree that is a surprise even to radio experts. You can tune in accurately, and tune out to change your station easily and completely. The De Forest pure tone quality—already famous—finds its best demonstration in this new set.

Strong claims? Yes, but a demonstration in your own home will bear them out. The simple technique of operation is easy to learn. You can get the essentials in a few minutes and have great fun in improving your skill as you go on.

Weigh the volume and clearness of De Forest tone in comparison with any other music-producing instrument—yes, even with the living artist standing before you on a distant stage. Once you have heard the De Forest D-17 Radiophone, you will understand why De Forest is the greatest name in radio.

Get the nearest authorized De Forest agent to bring one of these new D-17 Radiophones to your home for a demonstration. If no agent is available, write to us for information.

Prices, $125 to $200 f. o. b. factory.
(Without batteries)

DE FOREST RADIO COMPANY
JERSEY CITY, N. J.

Sit at ease in your own home. Cast out into a world-wide stream of ether waves, and catch—radio. You should know how simple the technique of radio has been made by the De Forest D-17 Radiophone.

Which will you choose?

A set that requires antenna wires, inside wiring fastened to the walls, ground wires, outside batteries, various insulating, switching and connecting devices, labor to hook up, and separate loud speaker.

or

De Forest D-17, a complete and self-contained instrument with tubes and places for batteries inside, built-in De Forest Loud Speaker, Loop Antenna—everything in one unit, compact, movable, efficient.

DE FOREST D-17 RADIOPHONE

REG. U. S. PAT. OFF.

D-17A Jan. 1925 $125

D-17L in leatherette (Fabrikoid) cabinet with speaker, $190 (later $170).

D-17M in mahogany cabinet with speaker, $200 (later $180).

Technical articles in *Popular Radio*, May 1925, pp.439–451, *QST*, Aug. 1925, pp.16–19, and *NY Herald-Tribune*, Jan. 25, 1925, pp.11,15.

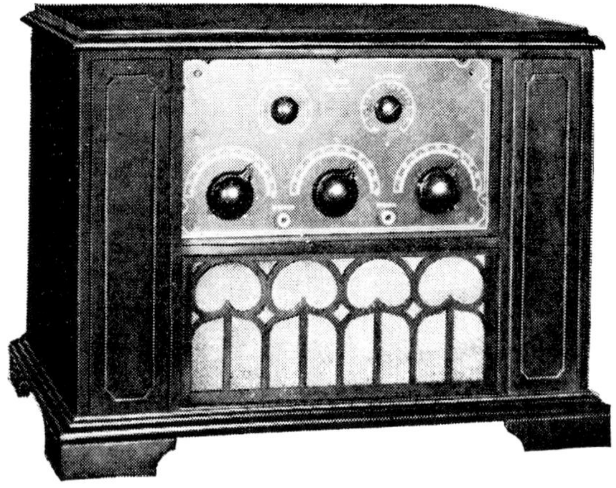

F-5-M Oct. 1925 $110
also F-5-L, leatherette.

1,720,453. MEANS FOR REDUCING INTERFERING VOLTAGES IN RADIO APPARATUS. FRANK M. SQUIRE, Hollis, N. Y., assignor to De Forest Radio Telephone & Telegraph Co., Jersey City, N. J., a Corporation of Delaware. Filed Feb. 26, 1924. Serial No. 695,286. 2 Claims. (Cl. 250—16.)

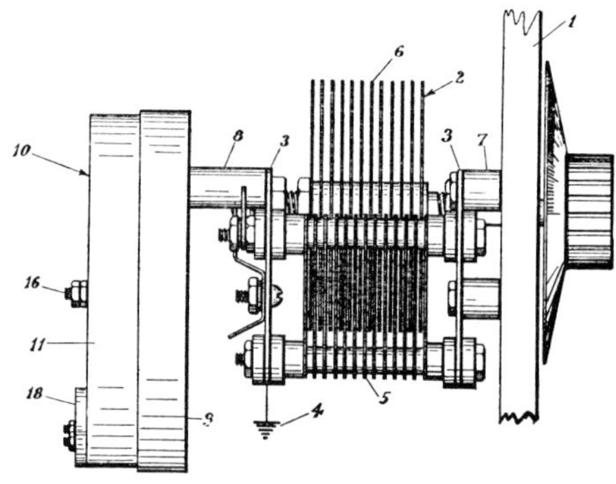

1. In combination with a radio transformer, a shield comprising two parts for supporting, prelocating and protecting said transformer, a condenser, and means for mounting the shield on the condenser.

Dave Crocker

D-12 July 1924 $161.20
with tubes and dry batteries, $180 with tubes and storage batteries, in Fabrikoid cabinet. Mahogany cabinet $15 more.
Technical article in *Radio Engineering* Oct. 1924, pp. 290–291.

D-14 Oct. 1924 $371.50

year ending	Dec. 1921	Dec. 1922	(9 mos.) Sept. 30, 1924	June 30, 1925	(6 mos.) March 31, 1929
gross sales	$108,439	$538,771	$563,240	$1,007,255	$976,486
net profit	($25,827)	$288,532	$116,482	$282,320	$176,829

year ending	March 31, 1930	March 31, 1931	March 31, 1932
gross sales	$2,767,505	$1,584,303	$744,589
net profit (loss)	($208,919)	($351,142)	($805,061)

F-5-AL or F-5-AW. July 1925 $90
leatherette or walnut. F-5-B also available, in a smaller cabinet, $85.

Radio Retailing (July 1925), p. 56

Five-Tube Portable Receiver

F-5 portable July 1925 $130

W-6-F Oct. 1925 $450
W-6-T table model (without base) $385

Popular Radio (July 1925), p. 1

Roy A. Weagant

Walt Curry

W-5-F Oct. 1925
$235 (with base)

DE FOREST

$22

This beautiful tone-recreator is winning tremendous success. Sets new standard of tone quality.

$55

A 2-tube coupled antenna circuit receiver utilizing regenerative detection and one stage A.F. Amplification. "Slow motion" low loss variable condenser. Etched metal panel. Mahogany cabinet. Price includes 2 Radiotron tubes, "Musicone" Headset and Grid Leak.

$84

A 3-tube receiver employing coupled circuit regenerative detection with 2 stages A.F. Amplification. Unusually selective. Cabinet in Adam Brown Mahogany two-tone effect. Ample accommodation for heavy duty "A," "B" and "C" dry batteries. Cushioned sockets, "slow motion" variable condensers. Price includes 3 Radiotron tubes, "Musicone" Headset, Phone Plug and Grid Leak.

A 4-tube receiver utilizing the new Super-Trirdyn circuit. Very selective—especially suited for work close to broadcasting stations. Cushioned sockets. Rich two-tone Adam Brown Mahogany Cabinet. Price includes 4 Radiotron tubes, "Musicone" Headset, Phone Plug and Grid Leak. $135

The New

Back of every DeForest & Crosley Radio lie the combined efforts of three of the most highly developed organizations in the Radio Industry. The inventive genius of Dr. Lee DeForest and his associate engineers has been combined with the fine production standards of Powell Crosley, Jr., in a product expressly designed and built to meet the exacting requirements of Canadian conditions.

DE FOREST RADIO CORPORATION

$135

Phonograph Journal of Canada (July 1925), pp. 78–79

The "Musicone" built into a fine Mahogany Cabinet that harmonizes with any setting. Acts like a sounding board—giving still mellower tone.

A 5-tube receiver utilizing two stages of Tuned R.F. Amplification, non-Regenerative Detector, and two stages of A.F. Amplification. Unique tone-recreating qualities. Incapable of howling or squealing. Readily logged. Beautiful Mahogany Cabinet with motifs in burl walnut. Ample battery accommodation. Price includes 5 Westinghouse Radiotrons, "Musicone" Headset, Phone Plug and Grid Leak.

Radio

After months of intensive study, searching tests and experiments, and weeks upon weeks of collaboration with the best cabinet designers, artists and fine metal craftsmen in the Dominion, we take pleasure in presenting the NEW "R" SERIES of DeForest & Crosley Radios.

In circuit they embody the latest thought of Radio engineering. In exterior design they represent, we believe, the ultimate in Radio ensemble. In dollar-for-dollar value they offer a standard never before available to the Canadian public.

LIMITED - TORONTO

$395

The same special type of 5-tube Tuned R.F. circuit as in the R.5 is here embodied in an elegant cabinet, fashioned in selected walnut with superb craftsmanship. A special type of loud-speaker unit is built in. Price includes 5 Westinghouse Radiotrons, "Musicone" Headset, Phone Plug and Grid Leak.

Same circuit as the R.5. In this elegant period design console, radio finds its ultimate expression — a circuit of unexcelled precision, beautiful tone properties and splendid simplicity, encased with loud speaker, in a cabinet piece of consummate artistry. Fashioned in selected walnut. Price includes 5 Westinghouse Radiotrons, "Musicone" Headset, Phone Plug and Grid Leak.

Phonograph Journal of Canada (Dec. 1924), p. 48

An Evidence of the Progress of the Radio Industry in Canada

Scenes Taken in the Factory of De Forest Radio Corporation, Ltd., Toronto, Where De Forest-Crosley Radiophones are Made

Section of one of the large assembly rooms showing line of workmen and girls engaged in assembling, wiring and soldering.

Section of Tube Testing Department, showing number of tubes being aged. Alll De Forest tubes are thoroughly tested before leaving the factory.

Section of the Testing Laboratory. All De Forest-Crosley Radiophones are thoroughly tested on actual reception before leaving the factory.

Section of the business office, showing Order and Accounting Departments.

Section of the D-12 Radiophone Room. Note the order in which the different parts entering into the construction of these Radiophones are kept.

A section of the Trirdyn Room, showing lines of workers engaged in the wiring of these Radiophones. On the left, note the inspector testing.

DeForest-Crosley Dealer Aid

It's MOST Important!

Quality, adaptability to merchandising conditions, style and price range—all are essentials of an effective Radio Franchise.

But, in considering the qualifications of a Radio Franchise, careful thought must be given to the degree of dealer support the manufacturer is prepared to extend —IT IS MOST IMPORTANT.

The records of DeForest-Crosley and DeForest-Crosley Dealers, made during the Fall and Holiday Seasons, firmly establishes the merchandising superiority of the DeForest-Crosley Franchise.

This, in turn, evidences the embodiment, in the De-Forest-Crosley Franchise, of the features essential to the establishment of pre-eminence — quality, adaptability and well-balanced style and price range—and that all-important requirement—co-ordinated co-operation, the brand of co-operation that gets results.

The Radio business at present is a veritable gold mine. Stake your claim and work it to the limit. The DeForest-Crosley Franchise offers you the most complete and modern equipment for this purpose.

Secure this Franchise. It's the most profitable thing you can do right now.

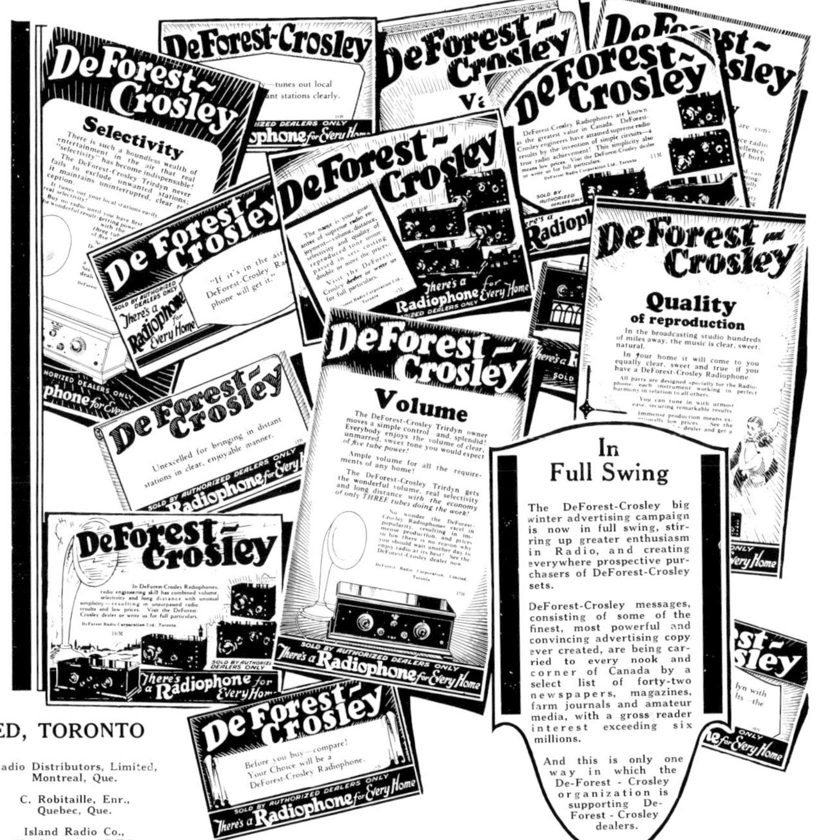

DE FOREST RADIO CORPORATION, LIMITED, TORONTO

EXCLUSIVE DE FOREST DISTRIBUTORS

Canada West Electric, Ltd., Regina, Sask.	Motor Car Supply Co., of Canada, Ltd., Calgary, Alta.	Radio Distributors, Limited, Montreal, Que.
Marshall Wells Alberta Co., Ltd., Edmonton, Alta.	Jas. S. Neill & Sons, Ltd., Fredericton, N.B.	C. Robitaille, Enr., Quebec, Que.
Marshall Wells B. C. Company, Ltd., Vancouver, B.C.	Phinney's Ltd., Halifax, N.S.	Island Radio Co., Charlottetown, P.E.I.
Marshall Wells Co., Limited, Winnipeg, Man.	Q.R.S. Music Co., of Canada, Ltd., Toronto, Ont.	

Phonograph Journal of Canada (Jan. 1925), p. 37

EAGLE

Eagle Radio Co.

Eagle is one of the shorter chapters in the history of Neutrodynes. It started in late 1922, advertising in November a portable loop on its own base "designed for . . . the Armstrong super-regenerative circuit and radio frequency" and called, reasonably enough, the Portabloop. By January 1923 Eagle also offered a Portabloop receiver, one of which seems to have found its way across the Atlantic to the Duke of York, as a wedding present.

In June 1922 the Wireless Specialty Apparatus Co. threatened legal action against crystal-set makers and dealers, for infringement of its Pickard patents, prompting them to form the Independent Radio Manufacturers, Inc. to fight WSA (see the Freed-Eisemann chapter). Eagle, so far as is known, did not make crystal sets, but it did join the IRM. WSA also owned two Pickard loop patents, and may have intimidated Eagle with these, accounting for Eagle's IRM membership.*

The IRM, by amazing good luck, arranged to build TRF radios under Prof. L.A. Hazeltine's newly-patented Neutrodyne circuit, and Eagle was fourth to get into production, after Fada, Freed-Eisemann, and Garod. In 1924 Eagle took in $226,327.16, not as much as the three early birds, but more than most of the late-comers. Assuming a wholesale price of $105 (list $175, less 40%, though it might have been 35% or 33⅓%), that would translate into just over 2000 sets, not a bad performance for a brand-new company.

And a hard act to follow, when the Neutrodyne's popularity waned. In spite of a procession of newer models, beautifully-made like the first ones, Eagle never got off the ground again. In February 1926 it had to tie up with Bamberger's department store to dispose of its sets; in mid-November it was thrown into receivership. Wurlitzer purchased the company (and its Neutrodyne license) in May 1927.

*On July 31, 1920 WSA's parent company, United Fruit, received a detailed report from its counsel Fish, Richardson & Neave on the validity of its Pickard detector and loop patents. From p.5, "We have no hesitation in saying that this Pickard loop patent No. 876,996 is an important element in the radio situation . . . it might be well worth while to try it out with the hope that it would be sustained as of controlling importance against competitors using the loop antenna . . ." (Clark Collection, Smithsonian Institution).

OSCAR O. KRAUSE
Vice-President, Secretary and Manager, Eagle Radio

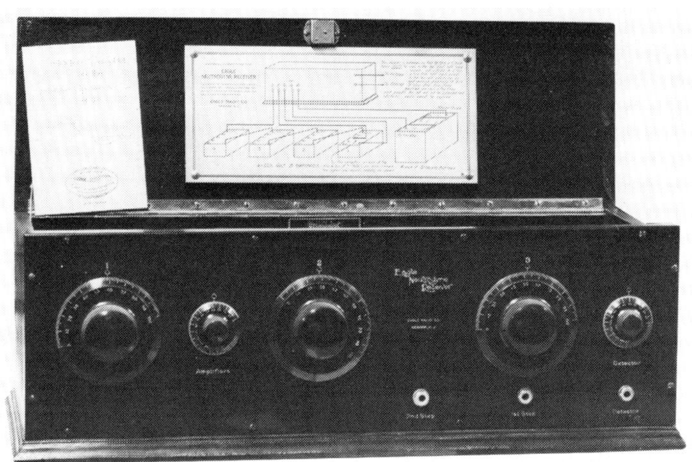

Wilson Norwood

A Nov. 1923 $175

Technical article in *Popular Radio*, Dec. 1924, pp. 568–576.

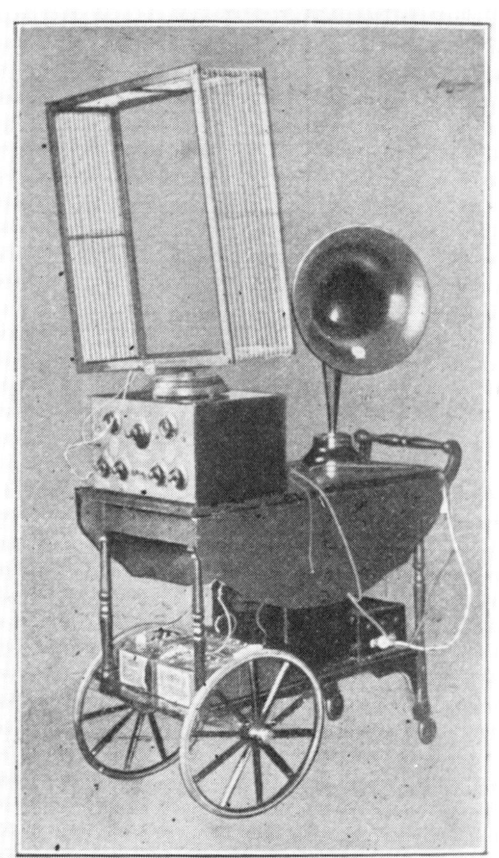

Photo: Vandyke.
The compact wireless receiving set owned by H.R.H. The Duke of York. It is installed at White ***Lodge, Richmond Park.***

Wireless World (Aug. 22, 1923), p. 698

Portabloop loop and receiver Jan. 1923

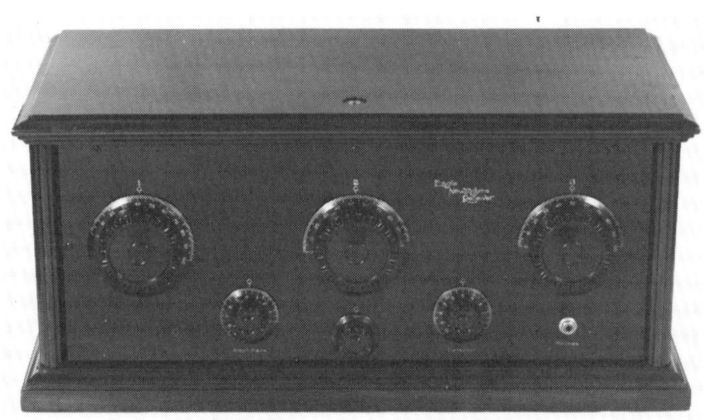

B Nov. 1924 $175
Console $275

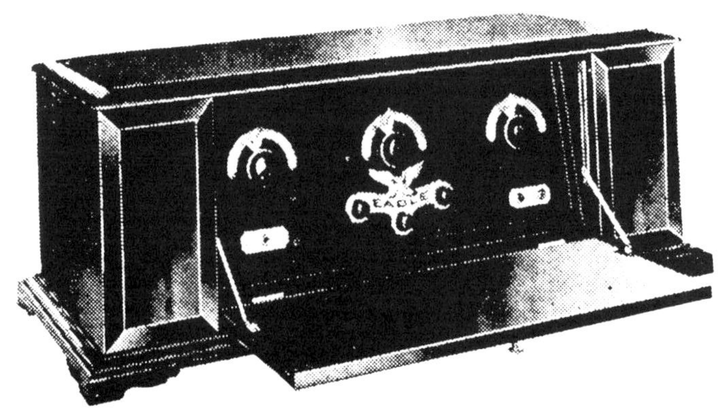

Radio Dealer (Sept. 1926), p. 72

K-2 Aug. 1926 $185
(not pictured) M Sept. 1926 $95

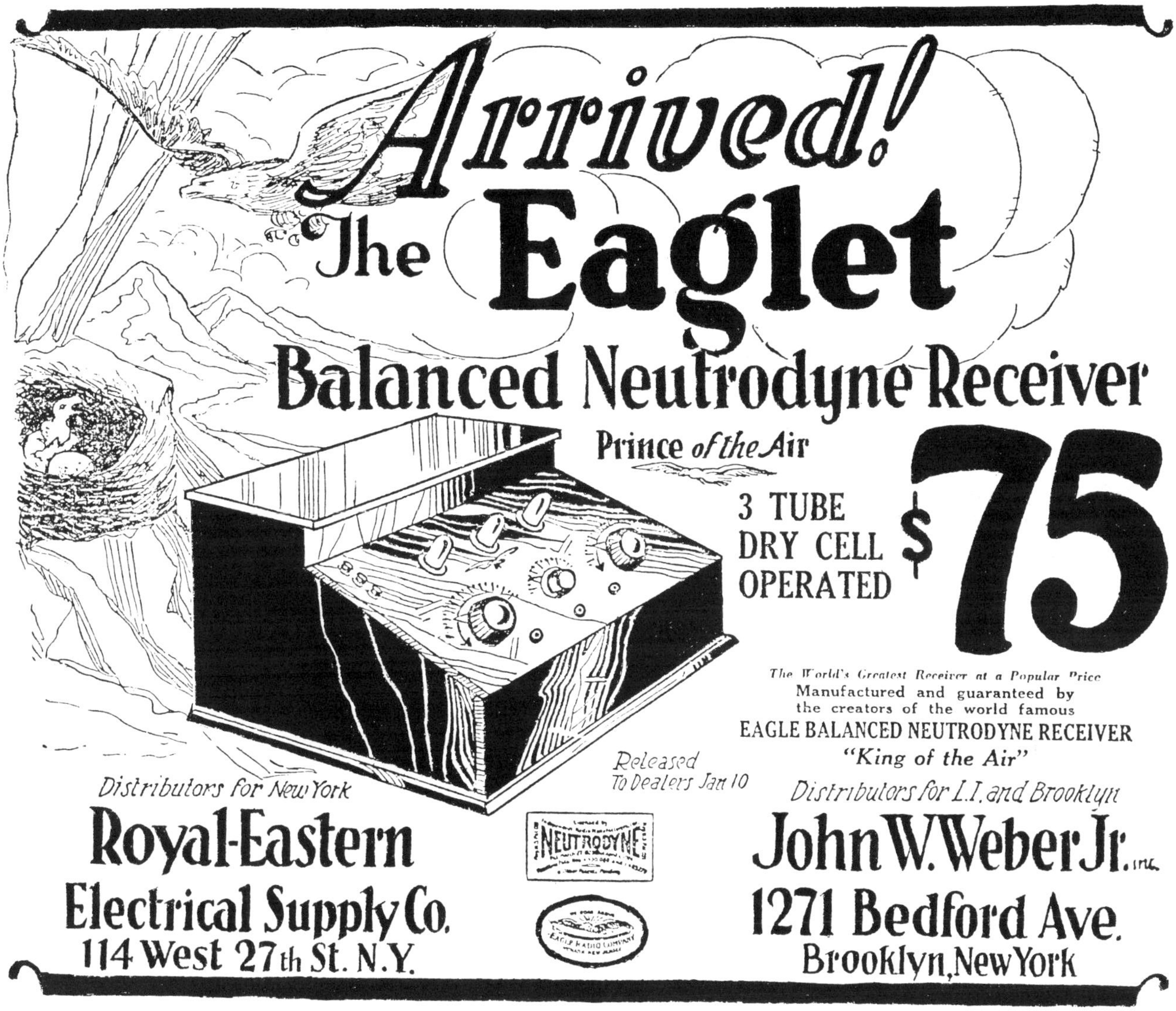

Arrived! The Eaglet

Balanced Neutrodyne Receiver

Prince *of the Air*

3 TUBE
DRY CELL
OPERATED
$75

The World's Greatest Receiver at a Popular Price
Manufactured and guaranteed by the creators of the world famous
EAGLE BALANCED NEUTRODYNE RECEIVER
"King of the Air"

Released To Dealers Jan 10

NEUTRODYNE

Distributors for New York
Royal-Eastern Electrical Supply Co.
114 West 27th St. N.Y.

Distributors for L.I. and Brooklyn
John W. Weber Jr. Inc.
1271 Bedford Ave.
Brooklyn, New York

N. Y. Herald-Tribune (Jan. 4, 1925)

EAGLE Model H RECEIVING SET. Manufactured by the Eagle Radio Co., 18 Boyden Place, Newark N. J. Five tube tuned radio frequency receiving set installed in quartered walnut finish cabinet measuring 11½" wide, 10½" high and 29½" long. Designed for use with power tube if desired. Three tuning dials, two filament controls, battery switch and loud speaker jack on front panel. List price $75.00.

Radio Dealer (Apr. 1926), p. 86

H Apr. 1926 $75
Not a Neutrodyne

EAGLE MODEL K RECEIVER with K-1 CABINET. Manufactured by the Eagle Radio Co., 18 Boyden Place, Newark, N. J. Five tube Neutrodyne receiving set enclosed in brown mahogany Duco finished console cabinet with following overall dimensions: 43¾ inches high, 40 inches long and 19½ inches wide. Art escutcheons on sloping panel. Compartments provided for speaker, batteries, charger, etc. Disappearing doors with drawers in center of table. List price $250.00.

Radio Dealer (Apr. 1926), p. 84

Apr. 1926 $175
K (later K-3)
Console Cabinet $75 extra

House Beautiful (Dec. 1925), p. 669

THE EAGLE

"All That is Best in Radio"

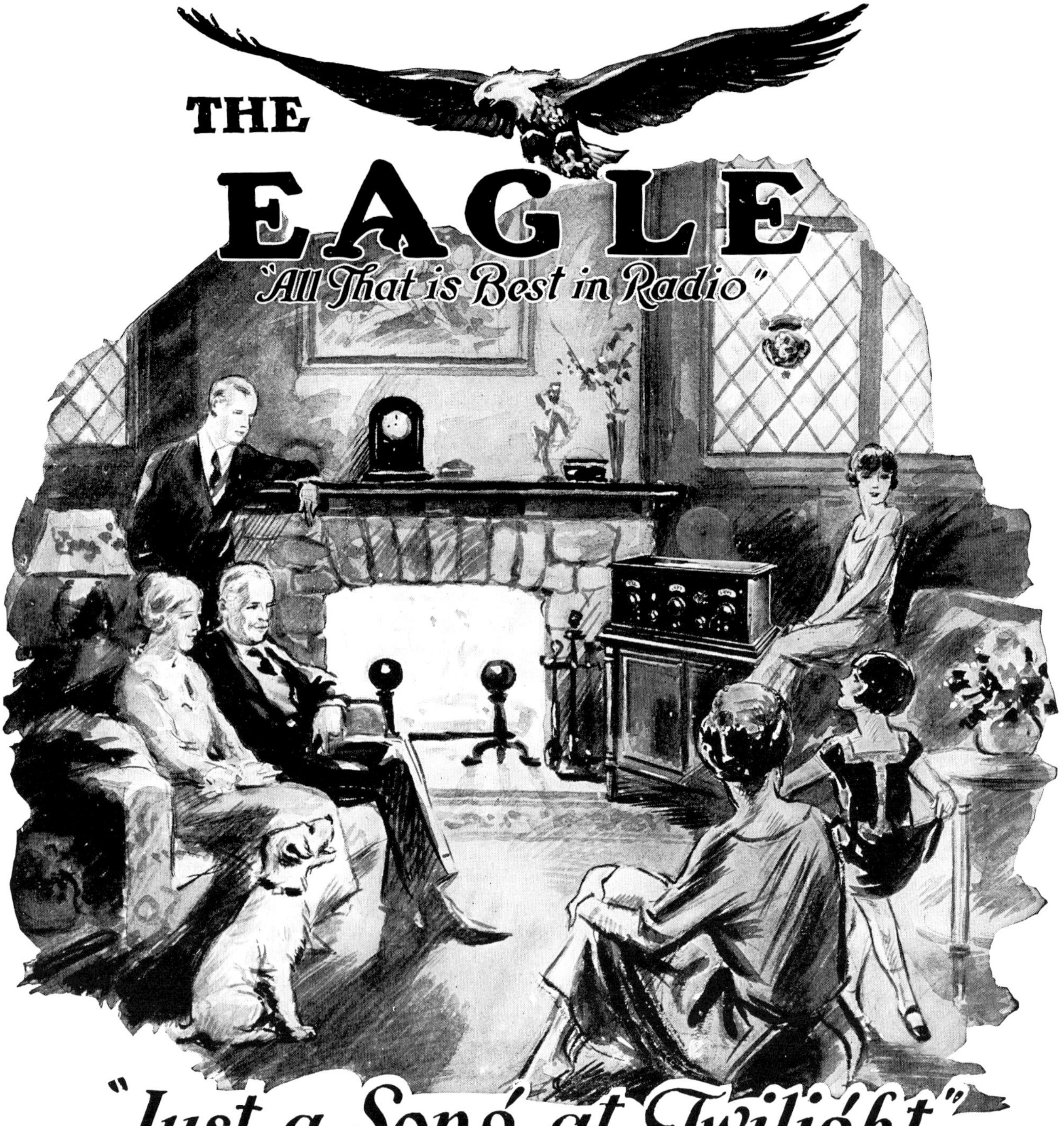

"Just a Song at Twilight"

THAT old song holds a wealth of sentiment not only for those in the twilight of life, but even the young folk, at the close of a busy day, will listen to it with keen appreciation.

You can truly enjoy this loved heart song! You can bring grand masterpieces played by world-famous orchestras, sermons, lectures, songs, dances, fun—right to your home with lifelike realism through an **Eagle Radio.**

Set shown is the table type.
Write for illustrated booklet of console models.

EAGLE RADIO COMPANY

16 BOYDEN PLACE, NEWARK, N. J.

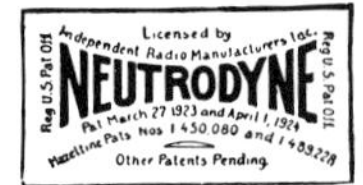

NEW

EAGLE MODELS

RADIO

MODEL D—$125.00

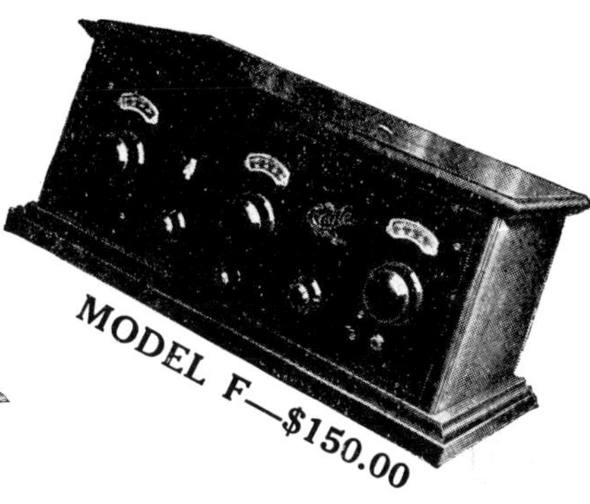

MODEL F—$150.00

CONSOLE MODEL C-1
WITH MODEL F
RECEIVER $235.00

AT ITS BEST

EQUALLED BY FEW

EXCELLED BY NONE

MODEL C-2
WITH MODEL F
RECEIVER—$250.00

The Trade Mark that stands for Dealer Co-operation and Profits

MODEL C-3
WITH MODEL F
RECEIVER—$275.00

Ask Any Eagle Dealer

Eagle Radio Company

16 BOYDEN PLACE NEWARK, N. J.

Radio Dealer (Sept. 1925), p. 89

Electrical Research Laboratories

George A. Pearson was a successful Chicago automobile distributor when he was bitten by the radio bug. Experimenting in his home workshop with all the new circuits and components as they came out, he decided to go into the manufacturing business himself, beginning in the fall of 1921 in part of a 5000 sq. ft. plant with a force of 20. Apparently his timing was good, and his products better, since in July 1923 he gave up his auto business, and a year later Erla filled a 50,000 sq. ft. building. Pearson presumably did not design the actual components and circuits himself, but hired engineers like William J. Schnell whose name appeared in magazine and newspaper articles, was later with Reichmann, then returned by 1930 as chief engineer.

Erla championed the reflex circuit beginning in April 1923, advertised parts for the Duo-Reflex in July, and by November was offering kits for one-, two-, and three-tube reflex sets. By April 1924 it had moved to a new plant at 2500 Cottage Grove Ave. and things looked so good that in November Pearson incorporated in Delaware with the intention of selling stock.

In June 1925 Erla made an operating merger with furniture maker Caswell-Runyan of Huntington, Indiana, claiming a combined capacity of 500,000 radio sets per year. Financial statements published for the "lambs" who bought stock showed sales of $22.2 and $24.2 million for 1923 and 1924, with profits of $2.9 and $4.1 million, but they didn't mention that these were only for Caswell-Runyan; Erla itself, from November 1924 to June 1925, sold only $608,091 and netted $43,321. Since nothing further was published concerning this merger, presumably Caswell-Runyan had the good sense to pull out.

From July 1925 to March 1926, Erla sold $1,835,966 worth of sets and parts, making $37,593 profit (less than the previous period, for three times the sales). Next season, it took out a license from RFL (Radio Frequency Laboratories, a Hazeltine competitor) and promoted several fancy console models. Quoting one of its New York jobbers, "the $285 (models were) supposed to have been ready for market early in September, but . . . were only delivered to us two days before Christmas." For the year ending in March 1927, Erla sold $1,208,287 and lost $79,695, at which point, with the better part of valor, it stopped publishing profit-and-loss statements, apparently doing even worse the next year.

Drastic action was in order; in September 1928 Pearson sold out to Greene-Brown, a large Chicago maker of B-eliminators who intended to combine the operations, building power packs for Erla's AC radios. Louis Frankel, formerly with Mohawk but now with a pocket full of cash, joined Burton Greene, who had a good idea but "had extreme difficulty in encompassing the 24 miles that intervened between the establishments, in his attempts to be in both places at once." Greene-Brown chief engineer Edwin Mraz joined Erla in the same capacity.

Erla succumbed to the inevitable in 1930, being reorganized as a private company in March 1931 by Frankel and another former official, Ernest Alschuler. Although the corporate name remained Electrical Research Laboratories until January 1945, it operated as Sentinel Radio Corp. from 1934, probably buying out the former company of that name. Sentinel also produced radios for tv enty to thirty private labels, some for export. Between 1938 and 1940 it moved to Evanston, Illinois, beginning TV production there in 1948. Sentinel's assets were sold to Magnavox in March 1956.

Mid-West Radio (Dec. 1925), p. 25

J.W. Caswell (left) and George A. Pearson (right) in 1925.

Radio Dealer (Oct. 1928), p. 29

At the left is a view of the machine parts room of the Erla Factory in Chicago, where modern equipment and excellent working conditions speed along production, and keep the workers happy. At the right we see a light corner of the factory where some of the preliminary work is done on Erla radio parts.

Radio Dealer (Oct. 1925), p. 125

Radio Dealer (Oct. 1928), p. 34

Burton Greene (left) and George A. Pearson (right) signing merger agreement in Sept. 1928.

Radio Engineering (Feb. 1926), p. 79

RADIO IN THE HOME

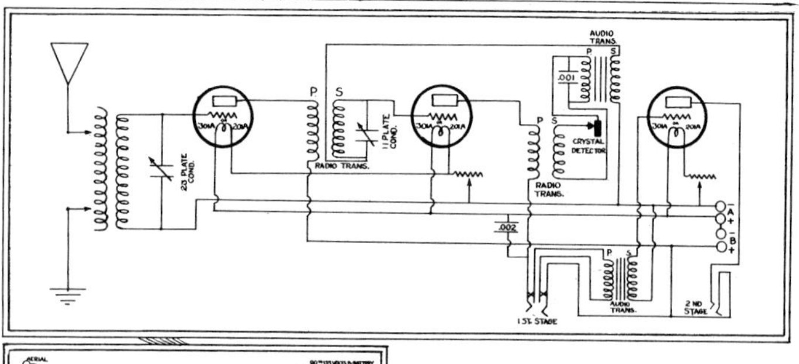

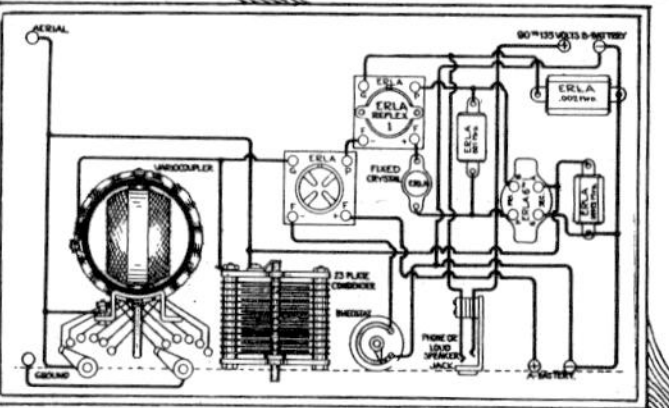

At the top of this page is the usual symbol diagram for the three-tube Erla hookup dealt with in this article
To the left is a pictorial diagram showing the hookup of the single-tube Erla reflex circuit

To the right is a pictorial diagram showing the hookup of the two-tube Erla reflex circuit, while at the bottom of this page is a picture drawing showing how to hookup the new three-tube Erla circuit which is dealt with in this article

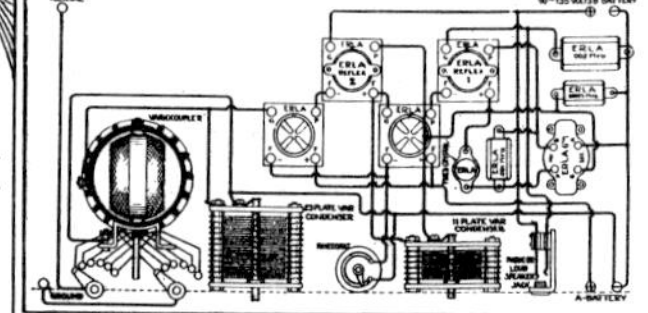

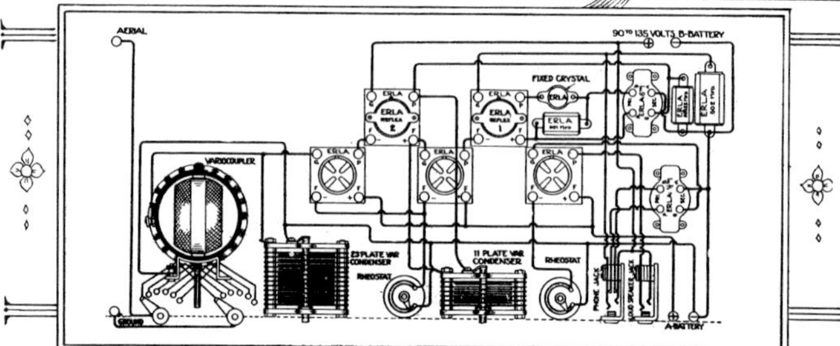

Nov. 1923

Radio Dealer (Oct. 1925), p. 191

"Erla Standard Cabinet"

Circloid Five (Standard Five) Sept. 1925 $69.50 DeLuxe $77.50. Console: $113.50. DeLuxe Console $142.50. Kit: $49.50.

Radio Dealer (Nov. 1925)

TRADE NAME: "Erla Portable Receiver."
MODEL: Portable built-in loud speaker.
TYPE: Five-tube SuperFlex.
TUBES: Five.
BATTERIES: Dry-cell "A" and "B." Not furnished.
CONTROLS: One.
AERIAL: Loop. Furnished with set.
PRICE: $145.00 without tubes or batteries.
MANUFACTURER'S NAME: Electrical Research Labs.

TRADE NAME: "Erla Floor Console."
MODEL: Cabinet built-in loud speaker.
TYPE: Five-tube SuperFlex.
TUBES: Five.
BATTERIES: Storage "A" and 60 to 90 volts "B." None furnished.
CONTROLS: Two.
AERIAL: Outdoor or loop.
PRICE: $270.00 without tubes or batteries.
MANUFACTURER'S NAME: Electrical Research Labs.

TRADE NAME: "Erla Table Cabinet."
MODEL: Cabinet.
TYPE: SuperFlex.
TUBES: Three, four or five.
BATTERIES: Storage batteries for three and four-tube sets, dry cells for five-tube set.
CONTROLS: Two.
AERIAL: Outside with three and four-tube sets, loop with five-tube set.
PRICE: Three-tube, $85.00; four-tube, $95.00; five-tube, $105.00 without accessories.
MANUFACTURER'S NAME: Electrical Research Labs.

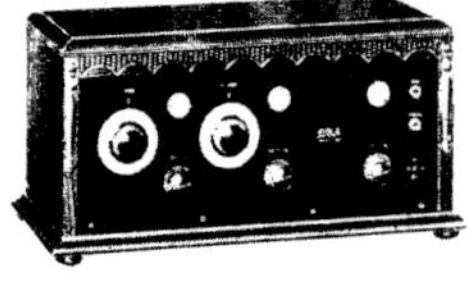

Radio News (Mar. 1925), p. 1653

Supereflex Feb. 1925 Advertised as kits since Nov. 1924. Construction of 3-tube model, *Radio Engineering*, Mar. 1925, pp. 153–160.

World's Best Radio

PEARSON

PEARSON De Luxe Floor Console Model PS-5

Cabinet 38¾ in. high, 21¼ in. wide, 16¼ in. deep, two-tone quartered and matched French Walnut. Scrollwork over a gold cloth panel conceals a fullsize, high quality loud speaker. Compartment for both charger and batteries. Contains the Pearson Navy license tuned radio frequency. Five 5-volt tubes—uses either indoor or outdoor aerial.

LIST **$146.50** without batteries or tubes.

These three sets rank with any on the market—AND THEY ARE PRICED ONE-HALF TO ONE-THIRD.

In range, volume, selectivity and tone there is no other set, regardless of price, which insures to the dealer so much turnover, profit and SATISFACTION.

Again, the Pearson Line gives more for the money. Again the Pearson dealer can sell with sensational values. Pearson challenges any dealer anywhere to read the following descriptions and not be convinced that the new Pearson 1926 Line affords his best chance for turnover and profit.

PEARSON PS-5 De Luxe

New style cabinet, French Walnut, bronze engraved panel, subpanel and dials, genuine Bakelite. Circuit licensed under Navy patent. Erla precision rheostats control the radio frequency tubes and detector. Automatic filament control for audio frequency tubes. Erla condensers.

LIST **$78.50** without batteries or tubes.

The PEARSON Five

LIST **$50** without batteries or tubes.

Every posted dealer and fan remembers the furore created by this $50.00 innovation. Like the watch that made the dollar famous, the Pearson Five was first to make $50.00 the price of the finest type of 5-tube performance. This genuine mahogany navy circuit set possesses every feature of the clarity, selectivity, volume—at any distance, and positive, easy control which have made Pearson one of the biggest names in radio.

All Pearson Circuits are Licensed under Navy Patents

Wire Today for the Slashed Prices which will enable you to run the most sensational sale in your history.

PEARSON Division, Electrical Research Laboratories, 1421 So. Michigan Ave., Chicago

The NEW ERLA RADIO

Radio Retailing (Nov. 1926), p. 161

Five-tube De Luxe Console—$146.50; Standard Console $129.50. All prices slightly higher west of Rockies.

Five-tube De Luxe Table Cabinet $95.50; Standard $89.50.

Erla six-tube De Luxe Table Cabinet, with Single Dial Control, $210.

Erla Omnitonic Cone Speaker $17.50.

Erla "Triple-Life" Tubes, Detector-Amplifier, 300 X I-A. $2.00. Power Tube, 300 X I-A, $4.50.

Hum-Free "B" Eliminator $45.00.

The beautiful new Erla "Super - Six" De Luxe Console, with Single Dial Control, $285.

You've wanted something new—something different to offer your customers. Something that will appeal to the novice as well as the expert and make propects of all. Here it is.

A new discovery is responsible for this improved radio—this new Mo-nod-ic Receiver that takes all the hocus-pocus out of radio—all the squeals and howls—and makes it simple enough for even the youngest member of the family to operate.

Super-Six (S-51?) Nov. 1926
Monodic S-5 (S-50) first advertised Aug. 1926

Ralph & Elinor Williams

'GRANADA" CONSOLE

By Electrical Research Laboratories ("Erla"), of Chicago. The finish is of dark antique walnut combined with birdseye maple. The drawer front is of satinwood with maple overlay. The chassis comprises the equipment for four r.f. stages (three of which are tuned), detector, and two transformer-coupled audio stages. Tuning is accomplished by means of a single dial, and there is a built-in loud speaker. Price $295.00. Furnished with an a.c. converter system, all tubes, and an output filter, price $395.00

Radio Broadcast (Jan. 1928), p. 226

Single-Six S-52	Sept. 1927	$90	Console	$170
Super-Seven S-61	Sept. 1927	$175	Consoles	$255, $295
75	Aug. 1928	$95	Consoles	$150, $175, $265
85	Aug. 1928	$265	Consoles	$325, $350

Saturday Evening Post (Mar. 14, 1925), p. 136

Shiny natives chasing glittering coins through clearest blue water—one of the delights of the West Indies

Erla Super-Five De Luxe Console, $225
Erla 3-tube De Luxe Superflex Console, $217.50

Erla Super-Five De Luxe Consolette, $125
Erla 3-tube De Luxe Superflex Consolette, $117.50

Erla De Luxe Super-Five Receiver, $75
Erla 3-tube De Luxe Supereflex, $69.50

More than *Wonders*

Just the wonder of hearing by air is no substitute for good taste and pleasing entertainment. A broadcasting station distributes the wonders of radio all around. But pleasure and pride in radio can come to you only from the right receiving set.

You want the clear, pure, true musical tone inherent in Erla principles. Erla clearness makes distance reception more than a stunt. At any distance, only real MUSIC or understandable natural speech is pleasing. And Erla, always rated more powerful, tube for tube, will give you any distance—*clearly*.

You want volume, certainly. But not mere loudness. Erla volume, super-abundant, intensifies clearness by enabling you to tune down for sharpest reception always.

And you like to be "boss," hand-picking your stations, instead of listening to something you cannot get rid of. Extreme selectivity not only is basic in Erla design, but simple control assures anyone of finest results.

These are the Erla betterments which now extend the whole field of radio. With distance AND clearness; volume AND clearness; selectivity AND clearness, Erla is captivating even those who have held aloof. Such advanced radio, sponsored by Erla, is so eagerly and widely received that it can be sold for less than the types it shades.

ELECTRICAL RESEARCH LABORATORIES, Chicago

Erla Super-Five Standard Model, $67.50
Erla 3-tube Standard Supereflex, $62.50

SENSATIONAL ANNOUNCEMENT

Mammoth new ERLA factory brings tremendous savings now passed on to you in a sensational NEW SCALE OF PRICES

ERLA
TRADE MARK REG

With 288,000 feet of floor space in Erla's new mammoth factory, all parts entering into the manufacture of Erla products will be produced under one roof . . . from parts to complete cabinets.

The resulting economies in manufacture are passed on to you. This sensational new scale of prices is effective at once. Your requirements can now be met immediately.

The 8-Tube Duo-Concerto Console, $119.50. This two-purpose instrument is Erla's latest triumph, engineered to give life-like reproduction of phonograph music by the simple attachment of the Erla electro-magnetic pickup. A snap of the switch instantly transports you from superb radio reception to your favorite phonograph record. The tones are all transmitted through the chassis of the Erla Duo-Concerto and through the Erla duo-dynamic speaker.

The 8-tube Duo-Concerto Radio and Phonographic Combination, $189.50. In this instrument, the phonographic reproducing mechanism is contained in the same cabinet with the radio receiving set.

Both models are GUARANTEED to provide selectivity, sensitiveness and range in excess of any other 8-tube receiver made. It is these two new Erla products which have been winning new friends for Erla dealers everywhere.

Write today for details of the Erla franchise.

Electrical Research Laboratories
2500 Cottage Grove Ave., Chicago

ERLA
Electro-Magnetic Pick-up

Another Erla triumph, the one electro-magnetic pick-up that assures perfect sound reproduction, uniform purity of tone without extraneous noises. Model P-16 with built-in volume control, replaces tone arm of any phonograph. **$13.00**

ERLA
Duo-Dynamic Cone Speaker

A major scientific advancement in sound reproduction. Actual comparative demonstration invariably proves its superiority and makes a profitable sale. Write for further information regarding 110-volt A-C cone in cabinet. **$42.50**

Radio Retailing (June 1929), p. 163

Technical Article, *Citizens Callbook* 11-1,(Jan. 1930), p. 77

Models C5F and C4F, or R-2/A-2

F.A.D. Andrea, Inc.

Frank Angelo D'Andrea's driving ambition was to get rich. From the time at age 11 when he stopped helping his father, a junk dealer, make his rounds collecting scrap, he tried a series of jobs: newsboy, prizefighter (using skills acquired as a newsboy), helper in an electroplating plant, and finally a tool-and-die maker. This last position, at the Frederick Pierce Co. who did experimental work for inventors, led him to radio when they were asked by Emil Simon to adapt a German-designed radio receiver for wartime production. After making a prototype for Simon, Pierce was unable to handle the resulting contract for the CE957 and so turned it over to De Forest for production, D'Andrea going along to supervise it.

In June 1920 D'Andrea went into business for himself, with his 16-year-old half-brother and a hired tool-and-die maker, in a little store at 1882 Jerome Ave. in the Bronx. Shortening his surname, he adopted his initials F.A.D.A. for his new company. When the radio boom hit in late 1921, Fada couldn't turn out crystal detectors fast enough, and soon was renting space in three more Jerome Ave. stores. Forty girls produced 1800 crystal detectors per day, detectors which cost 96¢ to make, and sold for $2.25. Fada's monthly receipts, at first $77.50, rose to as much as $50,000 in early 1922.

Although Fada made a line of vacuum-tube tuners, detectors, and amplifiers, its major business was its crystal detectors. Major enough to prompt the company to join the Independent Radio Manufacturers, Inc. to fight the Wireless Specialty Apparatus Co., which in June 1922 threatened legal action against all other crystal makers and dealers (see the Freed-Eisemann chapter). The IRM did succeed in enjoining WSA from publishing its misleading advertisements, but more importantly, it was the vehicle for introducing Prof. Alan Hazeltine and his newly-invented Neutrodyne circuit to a group of eager manufacturers. Fada was the first to get into production, with its model 160A, a four-tube reflexed design. This was soon joined by three kits: the $25 165A consisting of three tuned RF coupling transformer assemblies and two neutralizing condensers, the $64 166A four-tube reflex, and the $65.60 167A five-tube non-reflexed Neutrodyne.

RCA sued Fada around October 1923 for infringement of the Rice neutralization patent, but the suit was continually postponed, pending the outcome of an identical action against Garod (won by RCA in 1927).

1923 was an excellent year for Fada; sales figures were not reported, but ten years later Andrea claimed peak sales of $500,000 per month then. And Fada opened an additional plant at this time, near the original converted garage. But by March 1924 the old plant was closed down, and the new one running only until 2 PM. Business picked up again by November 1924, when Fada was behind in orders, taking in $1,103,003.91 for the year. Like many other radio companies, Fada's factory activity alternated between full employment and nothing: full during the last half of the year, in anticipation of winter sales, nothing during the remainder. Production for most companies was a matter of guesswork, since it was planned six months or more in advance. Particularly in the slow years of 1926–1927, overproduction was common, but Fada seems to have done it year after year.

Radio Merchant (May 1932), p. 71

F. A. D. Andrea

Radio Service Station

Complete facilities for panel engraving and model work for the experimenter.

Expert workmanship at prices amateurs will be glad to secure.

Watch for our next months announcement of "FADA" Radio Specialties

FRANK A. D. ANDREA

1882 Jerome Ave.
Bronx, New York

Andrea's first ad, Dec. 1920 *Radio News*

Fada had $400,000 worth of inventory in April 1925, and reportedly enough parts on hand to keep the factories busy for "the next two years." But the company was in good financial shape, and able to sit out the slow periods, without dumping merchandise. Neither was a $102,000 flood loss in April, caused by a water main break, the end of the world, especially as only overproduced inventory was damaged. Fada's reaction to the lack of business was to hire in April one of the best engineers in the business away from Western Electric, at $15,000 per year: Lewis M. Clement. He was given a blank check and the job of designing a top-notch TRF receiver. Meanwhile, the inexpensive 192A was brought out, the only new model for 1925.

Fada finally solved the problem of what to do with the mountains of excess sets by forcing its dealers to take a certain number of them along with the new models. And then proceeded to overproduce the 192A, finishing with 8500 of them in stock by January 1926, even after a strike by 500 of its 600-odd employees in November.

Fada formed a Canadian company in May 1925, first advertising to prospective dealers in July. For a time in January 1926, Fada planned a new factory, actually buying an acre of land in the Bronx, but thought better of it in June, leasing an additional plant of 10,000 sq. ft. instead in August. By October 1927 Fada had moved to Jackson Ave., Orchard & Queens St., Long Island City, New York, a plant vacated by the Rosenwasser Bros. shoe manufacturers.

In July 1926 Clement's new models were designed, and Fada began taking orders, claiming to be oversold by 2000 sets on October 15. By January 1500 were employed, and the company is said to have paid a larger royalty to Hazeltine for the last period of 1926 than any licensee had ever paid, and to have made nearly a million dollars in profits. Sales remained heavy even during the first half of 1927. Fada would have kept going on these same models, but in November 1927 was forced to discharge many employees and shut down while design of an AC set was rushed, to compete with the rest of the industry.

Clement left for a better offer at Kolster in November, F.X. Rettenmeyer taking over as chief engineer. By the following March, R.M. "Dick" Klein had quarreled with D'Andrea and had left also; he had been second in command and was as good a merchandiser as Clement was an engineer. Fada more or less fell apart, shutting down its plant by September for engineering redesign when its advertised models wouldn't work.

The company retrenched and managed to keep going until 1932, when it was sold to a group of Boston businessmen, the name changing to Fada Radio & Electric Corp. But after a 1932 loss of $266,216, and across-the-board salary cuts of 10% to 60% in mid-1933, Fada assigned for creditors in September 1934. It was revived by New York interests in November 1934, continuing in business into the 1940s. Meanwhile, Frank D'Andrea had formed Andrea Radio Corp. in 1934, taking 40 or 50 people from the defunct Fada company. He continued to run Andrea Radio until his death in 1965 at age 77, the company subsequently being headed by his son, F.A.D. Andrea, Jr. and daughter Camille.

The factory at 1581 Jerome Ave. was formerly a garage.

Fada Crystal Detector (Vertical Type).

$2.25 crystal detector

Outside view of Fada Tuner.

136-A	tuner	Oct. 1922	$21
137-A	one-tube receiver (not shown)		$27

Above: Part of Machine Shop.

Above—Office View. Right—A Corner of the Receiving and Stock Room.

Right: Punching Rheostat Pointers.

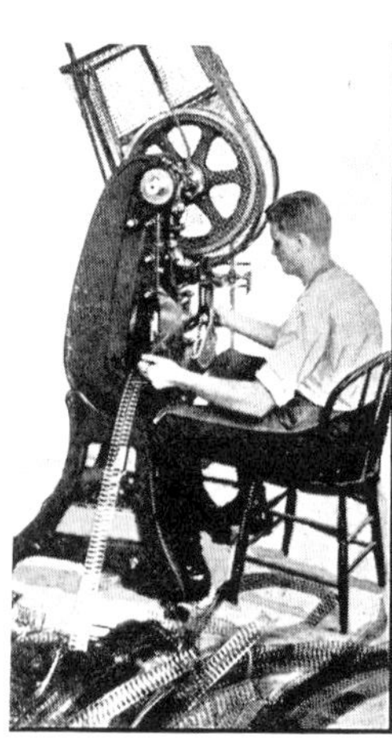

Catalog 140-A Oct. 1922 $65

Interior view of Fada Receiver-Amplifier, showing in detail the construction and perfection of every part.

NEUTRODYNE

The name of a marvelous new radio receiver circuit invented by Professor L. A. Hazeltine and used in the new FADA "One-Sixty" receiver amplifier.

Wireless Age (Apr. 1923), p. 4

Professor Hazeltine of Stevens Institute of Technology, Hoboken, N. J., after several years' work evolved the "neutrodyne" circuit which neutralizes the capacity coupling between the various parts of the circuit.

A broad license has been granted to FADA to manufacture radio equipment using this new "neutrodyne" circuit, and the "One-Sixty" receivers have been in production for several weeks.

The FADA "One-Sixty" is a four tube set incorporating tuner, two stages of tuned radio frequency amplification, vacuum tube detector and two stages of audio frequency amplification, one tube being reflexed.

From New York City, using only a 50 ft. aerial around the picture molding in a fourth floor apartment, the following broadcasting stations have been listened to, using in all cases, a loud speaker:

WSB	Atlanta, Ga.	WBAP	Fort Worth
KDKA	Pittsburgh	KSD	St. Louis
WFAA	Dallas	WGM	Atlanta
KYW	Chicago	KFI	Los Angeles
WDAF	Kansas City	WDAP	Chicago
WOAI	San Antonio	WHAS	Louisville
WOC	Davenport	PWX	Havana, Cuba

With the FADA "One-Sixty" receiver you will have the most modern receiver possible to design, and one that allows ultra-efficient reception of broadcasted concerts as well as long distance amateur 200 meter signals.

A four page bulletin has been published describing in detail the FADA "One-Sixty." We will be glad to send you a copy on request.

F. A. D. ANDREA
1581-C Jerome Avenue
New York City

The mechanical design and construction of the receiver has been given a great deal of attention and the workmanship is of the high grade class for which FADA instruments and parts are noted.

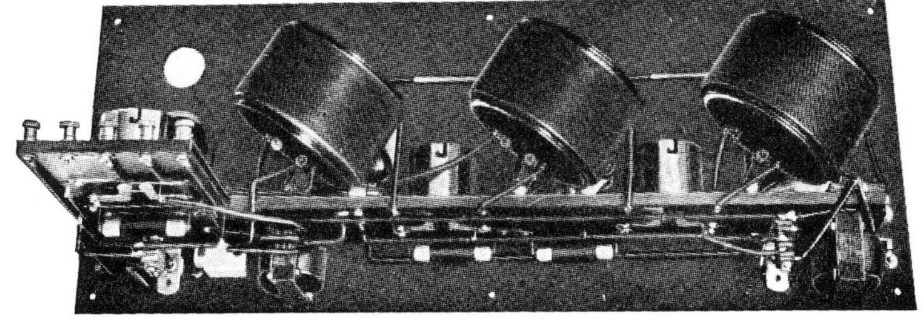

160-A Mar. 1923 $120

Andrea stated in 1929 that he had shipped his first radio in Feb. 1923.

FADA display at an amateur radio convention in New York, March 1-3, 1923

F. A. D. Andrea Had This One

Modulator (Apr. 1923)

Probably the best decorated booth of the lot was that of F. A. D. Andrea. An orange background was erected in the center of which a large map of the U. S. was framed showing the wonderful results obtained with the new circuit that F. A. D. A. is putting out. This set is built under license secured from Professor Hazeltine and is known as the Neutrodyne Circuit. It is brand new and if Stark is to be believed, it must be "the berries."

Mar. 1-3, 1923

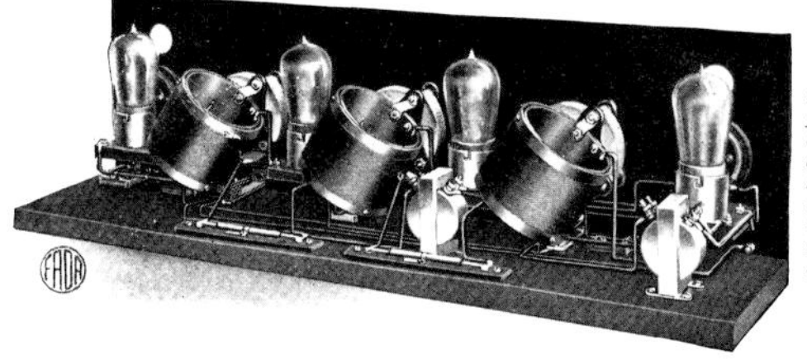

Rear view photograph of the four tube "Neutrodyne" receiving set. Note that the audio transformers are spaced well apart on the baseboard. This reduces the length of the wiring and reduces the capacity of the circuit.

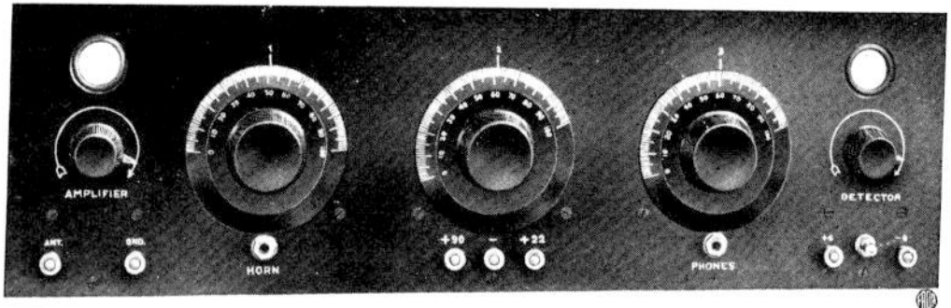

Photograph of panel front of home-made four tube "Neutrodyne" receiver giving two stages of tuned radio frequency amplification, vacuum tube detector, and two stages of audio frequency amplification.

166A kit $64

The rear view of the five tube home-made "Neutrodyne" receiving set. Audio frequency amplifying transformers on this set are both placed to the rear of the triple tube socket.

The front panel lay-out of the five tube "Neutrodyne" set is somewhat different, three of the vacuum tubes being placed at the right hand end of the panel.

167-A kit $65.50

The 169A (Aug. 1924, $72) is similar, with binding posts moved to the rear.

192-A Aug. 1925 $85

various consoles to $300.

EASTER RADIO SPECIALS

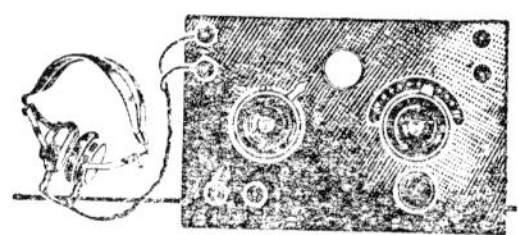

A beautifully engraved Remler audio detector control panel. A compact, good-looking, ideal piece of apparatus. List $8.50. Complete and ready to go to work **$3.85**

NATHANIEL BALDWIN PHONES

$7.29

This price good for Saturday only. One to a customer. No mail orders.

HAZELTINE'S NEUTRODYNE RECEIVER

The howl-less wonder of radio! The Fada "One-Sixty," especially constructed for local and long distance radio concert reception, is on demonstration here.

FADA No. 160 $120

ATWATER KENT

List	Item	Price
$ 8.00	Variocouplers	$ 5.75
8.00	Variometers	5.75
10.00	Mounted Variometer	7.75
13.00	Mounted Coupler	11.25
12.00	Couple Circuit Tuner	10.25
6.00	Detector Unit	3.65
13.00	Detector & 1-step	9.55
14.00	2-step Amplifier	9.85
	Mounting Boards	2.85

MISCELLANEOUS

LOUD SPEAKERS

List	Item	Price
$22.50	Bristol, Jr.	$15.95
$40.00	Bristol, Sr.	32.50
$1.85	Bradleystats	1.45
$7.50	Sleeper moulded variometers	5.65
$7.00	R.C.A.—U.V. 712 transformer	5.00
$5.00	Baldwin radio freq. transformer	2.28
$9.00	Original All Wave Coupler	5.85

$5.00 Sleeper Radio Freq. Trans. **$1.69**

Radiotrons U. V. 200, 201 and Cunningham 301 in stock.

J. L. LEWIS

132 WEST 32nd STREET

OPP. GIMBELS **OPEN SAT. TILL 9 P. M.**

CONVENIENT TO 33rd ST. "L", B. R. T., HUDSON TUBES AND PENN. STA.

The first Neutrodyne ad, from the *New York Globe* of Mar. 24, 1923

Radio Retailing (June 1926), p. 598

Here is one of the few photos of Frank A. D. Andrea that is in existence today. Mr. Andrea is reported to be extremely camera-shy, and enjoys having his picture taken as much as having a tooth pulled, which is nothing at all. However, on his return from a recent European trip, Mr. Andrea felt it necessary to doff the chapeau at Miss Liberty in New York Harbor and the cameraman sneaked up unawares.

Popular Radio (Jan. 1925)

The Best in Radio Equipment

The FADA Neutro-Junior

—a wonderful performer

MANY people prefer a Neutrodyne radio receiver but have deferred buying because they do not want a set that requires more than three tubes.

The FADA Neutro-Junior is a **three-tube** Neutrodyne. It is far superior to ordinary three-tube sets, for it possesses many of the desirable qualities of the larger Neutrodyne receivers. In volume, in selectivity, in ability to get distant stations and in tone quality the Neutro-Junior is an amazing performer. Enclosed in a solid mahogany cabinet it is a remarkable value at $75, enabling almost anyone to own a handsome, efficient radio receiver at a price they can afford to pay.

The Neutro-Junior has only two control dials, both of which read practically the same at all times. This ease of control simplifies your tuning.

The FADA full line of Neutrodyne receivers includes six models—three, four and five-tube sets in plain as well as in art-craft cabinets at a price range of $75 for the three-tube set to $295 for the five-tube Neutrola Grand.

Your dealer will be glad to demonstrate the Neutro-Junior or any of the other FADA models. Ask him.

F. A. D. ANDREA, INC., 1581 JEROME AVE, NEW YORK

FADA Neutrola Grand

No. 185/90-A The five-tube Neutrola 185-A mounted on FADA Cabinet Table No. 190-A. Price (less tubes, batteries, etc.) $295.

FADA Neutroceiver No. 175-A

Mahogany cabinet. Inclined panel and roomy battery shelf. 5 tubes. Price (less tubes, batteries, etc.) $160.

All apparatus advertised in this magazine has been tested and approved by POPULAR RADIO LABORATORY

175-A	Aug. 1924	$160	185-A	with speaker	$270
170-A	chassis only		195-A	Sept. 1924	$75

Radio Dealer (Dec. 1924)

NEW FADA PANELS

For Victor Models—List $110

The 197A Fits the No. 215 Victor
The 196A Fits the Nos. 400-405-410 Victor

Standard radio products, prompt deliveries and courteous dealings constitute the Triangle Pyramid of Service. Phone or wire your requirements to-day.

TRIANGLE RADIO SUPPLY COMPANY, Inc.
120 WEST 23rd ST., NEW YORK, N. Y.
Telephones: CHELSEA 4240-4241-4242

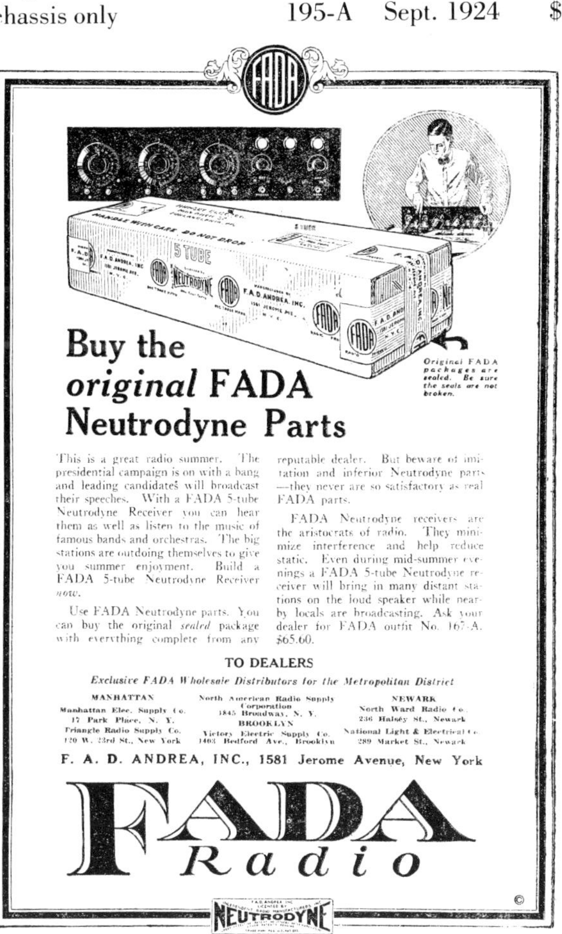

196-A, 197-A, Sept. 1924, $100.

National Geographic (1925)

FADA has set a standard of reception by which, more and more, all radio performance is being judged.

Learn what this standard offers *you*—by a complete demonstration in your own home, without obligation to buy.

The joy of pure, bell-like tone and full volume—the thrill of real distance—the ability to tune in with precision the stations you want as easily as you tune out those you don't want. And Fada Service that *guarantees* permanent performance!

Call up your dealer now and arrange for a demonstration tonight. You'll discover an entire new level of radio enjoyment.

Most Fada dealers will be glad to arrange convenient terms of payment. Send for book E, "Fada Radio—Standard of Reception".

F. A. D. ANDREA, INC.

CHICAGO NEW YORK SAN FRANCISCO
FADA RADIO, LTD.—TORONTO FADA RADIO, LTD.—LONDON

Manufacturers of TUNED RADIO FREQUENCY *receivers using the highly efficient* NEUTRODYNE *principle*

There is a Fada Radio model for every purse—all 5-tube Neutrodyne sets for dry cell or storage battery tubes, from $85 to Art Cabinet models up to $400. Illustrated is the Neutrola-Grand at $225.

LICENSED BY
INDEPENDENT RADIO MANUFACTURERS, INC.
REG U S PAT OFF
NEUTRODYNE
REG U S PAT OFF
PAT MARCH 27 1923 AND APRIL 1, 1924
HAZELTINE PATS. NOS 1450 080 AND 1489 228
OTHER PATENTS PENDING
F A D ANDREA INC., 1581 JEROME AVE, N.Y.C.

Fig. 4 An eight-tube tuned-radio-frequency receiver which is completely shielded, and is extremely easy to tune and control.

Al Monroe

480A	July 1926	$300
SF 50/80	console	$400

Technical article in *Popular Radio*, Sept. 1926, pp.429, 462–465.

Fig. 3. The coils and condensers of the R.F. stages are here shown exposed by the removal of the top sections of the shields.

Radio News (Dec. 1926), p. 651

460A (R60 chassis) July 1926 $150

In same consoles as 170A in 1925:

SF 10/60	Davenport table	$225
SF 20/60	Beethoven Grand	$250
SF 30/60	Queen Anne Desk	$300
SF 40/60	Console	$275

Lewis M. Clement

Ray Chase

265A Special	June 1927	$95
AC Special	Feb. 1928	$160

Technical article in *Radio Broadcast*, Dec. 1927, pp.128–130.

Window Display of John M. Norr, Jr., Elizabeth, N.J. 1927 (Fada-Sales 5-1)

The Fada 480-B

A VIEW OF THE CABINET

THE SHIELDING ARRANGEMENT

480B Sept. 1927 $300

Short technical article in Radio Broadcast, Aug. 1928, p.221.

Radio Broadcast (Aug. 1928), p. 221

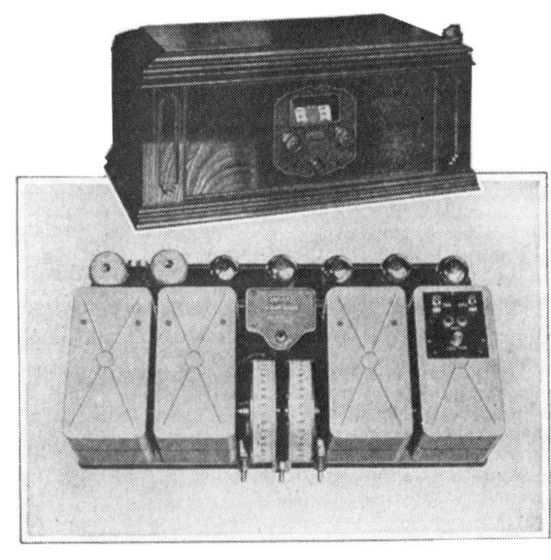

THE FADA 7

Another receiver employing four stages of r.f., but two tuning controls are used with this model. A loop is supplied with the Fada 7 although an outside antenna may be used successfully. The loop fits into a special clamp on the side of the cabinet. The two audio stages are transformer-coupled. A special arrangement in the detector circuit reduces the possibilities of overloading. Price, $185.00

Radio Broadcast (Jan. 1928), p. 227

475A Seven	Sept. 1927	$185
SF 45/75 console		$285
AC Seven	Feb. 1928	$250
AC console		$350

10 June 1928 $110
12 (DC) Aug. 1928 $120

Fada-11

Fada-Sales 5-6

11 June 1928 $135
17 Jan. 1929 $135

Short technical article in *Radio Broadcast*, Dec. 1928, p. 122.

Fada-Sales 5-6

Fada-Sales 6-4

Fada-Sales 5-6

30 June 1928 $187.50

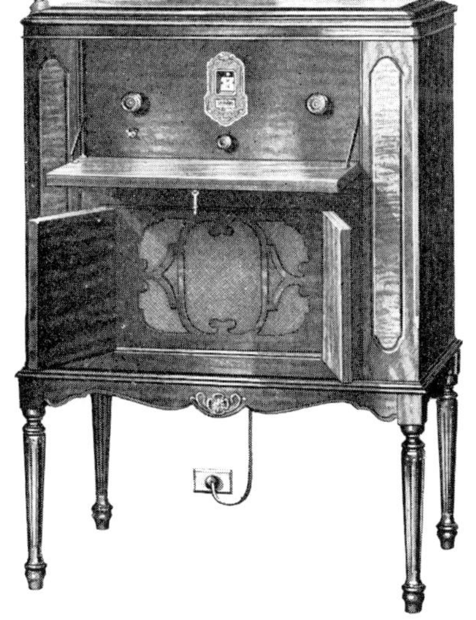

Fada-Sales 5-6

70 July 1928 $340
Two types, with 71s or 10s in audio output.

Fada-Sales 5-6

31 June 1928 $235.

Fada-Sales 7/8

Frank A.D. Andrea and chief engineer F.X. Rettenmeyer (later with RCA) in the Fada laboratory about July 1929.

Radio Retailing (June 1929), p. 50

72 radio-phone Nov. 1928
77 June 1929 $675

Fada-16

Fada-Sales 6-1, p. 514

16 Jan. 1929 $110
20 June 1929 $99.50
18 Jan. 1929 $120 DC
22 Oct. 1929 battery

Radio Retailing (June 1929), p. 50

75 June 1929 $360

35 June 1929 $245
35B Nov. 1929 $255
35C Nov. 1929 $220

Technical article in *Radio Broadcast*, July 1929, pp.171–173.

Fada-32

Fada-Sales 6-1, p. 515

32 Jan. 1929 $225

Radio (Aug. 1929), p. 43

25 June 1929 $165
Technical article in *Radio*, Aug. 1929, pp. 43–44.

FEDERAL

Federal Telephone & Telegraph Co.

The Federal Telephone & Telegraph Co. was incorporated on December 28, 1908 to consolidate a number of independent telephone companies in western New York State. On March 1, 1918 it disposed of its operating interests to the New York Telephone Co. (wholly owned by AT&T) and was left with the manufacturing portion: the plant formerly run by the Century Telephone Construction Co. in Buffalo, a branch in Bridgeburg, Ontario, and the plant of the General Drop Forge Co. Federal made telephones, switchboards, and accessories, and soon, radio parts; many of its personnel were amateur radio operators.

When the public discovered radio in late 1921, Federal had the obvious idea of making complete radio sets to cash in on the boom. It started with a detector-amplifier, then a two-stage amplifier, in November and December of 1921. Since Federal did not make rheostats and tube sockets at this time, it arranged to swap headsets and transformers to Adams-Morgan in return for Paragon rheostats and sockets. Paragon was too busy with its own orders to fill Federal's on time, so Federal eventually cancelled and switched to another maker (possibly De Forest); meanwhile it had already delivered the headsets and finally had to go to court to get paid for them.

The upshot of all this was that Federal was late in getting the units to market. When it displayed its line at the New York Amateur Show on March 7 to 11, 1922, in setting up a dummy station for a series of publicity shots, it used two detector-amplifier units rather than one, and the two had different knobs, suggesting that Federal had scavenged all its prototypes to outfit the booth. And these units were not nationally advertised until April, when business was essentially nil.

Federal did better with its crystal set, which used only its own parts. Appearing on a blueprint dated January 24, 1922, it was being advertised by March, and no doubt had substantial sales, so the season was not a total loss. Later that year Federal added a matching two-stage amplifier but relatively few were sold and it lasted less than a year in the line, whereas the Jr. went on into 1925.

While business was in the doldrums in mid-1922, Federal's engineers used the time to design a new line of RF amplifier units and two complete receivers. The amplifiers, nos. 55 and 56, are so rare that it is questionable how many were actually marketed, but no doubt exists about the 57 or 58. The 57 was the simpler, single-circuit set, not as selective but well-suited to broadcast conditions at the time when all stations were on two frequencies. The 58 with its double-circuit tuner was intended for distance reception. Later in the season Federal offered the type 60 double-circuit tuner for use with the earlier detector and amplifier units.

Thus far, Federal had done rather well with its high-priced (double the industry average) quality line, so for 1923–1924, it was more of the same. As stations were now spread over the broadcast frequency band, the new sets had to have double-circuit tuners. For the simpler set, type 59, Federal added a few frills to the old 58, increased the panel height, and built it into an imposing mahogany cabinet. For the 61, so that it would work with a loop antenna, two more RF stages were added (plus several more front-panel controls to run them), and not only was the radio imposing, so was the price, $223.

Federal was lucky. In those days when all radios were technical-looking black boxes with frightening arrays of knobs and dials, the 61 didn't scare the fans away. In fact the number of dials was a status symbol for the owner who knew how to manipulate them, and the Federals could perform superbly. Furthermore, there was little competition for this high-performance part of the market. RCA, still trying to get its new superheterodyne into production, could only offer the $162.50 Radiola VI, an antiquated beast. Grebe's top-of-the-line CR12 had most of the bells and whistles, and was a far better value at $175, but few customers ever got a chance to compare them head-to-head. Even the popular Neutrodynes weren't cheap, at $150 or more. So Federal made hay.

In July 1924 Federal's plants were transferred to the Federal Telephone Manufacturing Co., which had been organized in November 1923 for the purpose.

The sun couldn't shine forever. RCA finally brought forth its superheterodyne which sold in record numbers, taking much of Federal's market. As public tastes moved away from complicated sets crowded with controls, Federal's business went rapidly to pieces. When millionaire owner Burt G. Hubbell died on January 24, 1925, possibly as a result of business pressures, his successor lost no time in salvaging what he could. Overproduced 61s were dumped to Bloomingdales where they appeared at ⅔ off list.

Stocks of parts were used up in making variations of the older models, such as the 161, a gold-trimmed 61. Finally in May the company was put up for sale, at $2 million.

Lawrence C.F. Horle had been hired as chief engineer in 1924, and when the company now failed to sell, put his new Ortho-Sonic models into production. About this time, September 1926, the Federal Radio Corp. was formed, probably as a sales organization since the two older names continued to exist. No sales figures are available for the Ortho-Sonics, but they were successful in keeping the wolf from the door for several years. On July 3, 1929 however, Federal went into receivership; broadcast station WGR was sold to a Buffalo group, while Federal was bought by the Acoustic Products Co., owner of Sonora.

No. 8 Detector and One Stage Amplifier

Catalog, p. 15

8 April 1922 $52

A New Two-Stage Amplifier

This welcome new addition to the Federal line is meeting with the same popular approval that has so consistently distinguished all

RADIO APPARATUS

Metal shielding; metal diaphragm between each stage, tube sockets and transformers of latest design and mounted on metal brackets; metal brackets grounded to cabinet and one side of filament battery; eliminates howling; highest degree of amplification; rheostat of ample capacity to prevent heating; sturdy construction; beautifully engraved panel with binding posts marked for ease in connection.

Federal Telephone & Telegraph Co., Buffalo, N. Y.

Modulator (Jan. 1922)

Note the Paragon knobs.

No. 9 Two Stage Amplifier

9 Jan. 1922 $58

DEMAND

FEDERAL JUNIOR

RADIO RECEIVING SET.

Price, complete, $25.00.

For more than ten years FEDERAL Radio Apparatus has been Standard throughout the World.

FEDERAL TELEPHONE & TELEGRAPH CO.

BUFFALO, N. Y.

N. Y. Globe (Feb. 25, 1922), p. 22

Herb Parsons

Jr.	Feb. 1922	$25 (later $20).
Jr. Amp.	Aug. 1922	

No. 55 Two Stage R. F. Amplifier

Catalog, p. 14

55 Oct. 1922 $58

Lou Lindauer

57 Oct. 1922 $98

No. 56 One Stage R. F. Detector

Catalog, p. 14

56 Oct. 1922 $52

Lou Lindauer

58 Oct. 1922 $116

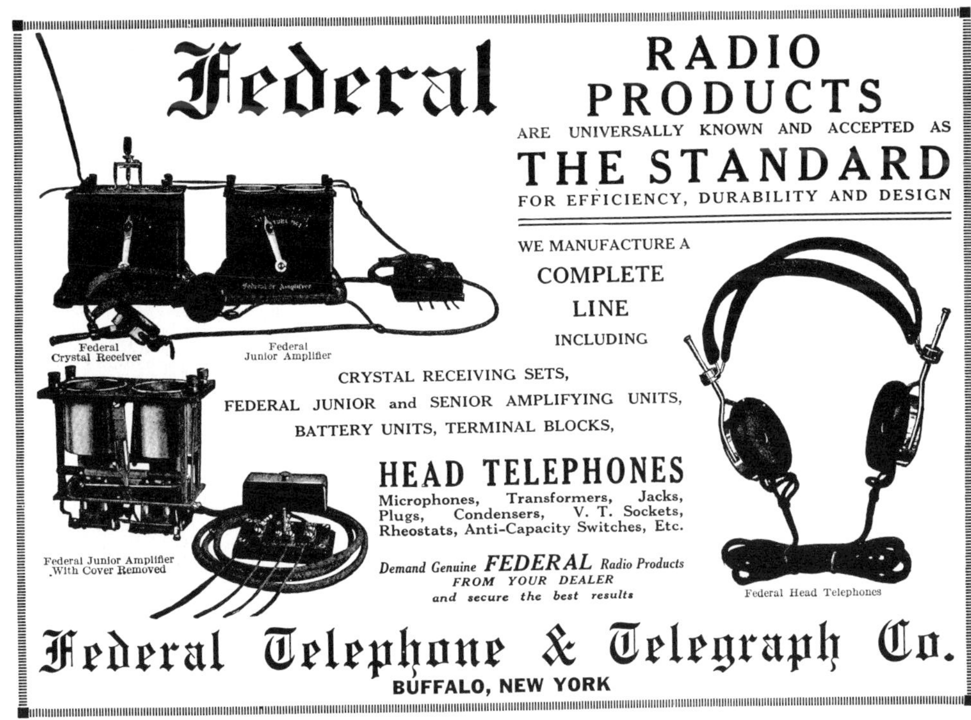

Radio News (Oct. 1922)

Lou Lindauer

110 Sept. 1923 $87

The 110 used a less-selective single-circuit tuner, the same as the older 57.

FEDERAL insures to its users only the highest refinement of the art. Each and every manufacturing necessity to produce a harmonious radio set is known to Federal Engineers, and no Federal set is produced without them.

Federal radio sets are built with two thoughts in mind—first: simplicity of operation, but three controls being necessary—second: special controls for refinement for "lovers of good music" interested in reproducing all the beauty of tones that fill the air.

FEDERAL TELEPHONE & TELEGRAPH CO.
Buffalo, N. Y.

Boston, New York, Philadelphia, Chicago, Pittsburgh, San Francisco, Bridgeburg, Canada.

Popular Radio (Aug. 1924), p. 12

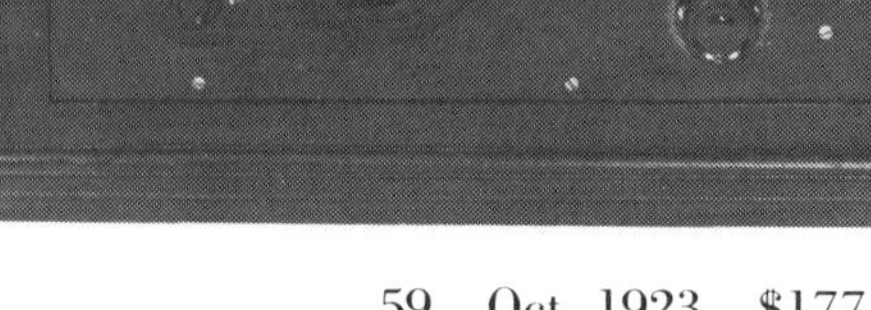

59 Oct. 1923 $177

61 Oct. 1923 $223

60 Oct. 1923 $42

141 Dec. 1924 $150

Lou Lindauer

102 May 1924 $140

TRADE NAME: Federal.
MODEL: 142.
TYPE: Built-in loud speaker.
TUBES: Five.
BATTERIES: None furnished.
CONTROLS: Two.
AERIAL: Outdoor or indoor.
PRICE: $230.00 without accessories.
MANUFACTURER'S NAME: Federal Telephone & Telegraph Corp.

TRADE NAME: Federal.
MODEL: 143.
TYPE: Built-in loud speaker.
TUBES: Five.
BATTERIES: None furnished.
CONTROLS: Two.
AERIAL: Outdoor or indoor.
PRICE: $330.00 without accessories.
MANUFACTURER'S NAME: Federal Telephone & Telegraph Corp.

TRADE NAME: Federal.
MODEL: 144.
TYPE: Built-in loud speaker.
TUBES: Five.
BATTERIES: None furnished.
CONTROLS: Two.
AERIAL: Outdoor or indoor.
PRICE: $330.00 without accessories.
MANUFACTURER'S NAME: Federal Telephone & Telegraph Corp.

142 Mar. 1925 $230

Radio News (Mar. 1925), p. 1655

GIMBELS Presents for the First Time in New York: the

PhonoradiO

The President Speaks---Then You Touch a Lever--- And Play Your Favorite Song!

That's how simple the Phonoradio works! The perfect union of Radio and Phonograph. A fitting offering for Gimbels, an organization that has performed marvels in the popularization of Music; a climax in the career of New York's leading retailer of Radio equipment. Phonoradio! The combination of one of the world's greatest Phonographs, the Emerson, with the four tube Federal Radio, designed and built specially for the Phonoradio by one of the pioneers in the industry, the Federal Telephone and Telegraph Co., of Buffalo, New York. So simple that a child can operate it. No sound box to change. No loud speaker to adjust. So wonderful that with one little touch, presto! the Radio becomes a Phonograph or the Phonograph is transformed into a Radio! And both of them satisfactory as separate units and as an harmonious whole.

The reproducing unit of the Radio as well as the Phonograph utilizes the famous Music Master Reproducing Horn, well known for its clear, mellow tones, with no tinny sounds or throaty distortions. The loud speaker unit is connected permanently inside the sound chamber and a slight push or pull of the lever (see illustration) throws one or the other into use. All batteries and other accessories are housed in the left hand side of the console, under the radio panel. No unsightly wires! The top may be lowered while either the radio or the phonograph is operating!

From Phonograph to Radio and Back in a Flash!

Famous Federal Radio Equipped With Storage Batteries

One of the simplest to operate, the Federal 4 tube Receiving Set is also one of the best. It employs one step of radio frequency amplification, a detector tube and two steps of audio frequency amplification. The circuits are so arranged that absolute freedom of interference from unwanted stations is assured. Storage Batteries! Which mean great power, efficiency and volume. All instruments are concealed; only the three tuning controls are exposed. This receiving set has remarkably clear tonal quality.

And

$280 Complete With All Accessories

Wherein This Newest Development of the Phonoradio Unit Differs from All Others—

The Phonoradio is not just a "put together" instrument. The radio has been carefully designed to harmonize in every particular with the phonograph unit—and the entire instrument perfected by one of the most famous phonograph makers in America. No sound box to change. No loud speaker to adjust. The famous "Music Master" Horn is built in and serves both as loud speaker and phonograph reproducer. No handle to pull off—you just touch a lever and you have a radio or a phonograph at will.

Famous Emerson Period Phonograph: With Music Master Horn

Encased in an artistic Cabinet of the "Queen Anne" Period, the Phonoradio possesses a simple architectural charm. It is brown mahogany finish, with nickel plated trimmings, improved Emerson throw-back tone arm, special Emerson double spring motor, 12 inch turn table, automatic stop, tone modifier and album filing system. The big feature, of course, is the marvelous reproducing "Music Master" Horn of straight-grained spruce, the wood of which the best loud speakers are made, producing a wonderful velvety richness of tone. This is the mouthpiece of both Phonograh and Radio.

Convenient Terms of Payment May Be Arranged

Included at the Price

1. Beautiful Emerson Period Console Phonograph.
2. Federal Four Tube Radio Receiving Set.
3. Famous "Music Master" Horn (Built in).
4. Four UV201A Tubes.
5. One large size Philco 6 volt Storage Battery.
6. Two large size 45 volt French Ray-O-Vac B Batteries.
7. Aerial Equipment.

Come In and See the Phonoradio Demonstrated

OPEN TONIGHT

5:30 to 9 o'Clock

After 5:30 use the 33rd Street Entrance, near Broadway.

GIMBELS PHONORADIO SHOP—Eighth Floor.

If You Cannot Call Mail this Coupon for Complete Details

COUPON

GIMBEL BROTHERS, B'way and 33rd St.
PHONORADIO SALON
EIGHTH FLOOR

GENTLEMEN—
Kindly send me complete information about the wonderful PHONORADIO.

NAME..

ADDRESS..

....

EIGHTY-ONE YEARS OF FAITHFUL SERVICE

GIMBEL BROTHERS

32ND STREET - BROADWAY - 33RD STREET NEW YORK CITY

N. Y. Sun-Globe (Mar. 1, 1924), p. 23

140 Mar. 1924

Radio Retailing (July 1925), p. 92

145 July 1925 $150 (table model).
Advertised in July through Sept. 1925 in this console by The Harponola Co. of Celina, Ohio.

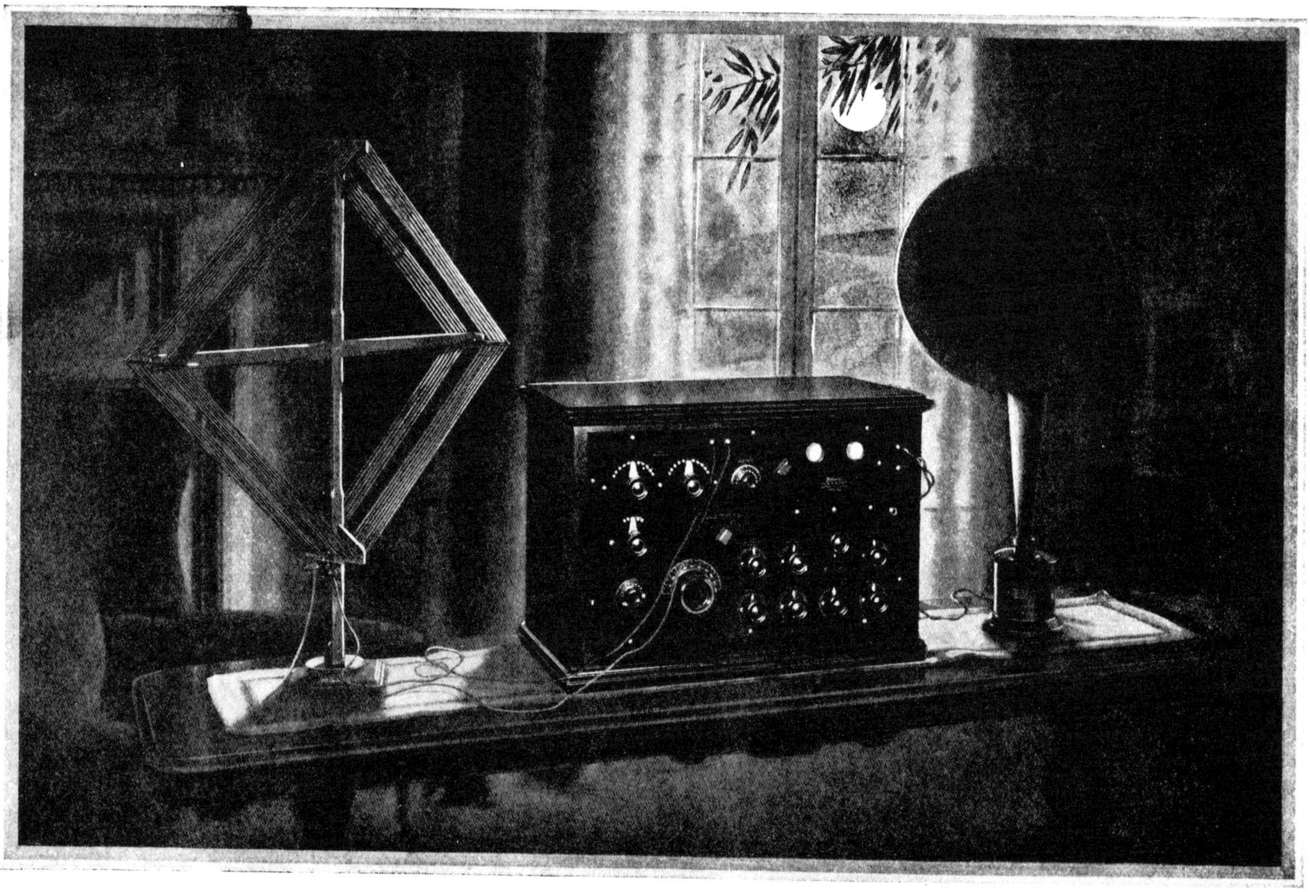

A Midsummer Night's Dream

A WHISK of the dial and the Federal whispers its secrets of the night. Another turn and wondrous music fills the room with tones of bell-like clearness. No reading from the master's pen could rest you—thrill you—as can this gripping romance from your Federal.

There is the joy of mastery in its positive performance, the zest of accomplishment in its exceptional selectivity and a constant inspiration in the beauty of its tone.

And best of all is the satisfaction of knowing that your Federal will give you this same joy—these same gripping thrills—today, tomorrow and for time to come.

Besides the complete Federal Sets—for those who wish the fun of making their own—there are over 130 standard radio parts backed by Federal's quarter-century experience and bearing the Federal iron-clad performance guarantee.

FEDERAL TELEPHONE AND TELEGRAPH CO.
BUFFALO, N. Y.

The three most popular Federal Sets are:

Set No. 59 . . .	$177.00
Set No. 61 . . .	223.00
Set No. 102 . . .	140.00

Prices include Head Sets

Look for this sign

BOSTON NEW YORK PHILADELPHIA CHICAGO PITTSBURGH SAN FRANCISCO BRIDGEBURG, CANADA LONDON, ENGLAND

Phonograph Journal of Canada (Dec. 1924), p. 51

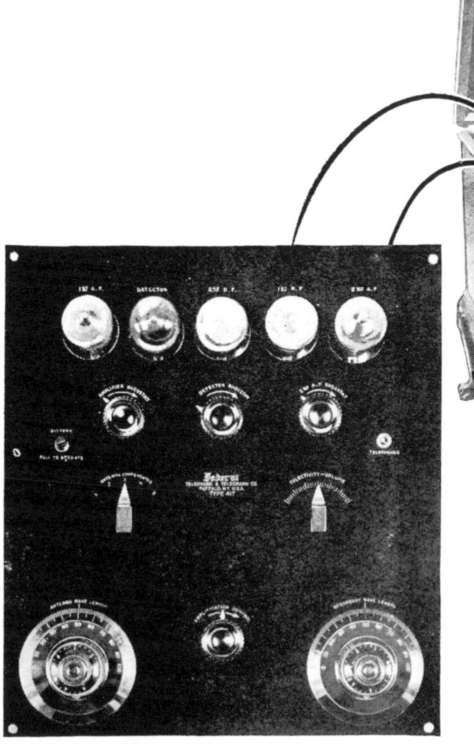

No. 417—Gold finish for Victor Art Model Consoles.
No. 200—Nickel finish for Victor No. 215 Special.

New Federal Radio Panels for Music Merchants

SOME time ago Federal turned its vast engineering abilities to the construction of radio panels that would be in keeping with the latest Victor phonograph creations.

Out of all this effort has been developed the Federal panels No. 200 and No. 417 designed for Victor Console Art Models and the Victrola No. 215 as well as the Federal No. 135 panel for ***all*** upright phonographs.

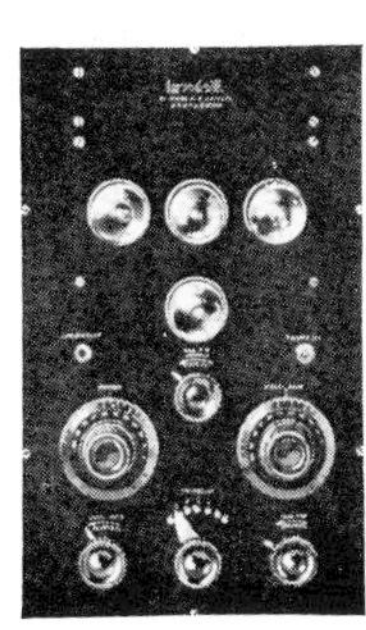

No. 135
For all
Upright Phonographs
and
Victor No. 210 Console

These new Federal panels are built upon the same basic principles and traditions which have maintained Federal leadership in the electrical world for over a quarter century.

They incorporate advantages of tone beauty, selectivity and distance range that will astonish even the experienced radio enthusiast. They will hold the good will of all who buy them.

Music merchants are invited to write Federal for literature and for the name of the nearest distributor of these new panels.

Century Telephone Construction Company, Inc.
Bridgeburg, Ontario
Canadian Distributors for Federal Radio Equipment

135 Mar. 1924 200 Dec. 1924 417 Dec. 1924
The 135 and 140 differ only in panel layout; instruction manuals for both are dated Dec. 1923.

Once again you are faced by the Christmas gift problem. May we offer this suggestion?

There is a dear friend. There is a little child, bedridden. There is a wife tied closely to home by her duties. There is a pair of silver haired lovers out on the farm. There is a home where youth is just coming into its own. There are others whom thoughts of Christmas bring to mind.

Give RADIO—the greatest gift in the world—to at least one of these.

Give the Magic Box that brings the nation's finest singers, musicians, bands, orchestras, fun makers, speakers, preachers—market reports, weather reports—anything and everything into the home *on demand.*

Give Federal Ortho-sonic!

This, we claim, is the *greatest gift in the world*—costing the giver *less* and giving the recipient *more* that is worth while than even the jewels of a queen.

FEDERAL RADIO CORPORATION, *Buffalo, N. Y.*
(Division of Federal Telephone and Telegraph Company)
Operating Broadcast Station WGR at Buffalo

A-10 Five tubes — 199's or 201-A's. Balanced tuned radio frequency. Cabinet has rich brown mahogany finish. Federal standard parts used throughout. Without accessories $75

B-35 Five tubes — 199's or 201-A's. Balanced tuned radio frequency. Beautiful cabinet of selected mahogany, finished in rich two-tone effect. Enclosed speaker. Ample space for batteries Without accessories $250

C-35 Seven tubes — 199's or 201-A's. Balanced tuned radio frequency. In handsome highboy cabinet. Superior quality built-in Federal Speaker. Without accessories . . . $300

C-20 Seven tubes — 199's or 201-A's. Balanced tuned radio frequency. Genuine mahogany cabinet, finished in rich brown. Operates on self-contained loop. Without accessories $165

This Federal Designator is displayed by the Retailer who handles genuine Federal Ortho-sonic Radio Sets. Look for it

ORTHO-SONIC

Of, pertaining to, or producing tone values in sound reproduction corresponding exactly to the natural tones

A,B,C models first advertised Sept. 1925

Why not have the latest?

B-20 Five tubes — 199's or 201-A's. Balanced tuned radio frequency. Genuine mahogany cabinet with rich finish. Micrometer tuning controls. Space for batteries. Without accessories $100

C-30 Seven tubes — 199's or 201-A's. Balanced tuned radio frequency. Genuine mahogany cabinet, with rich brown finish. Two micrometer tuning controls. Space for batteries. Operates on self-contained loop. Federal enclosed adjustable speaker. Without accessories $200

B-36 Five tubes — 199's or 201-A's. Balanced tuned radio frequency. Floor cabinet of choice mahogany woods with rich, two-tone brown finish. Enclosed Federal Speaker. Space for batteries. Without accessories $250

C-40 Not illustrated. Seven tubes — 199's or 201-A's. Balanced tuned radio frequency. In specially designed custom-built hand-carved cabinet. Every detail and feature is provided in this model to make it the last word, not only as a radio receiving set, but also as a beautiful furniture piece for the select home. Without accessories $350

B-30 Five tubes — 199's or 201-A's. Balanced tuned radio frequency. Extremely sensitive and selective. Mahogany cabinet finished in rich, lustrous brown. Micrometer tuning controls. Ample space for batteries. Federal enclosed adjustable speaker. Exceptional volume and tone. Without accessories $130

You read many *claims* about "tone." Why not make this test? Why not let your own ears judge the difference between mere *radio tone* and *Ortho-sonic lifelike presentation?*

Go to any Federal Retailer. He will tune in a Federal Ortho-sonic receiver on any station. Then "*Listen with Closed Eyes.*" This is the supreme test of tone.

Note the startling *reality* of the presentation. There is no sense of distance. The instrument or the voice is *in the room*. Tone, volume, personality are *real*. This is *Ortho-sonic* radio—"*Rivaled only by Reality.*"

Yet Federal Ortho-sonic Radio Sets, in cabinets of modern design and exquisite finish, cost no more than others. There is a type and a price for every home.

See the Federal Retailer in your vicinity. If you do not quickly find him write for his name and a free copy of our booklet, "Radio Reality."

FEDERAL RADIO CORPORATION, *Buffalo, N. Y.*
(Division of Federal Telephone and Telegraph Company)
Operating Broadcast Station WGR at Buffalo

Demand for Radio Increases Daily

Radio Topics (Mar. 1924), pp. 23–24

A MONUMENT to the public demand for radio apparatus and American industry has sprung up in Buffalo.

The immense plant of the Federal Telephone and Telegraph Company, Buffalo, N. Y., has become a vast community of workers, striving to feed the appetite of the radio-hungry American people.

At the outbreak of the World War this manufacturing plant was engaged mainly in the manufacture of high grade telephone equipment.

Like many other industries, radio was spurred on by the needs of our country during the war, until now it stands alone as the one which commands the greatest attention of the nation's masses.

Today the Federal Telephone and Telegraph Company makes radio equipment almost exclusively.

Drilling and engraving panels for Federal 61 sets.

An Immense Plant

An idea of the magnitude of its organization may be obtained from the following facts:

The factory at present employs over 800 operatives. A 50 per cent increase is contemplated during the coming month. It is expected that 30,000 complete sets of every kind will be turned out in the next few months. Completed sets have yet to become as popular as the home built set. Many more parts are sold as units than are used in the assembly of complete sets. The demand for radio is so great that Federal alone expects to produce over 175,000 headsets this year.

Any one who is capable of work is hired and given a chance, but the rigid inspection of each one's work is such that a great many employes become discouraged over their failure to produce parts which do not come up to the specifications.

Modern Machinery Used

Upon entering the factory, visitors are noticeably impressed with the immensity of the plant and the vast number of persons engaged in making and assemblying the various parts.

The most complete and modern machinery has been installed to insure accuracy and precision.

A person who purchases Federal radio, little realizes the large number of machines and operations needed to produce even the most insignificant looking part.

There are at present eleven engraving machines, used solely for the cutting of the lettering into the panel fronts of the sets.

One of the most impressive sights is the new multiple drill machines. These marvelous machines drill 20 holes at the will of the operator in just one pull of the lever.

Young ladies employed to test sets. Each part passes rigid inspection.

Shop Never Closes

There are so many different kinds of nuts and screws and other miscellaneous parts used in the manufacture of radio that it is necessary to run the machine department night and day to keep up with the requirements of assembly. All of the output of the various departments is used solely in Federal equipment. No work is done for outside concerns. Millions of screws and punchings must be purchased each month to supplement the output of what is probably the largest machine shop which, practically speaking, is engaged exclusively in radio production.

All of the bakelite moulding is done outside of the factory. The entire capacity of a large punching plant is contracted for as a supplement to the twenty punching presses in the plant proper.

Despite the fact that there are over 40 hand and automatic screw machines in the plant, it is necessary to have done in outside places more work of this kind than is actually done in the factory.

The tool making department alone employs thirty men who do nothing but make special fixtures, dies and punches for machines.

The storage warehouse, packing and shipping departments and other buildings, all in process of expansion, require much greater floor space than the factory building proper.

Countless Operations

There are countless operations which never occur to the prospective owner of a Federal set, such as the nickel, and copper plating of metal parts, the enameling of metal cases, the assembly of the transformers, the testing of every part and set.

A large office force is required to handle the mass of correspondence and orders.

A whole section is devoted to answering persons who write in and say they heard WGR, the big Federal station. An excellent method of logging such testimonials discloses the fact that the average reliable distance which WGR covers is 550 miles.

The personnel comprising the radio sales force in the United States only, numbers over 100.

All this in answer to the demand of the public for more and better radio.

Two New Stations on the Air

THE largest and most powerful radio broadcasting station in Canada, operated by Canadian National Railways, was thrown open February 27. The call of the new station is CKCH, and is the first of a chain of stations to extend across Canada. A musical program and talk by Sir Henry W. Thornton, chairman and president of the board of directors of the Canadian Railway, opened the station.

The new station expects to have a range beyond that of any station in the dominion, due not only to its up-to-date equipment, but also to the height of its aerial, which stands on the roof of the Jackson building and reaches 200 feet above the ground.

It will broadcast on a wavelength of 435 meters. The initial program was relayed by station CHYC, Northern Electric Company, Montreal, on a wavelength of 341 meters, so that radio listeners everywhere in Canada and the United States should have no difficulty in receiving the program.

Boston Has Fine Station

Another new station is WBZ, located on the Brunswick Hotel, Boston, Mass.

Programs of classical and popular music, jazz music, lectures and theatrical features will be broadcast. On the top of the Brunswick Hotel has been built a studio that equals anything that has been done in the past for convenience and beauty. In the building that has been constructed on the roof of the hotel is the studio, 30 by 40 feet in size, and a reception room with a ladies' parlor.

In order to make the project successful, it was necessary to have constructed an entirely new line from Boston to Springfield, by the Western Union Company. This line is slightly over one hundred miles in length and connects the studio at the Brunswick with the radio station at East Springfield, Mass.

At Springfield this telephone line goes to Station WBZ of the Westinghouse Company, located at this company's plant. For the past two months the radio station has been undergoing a complete renovated so that when the studio opens, everything will be in first class condition. The signals will radiate from the long antenna 210 feet above ground and supported from massive steel towers.

The three interests, the Boston Herald-Traveler, the Westinghouse Electric & Manufacturing Company and the Brunswick Hotel of Boston, have been working day and night for the past two months to get everything ready for the opening date. A new era in remote studio broadcasting will be started and the radio fans will have a new station that will equal and perhaps surpass anything that has been done in broadcasting.

A corner of the assembling room where Federal 61's are made.

Calls Troops by Radio

Radio was employed for the first time in mobilizing the national guard when the 132nd Infantry of Chicago, commanded by Col. William E. Swanson, was notified via the ether to report at the armory at 2640 West Madison street. As soon as the order was received from Springfield directing the mobilization of the regiment to be held in readiness to move to Herrin, the adjutant relayed the information to station KYW and it was broadcast.

It is estimated that more than 50 per cent of the soldiers either received the message themselves or were notified of it by neighboring radio fans.

ENGLISH ART MODEL
A custom-built Ortho-sonic set
Without accessories, $600.00

ORTHO-SONIC ANNOUNCEMENT

New beauty in the Federal line

WHEN the Federal Radio Corporation introduced Federal Ortho-sonic Radio, a real sensation was created.

Ortho-sonic tone, brought in by the exclusive patented Federal Ortho-sonic circuit, was unlike anything hitherto produced in radio. For it was unusually true to life. Music critics and radio experts alike marveled at it and quickly accepted it. It at once set a new standard of radio-tone perfection.

Now comes another Federal achievement. Sets enhanced by designs whose graceful contours, handsome inlays and hand carvings, and duo-tone wood effects have united in a new standard of beauty and artistry.

Here, we believe, is the ultimate in radio. Here, with beauty and individuality, you get the super-selectivity of sets shielded by the improved Federal method and giving remarkable freedom from interference.

You get super power and super range—the surety of bringing in far distant stations clearly.

You get the ideal one-dial control perfected by Federal engineers and enabling you to bring in clearly any station you want, by the mere twist of the wrist—a control remarkable for its simplicity and effectiveness.

And you get also extreme durability, with dependability. In reality every Federal Ortho-sonic Set is a precision instrument built by craftsmen whose reputation for wireless, telephone, and navy radio reaches back over a quarter of a century. Radio built by makers who are responsible and here to stay.

No furniture you can put in your home will ever be more decorative, more enduring, more fruitful of pride and joy.

No radio you can buy will ever serve longer or more satisfactorily. Yet Federal prices are conspicuously moderate. And there's a model for your own particular pocketbook.

Note on opposite page Federal's 14 Points. Then go to your Federal retailer or phone him for a free, non-obligatory, home demonstration—listen in with closed eyes. You will then be thrilled into a complete understanding of the enthusiasm of every Federal owner. And Federal owners are legion.

THE Greater Federal Line includes many models, ranging in price from $75 to $400, and four de luxe custom-built models retailing at $600 to $1000. B and C models have space for all batteries, also for current supply devices which operate from electric light sockets. C models are completely self-contained and portable.

FEDERAL RADIO CORPORATION, *Buffalo, N.Y.*
(Division of Federal Telephone and Telegraph Company)
Operating Broadcasting Station WGR at Buffalo

Federal ORTHO

★ The fundamental and exclusive circuit making possible Ortho-sonic reproduction is patented under U. S. Letters Patent No. 1,582,470.

E-10

E-10—Six tubes. Wet or dry battery type. Balanced, tuned radio frequency. Single control. Extremely selective. Maximum receiving range. All-metal construction. Illuminated scale. Perfect control of volume. Beautiful brown mahogany cabinet—rosewood inlay. Well finished. Satin texture. Without accessories . . . $150.00

E-5—Console as Illustrated. Specially designed and finished to match E-10 Ortho-sonic. Ample battery space. Equipped with gliders . . . $40.00

A-10

A-10—Five tubes. Wet or dry battery type. Balanced, tuned radio frequency. Rich brown mahogany finish. Without accessories . . . $75.00

E-40—Six tubes. Wet or dry battery type. Balanced, tuned radio frequency. Single control. Extremely selective. Maximum receiving range. All-metal construction. Illuminated scale. Built-in speaker. Receiver compartment slides out. Walnut cabinet, antiqued and inlaid with rosewood. Ample battery space. Without accessories . $300.00

E-40

F-10—Seven tubes. Wet or dry battery type. Balanced, tuned radio frequency. For use with loop only. Single control. Maximum selectivity—long range reception. All-metal construction. Illuminated scale. Perfect control of volume. Beautiful mahogany cabinet. Vermillion inlay. Finish rich brown—satin texture. Without accessories . . $250.00

F-5—Console. As illustrated. Specially designed and finished to match F-10 Ortho-sonic. Ample battery space. Equipped with special rubber-tired casters . . . $50.00

F-10

Federal's 14 Points

1. Orth-sonic receiving sets made complete in Federal factories assures perfect matching of parts.
2. Built around Federal's own patented circuits.
3. Ortho-sonic tone quality—the resu'tofyears of acoustical research.
4. Rugged, all-metal construction—will last a lifetime.
5. Cabinets of carefully selected mahogany and walnut.
6. Multi-shielding. Provides isolation of all circuits from one another and from extraneous influences.
7. Simplified control with the maximum efficiency.
8. Razor-edge selectivity; allowing reception of distant stations through locals.
9. Illuminated, self-indicating scale, acting as pilot light.
10. Uses dry battery tubes, also standard tubes, without adjustment.
11. Re-radiation proof—does not interfere with your neighbor's receiving set.
12. Includes a model for every purse and a design for every setting.
13. A precision instrument built by Federal telephone and radio experts.
14. Backed by a solid, substantial company—can never become an "orphan."

D-40

D-40—Five tubes. Wet or dry battery type. Balanced, tuned radio frequency. Centralized control. Built-in speaker. Deep, rich tone quality. Receiver compartment slides out. Walnut cabinet. Ample battery space. Without accessories . . . $200.00

D-10

D-10—Five tubes. Wet or dry battery type. Balanced, tuned radio frequency. Centralized control. Very selective and sensitive. Mahogany cabinet, mahogany lined. Finish, rich brown. Without accessories . $100.00

D-5—Console. As illustrated. Specially designed and finished to match D-10 Ortho-sonic. Ample battery space. Equipped with gliders. . . $30.00

The sign displayed by all designated Federal retailers. Wherever you see it feel free to enter, and listen in.

F-40

F-40—Seven tubes. Wet or dry battery type. Balanced, tuned radio frequency. For use with loop only. Single control. Maximum selectivity—long range reception. All-metal construction. Illuminated scale. Extra large built-in speaker. Walnut cabinet, artistically antiqued. Inlaid with Vermillion. Ample battery space. Without accessories . $400.00

SONIC* Radio

Reg. U. S. Pat. Off.

D, E, F models Aug. 1926

1927 models, first advertised in June, same as D, E, F 1926 models, with these AC models added: D-10-60, $185 E-10-60, $275. F-10-60, $360. D-40-60, $285. E-45-60, $460. F-45-60, $600. Plus 8 custom-built models, to $1225.

B20 Ralph & Elinor Williams

E10 Henry Harder

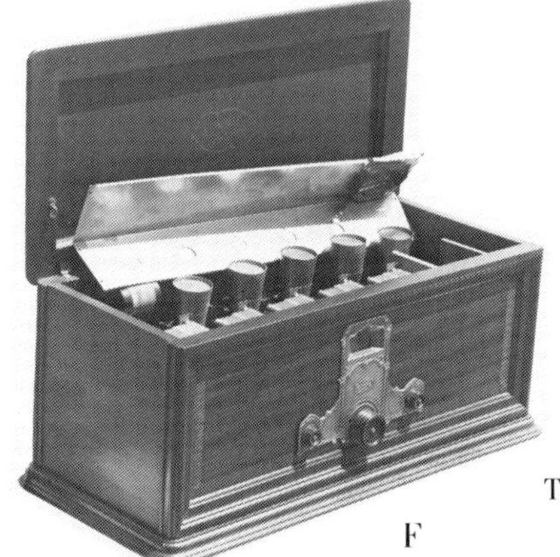

Tim Keener

F

Radio Dealer (Aug. 1928), p. 6

Electrical Record Vol. XLII, No. 2

FEDERAL *Ortho-sonic*

Custom-Built Models—

For those seeking the ultimate in radio cabinet work and design, we offer these four splendid examples of furniture craftsmanship.

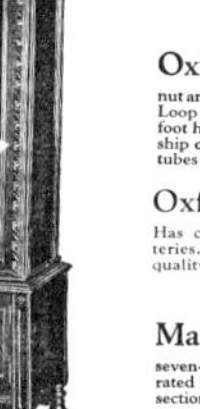

The Oxford

Oxford Seven tubes. Single control. Illuminated scale. Custom-built cabinet of genuine walnut and maple finished in deep brown, carefully antiqued. Loop concealed in door. Contains famous Federal seven-foot horn. The cabinet is characteristic of the workmanship of the early English artisan. Without tubes or accessories (but including loop).... $650.00

Oxford 60 Oxford receiving set with circuit adapted for light-socket operation. Has current supply equipment, eliminating all batteries. Built-in coupler gives finest tone quality. Without tubes $775.00

The Mandarin

Mandarin Seven tubes. Single control. Illuminated scale. Contains famous Federal seven-foot horn. Loop concealed in door. Cabinet decorated in an authentic Chinese Chippendale, the upper section being beautifully finished with hand-applied Chinese decorating of the raised type. Lower section is of selected walnut finished in a dark brown tone with a tracery of maple. Without tubes or accessories (but including loop) $1000.00

Mandarin 60 Mandarin receiving set with circuit adapted for light-socket operation. Has current supply equipment, eliminating all batteries. Built-in coupler gives finest tone quality. Without tubes. $1125.00

The Louvain

Louvain Seven tubes. Single control. Illuminated scale. Custom-built cabinet of walnut and cherry burl. Semi-arched top, characteristic of later English design, finished in a dull wash. Loop concealed in door. Contains famous Federal seven-foot horn. Without tubes or accessories (but including loop) $775.00

Louvain 60 Louvain receiving set with circuit adapted for light-socket operation. Has current supply equipment, eliminating all batteries. Built-in coupler gives finest tone quality. Without tubes....... $900.00

The Milan

Milan Seven tubes. Single control. Illuminated scale. Custom-built cabinet of genuine walnut and maple finished in deep brown. Loop concealed in door. Contains famous Federal seven-foot horn. Splendid example of cabinetmaker's art in Italian Renaissance. Wood of selected walnut with all decorations hand carved and raised. Without tubes or accessories (but including loop)............. $1100.00

Milan 60 Milan receiving set, with circuit adapted for light-socket operation. Has current supply equipment, eliminating all batteries. Built-in coupler giving finest tone quality. Without tubes $1225.00

FEDERAL RADIO CORPORATION, *Buffalo, N. Y.*
(Division of The Federal Telephone Manufacturing Corp.)
Operating Broadcast Station WGR at Buffalo

Electrical Record (Aug. 1927)

H John Bayusik

July 1928 G-10-60 $130 G-40-60 $220

Aug. 1928 models:

(F-11 series, wire antenna):				F-11	$250	F-11-60	$360			
(F-10 series, loop antenna):										
		F-50	$650	F-60	$775	F-70	$1000	F-80	$1100 (battery)	
F-40-60	$460	F-50-60	$775	F-60-60	$900	F-70-60	$1125	F-80-60	$1225 (AC)	
F-41-60	$595	F-51-60	$810	F-61-60	$935	F-71-60	$1160	F-81-60	$1260 (AC with dynamic speaker)	

not illustrated:
Dec. 1928: F-43 $295 F-43-60 $370

Short technical article in *Radio Broadcast*, Aug. 1928, p.222.

Complete
$155.50
Less Tubes

Federal-Findlay Combination

FINDLAY Metal Console Tables are now available for Federal H10-60 Models. This specially designed unit makes an ideal inexpensive installation worthy of any home.

WM. T. BAXTER CORPORATION

4234 Broadway, near 180th St.

Tel. WAShington Heights 3139

N. Y. Sun (Dec. 15, 1928)

Federal Announces The New H-Series

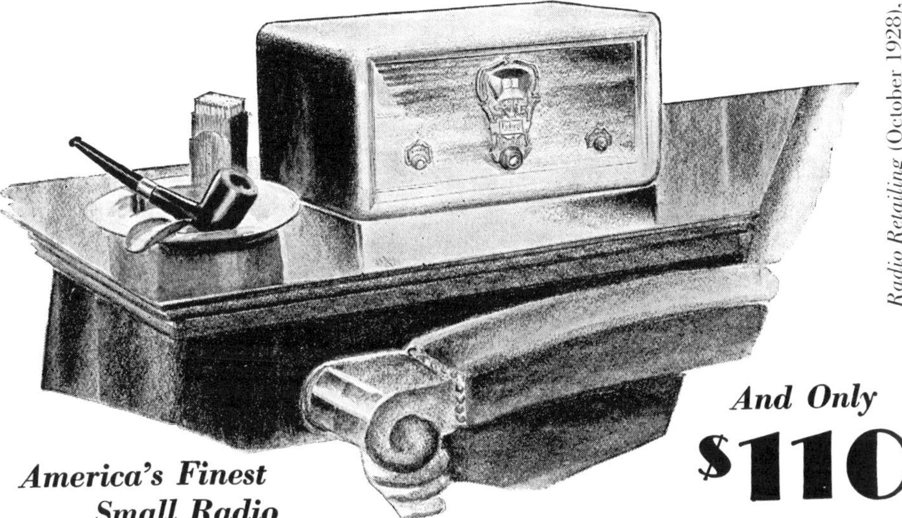

Radio Retailing (October 1928), p. 131

And Only

$110

America's Finest Small Radio

FEDERAL presents a new achievement—the new H models – table and console receivers of remarkable beauty and compactness.

The table model easily tucks away in any of dozens of places in the home wherever there's a few square inches of unused room. The console model fits into the decorative scheme in the smallest nook or corner without rearranging the room or disturbing the furniture. This beautiful compactness appeals particularly to women. It alone will mean thousands of sales for designated Federal retailers this year.

The console model is provided with built-in speaker specially designed to respond to low frequencies. It is also available with built-in dynamic speaker.

All Metal Chassis · Wood Cabinets · Two Way Selectivity · Push-pull Amplification · Ortho-sonic Tone · Single Dial, electrically illuminated · Unusually Compact

7 A. C. tubes (including rectifier)

Prices (without tubes or accessories)

Oct. 1928: H-10-60 $110 H-40-60 $185 H-41-60 $210

Technical article in *Citizens Callbook*, vol.10 no.4, Nov. 1929, p.89.

Radio (July 1929), p. 92

"K" TABLE MODELS
K 10-60 60 cycle $127.50
K 10-25 25 cycle $137.50

"K" CONSOLE MODELS
K 40-60 60 cycle $197.50
K 40-25 25 cycle $207.50
with DYNAMIC SPEAKER
K 41-60 60 cycle $227.50
K 41-25 25 cycle $237.50

Prices do not include tubes and are slightly higher in west.

Jan. 1929

The Federal new L chassis is similar in general to the M chassis, excepting that screen-grid tubes are used in the three r.f. stages. It is made in two models, L36 and L46, with built-in dynamic speaker and timbre control.

Technical article on M and L in *Radio*, July 1929, p.92.

July 1929: L-36 $149.50 L-46 $179.50
Aug. 1929: M-10 $175 M-36 $245 M-41, M-42, M-46 $295

FERGUSON

J.B. Ferguson, Inc.

Joseph B. Ferguson worked for Ship Owners Radio Service, Inc. for several years, as chief engineer and after May 1923 as general manager. In December he struck out for himself, forming J.B. Ferguson, Inc. along with J.B. Jr., with $10,000 capital. He began in a small way, building relatively expensive four-tube sets and getting writeups in the New York City newspapers. In April 1925 Raymond J. Ketcham, with Ferguson four years at Sorsinc, joined him again as chief engineer, and probably contributed a good deal to their most innovative model, the Eight, which appeared several months later.

In November 1925 Ferguson found his financial angel, Thomas B. Wickwire, Jr., formerly of the Wickwire-Spencer Steel Corp. (which he had just left, probably with a push, as the company shortly went into receivership and closed a few of its many plants). Ferguson moved to a new factory in Long Island City, New York that was undoubtedly financed by Wickwire. Wickwire considered taking over the ailing De Forest company around February 1926, but decided against it and left De Forest after a month or two. In August Ferguson opened a second plant near the first, and by the end of the year was said to be well behind in orders.

J.B. Ferguson left his company in June 1927 for another venture: building AC sets in Kingston, Pennsylvania ("I do not make receivers until I get non-cancellable orders for them; no overproduction in our plant"). Meanwhile, Wickwire took over the old company entirely, making, besides the Ferguson models, two "Homer" models under the Technidyne patents (tuner followed by a multi-stage untuned RF amplifier, avoiding the need of an RCA license). But he lost money the whole time: $128,652 on the factory and $145,442 on his jobber, whom he financed, and assigned for creditors in February 1928.

J.B. Ferguson, after returning from Kingston, joined a syndicate promoting Baird's television system, ran a New York factory assembling Rola speakers, then at the end of 1929 built sets under the Balder name. In 1932 his Ferguson Radio Corp., incorporated in March or April, was offering six-tube midgets for the export trade. He started making short-wave sets in England by 1937, his company within four years becoming a part of Thorn.

T. H. Wickwire, Jr., Vice-President

J. B. Ferguson, President

Interior of the four tube receiver.

N. Y. Sun (Feb. 16, 1924), p. 1

This chassis, probably an early form of the TRF 3, was described in the *NY Sun*, Feb. 16, 1924 and the *NY Evening World*, Apr. 12, 1924. A "4-tube Ferguson" listed at $125 in May 1924.

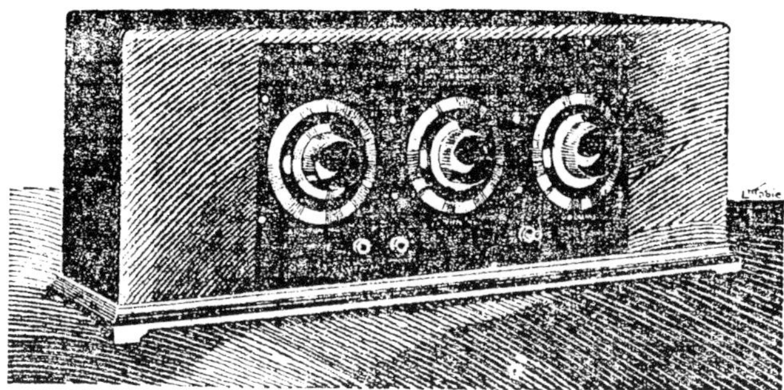

A *Home* Receiver *All* Can Operate

THE Ferguson Type TRF 3 is just as sensitive—just as selective as most receivers costing two or three times the price. And of course there's plenty of volume for loud speaker operation.

***Anyone* can get peak results from a Ferguson TRF 3**
—ask any of the following dealers to prove it.

Adams Flanigan Company
149th St. and 3rd Ave.
New York City

Ludwig Baumann & Co.
New York and Newark Stores

Deutsch Bros.
121st St. and 3rd Ave., N. Y. C.
58 Avenue A, cor. 4th St., N.Y.C.
6th Ave and 20th St., N. Y. C.
3rd Ave. and 123rd St., N.Y.C.
Broadway and Saratoga Street,
Brooklyn, N. Y.

Effanel Elec. & Radio Corp.
3604 Broadway
New York City

O. W. Gellentien
2027 Third Ave.
New York City

Sol Lazarus
216 East 59th St.
New York City

W. P. Plummer
29 Broadway
New York City

Throg's Neck Elec. & Supply Co.
3432 East Tremont Ave.
New York City

A. & R. Radio
540 West 125th St.
New York City

Broadway Radio Corp.
148 Pearl Street
New York City

Charles W. Down
321 West 44th St.
New York City

Fred J. Dreher
286 Amsterdam Ave.
New York City

A. Finkenberg's Sons
124th St. and 3rd Ave.
New York City

Kranz Bros. (Mike's)
Greenwich and Dey Sts.
New York City

Merchants Radio Stores
381 Broadway
New York City

Tremont Music Shop
452 East Tremont Ave.
New York City

Matthew Conroy
147 East 76th St.
New York City

Bigelow-Halstead, Inc.
16 Division Street
New Rochelle, N. Y.

Namsel Radio Co.
262 Huguenot St.
New Rochelle, N. Y.

Earle Radio
272 Main St.
Hackensack, N. J.

Halpin Edelson & Co.
17 East 8th St.
Yonkers, N. Y.

Wireless Equipment Co.
7 Union Place
Stapleton, S. I.

Kelly's Radio Store
1104 Lafayette Ave.
Brooklyn, N. Y.

R. & S. Electric
109 Ralph Ave.
Brooklyn, N. Y.

Rex Radio Shop
238 Reid Ave.
Brooklyn, N. Y.

John Streiff
73 Flatbush Ave.
Brooklyn, N. Y.

Manufactured by
J. B. FERGUSON, INC. 80 Beaver St.

Distributed by
THE RANCE CORPORATION 86 Church St.

N. Y. *Herald-Tribune* (Dec. 21, 1924), p. 14

Ferguson Six—*Console Model*
Complete with special built in loud speaker, with Amplion unit, compartments for "A" and "B" batteries and battery charger. Cabinet work the finest, genuine walnut, finished in latest high light styles. Plenty of room for books and radio literature. Top panel drops down when receiver is in operation, providing a handy desk arrangement.
Price, $290
Batteries, tubes, and charger extra

Ferguson Six—*Cabinet Model*
Employs two stages of balanced tuned radio frequency oscillating detector (equal to regeneration for added volume, but without radiation or effect on tuning); three stages of perfectly matched audio frequency. De Luxe high light walnut or mahogany cabinet with separate compartments for "B" batteries.
Price, $180
Tubes and batteries extra

Radio Broadcast (June 1925), p. 285

Six Feb., 1925 $180

Fig. 1. A real radio machine, designed to be put together quickly and to stay put forever

Radio Engineering (Apr. 1926), p. 160

Eight Chassis

Dick Doremus

TRF3 Dec. 1924 $130
3 models by Mar. 1925: 3-A (same, 30 to 120 meters) $130. "Traveling" model $120. 3-V Victor phono panel $110.

Radio Dealer (Sept. 1925), p. 139

Eight Sept. 1925 $226

After Feb. 1926, in a shorter 28½″ cabinet. Technical articles in *Popular Radio*, June 1926, pp. 131–136, and *Radio Engineering*, Apr. 1926, p. 160.

Four Sept. 1925 $95

FERGUSON MODEL FOUR. Manufactured by J. B. Ferguson, Inc., 41 East 42nd street, New York City. Four tube receiving set employing one stage of balanced tuned radio frequency, non-radiating detector, and two stages of audio frequency. Sloping Bakelite panel. Mahogany finished cabinet with space provided for "B" batteries. List price $95.00.

Radio Dealer (Mar. 1926), p. 92

Radio Dealer (Apr. 1926), p. 5

Ten Apr. 1926 $110
By Sept. 1926, used a dial like the Fourteen. Described in *Radio News*, Sept. 1926, pp. 220–221, and *Popular Radio*, Dec. 1926, pp. 787, 790–792.

Radio Dealer (Oct. 1926), p. 3

Twelve Oct. 1926 $75

THE FERGUSON MODEL 14 RECEIVER

Radio Broadcast (June 1927), p. 122

Fourteen Apr. 1927 $235
(not shown) Eighteen Sept. 1927 $195

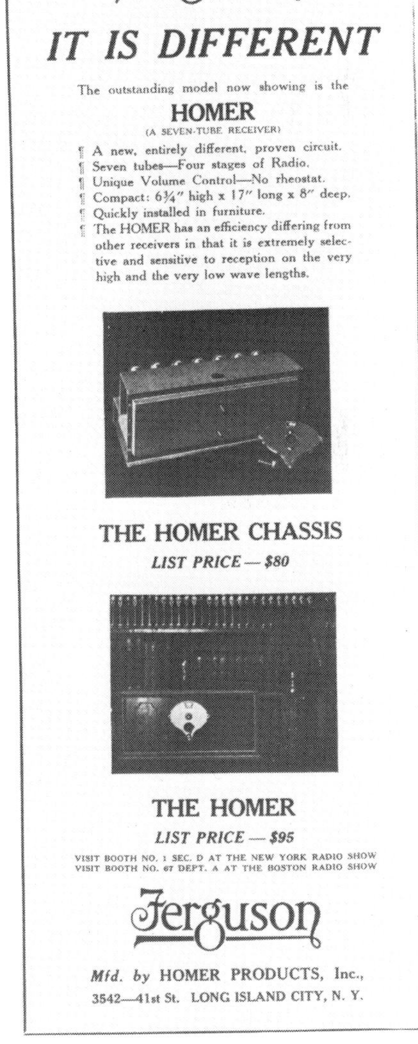

Radio Retailer & Jobber (Sept. 1927)

Homer Sept. 1927 $95.
Also Homer Sr., $150

Index